Principle of the Beginning (229) The list of stakehol_ _prise) that are included in the early stages of product de_ the architecture.

Principle of Balance (244) Many factors influence and act on the conception, design, implementation, and operation of a system. One must find a balance among the factors that satisfies the most important stakeholders.

Principle of the System Problem Statement (247) The statement of the problem defines the high-level goal and establishes the boundaries of the system. Challenge and refine the statement until you are satisfied that it is correct.

Principle of Ambiguity and Goals (252) The architect must resolve this ambiguity to produce (and continuously update) a small set of representative, complete and consistent, challenging yet attainable, and humanly solvable goals.

Principle of Creativity (267) Creativity in architecture is the process of resolving tensions in the pursuit of good architecture.

Principle of Apparent Complexity (289) Create decomposition, abstraction, and hierarchy to keep the apparent complexity within the range of human understanding.

Principle of Essential Complexity (292) Functionality drives essential complexity. Describe the required functionality carefully, and then choose a concept that produces low complexity.

Principle of the 2nd Law (295) The actual complexity of the system always exceeds the essential complexity. Try to keep the actual complexity close to the essential complexity.

Principle of Decomposition (296) Decomposition is an active choice made by the architect. The decomposition affects how performance is measured, how the organization should be set up, and the potential for supplier value capture.

Principle of "2 Down, 1 Up" (298) The goodness of a decomposition at Level 1 cannot be evaluated until the next level down has been populated and the relationships identified (Level 2).

Principle of Elegance (300) Elegance is appreciated internally by the architect when a system has a concept with low essential complexity and a decomposition that aligns many of the planes of decomposition simultaneously.

Principle of Robustness of Architectures (349) Good architectures can respond to change by being robust (capable of dealing with variations in the environment) or by being adaptable (able to adapt to changes in the environment).

Principle of Coupling and Organization of Architectural Decisions (354) The sequence of architectural decisions can be chosen by considering the sensitivity of the metrics to the decisions and the degree of connectivity of decisions.

SYSTEM ARCHITECTURE

Strategy and Product Development for Complex Systems

Edward Crawley   Bruce Cameron   Daniel Selva

Boston Columbus Indianapolis New York San Francisco Hoboken
Amsterdam Cape Town Dubai London Madrid Milan Munich Paris Montréal Toronto
Delhi Mexico City São Paulo Sydney Hong Kong Seoul Singapore Taipei Tokyo

Vice President and Editorial Director, ECS: *Marcia J. Horton*
Executive Editor: *Holly Stark*
Field Marketing Manager: *Demetrius Hall*
Senior Product Marketing Manager: *Bram van Kempen*
Marketing Assistant: *Jon Bryant*
Senior Managing Editor: *Scott Disanno*
Global HE Director of Vendor Sourcing and Procurement: *Diane Hynes*
Director of Operations: *Nick Sklitsis*
Operations Specialist: *Maura Zaldivar-Garcia*
Cover Designer: *Black Horse Designs*
Production Project Manager: *Rose Kernan*
Program Manager: *Erin Ault*
Manager, Rights and Permissions: *Rachel Youdelman*
Associate Project Manager, Rights and Permissions: *Timothy Nicholls*
Full-Service Project Management: *George Jacob, Integra*
Composition: *Integra*
Printer/Binder: *Edwards Brothers*
Cover Printer: *Phoenix Color/Hagerstown*
Typeface: *10/12, Times LT Std*

Copyright © 2016 by Pearson Higher Education, Inc., Hoboken, NJ 07030. All rights reserved. Manufactured in the United States of America. This publication is protected by Copyright and permissions should be obtained from the publisher prior to any prohibited reproduction, storage in a retrieval system, or transmission in any form or by any means, electronic, mechanical, photocopying, recording, or likewise. To obtain permission(s) to use materials from this work, please submit a written request to Pearson Higher Education, Permissions Department, 221 River Street, Hoboken, NJ 07030.

Many of the designations by manufacturers and seller to distinguish their products are claimed as trademarks. Where those designations appear in this book, and the publisher was aware of a trademark claim, the designations have been printed in initial caps or all caps.

The author and publisher of this book have used their best efforts in preparing this book. These efforts include the development, research, and testing of theories and programs to determine their effectiveness. The author and publisher make no warranty of any kind, expressed or implied, with regard to these programs or the documentation contained in this book. The author and publisher shall not be liable in any event for incidental or consequential damages with, or arising out of, the furnishing, performance, or use of these programs.

Library of Congress Cataloging-in-Publication Data On File

2 17

ISBN-10:   0-13-397534-7
ISBN-13: 978-0-13-397534-5

Contents

Foreword vii
Preface ix
Acknowledgments xi
About the Authors xii

PART 1 System Thinking 1

Chapter 1 Introduction to System Architecture 2
Architecture of Complex Systems 2
The Advantages of Good Architecture 2
Learning Objectives 5
Organization of the Text 6
References 7

Chapter 2 System Thinking 8
2.1 Introduction 8
2.2 Systems and Emergence 8
2.3 Task 1: Identify the System, Its Form, and Its Function 13
2.4 Task 2: Identify Entities of a System, Their Form, and Their Function 17
2.5 Task 3: Identify the Relationships among the Entities 26
2.6 Task 4: Emergence 28
2.7 Summary 33
References 34

Chapter 3 Thinking about Complex Systems 35
3.1 Introduction 35
3.2 Complexity in Systems 35
3.3 Decomposition of Systems 39
3.4 Special Logical Relationships 43
3.5 Reasoning through Complex Systems 44
3.6 Architecture Representation Tools: SysML and OPM 45
3.7 Summary 48
References 50

PART 2 Analysis of System Architecture 51

Chapter 4 Form 53
4.1 Introduction 53
4.2 Form in Architecture 53
4.3 Analysis of Form in Architecture 58
4.4 Analysis of Formal Relationships in Architecture 63
4.5 Formal Context 75
4.6 Form in Software Systems 77

4.7 Summary   82
References   82

Chapter 5   Function   83
5.1 Introduction   83
5.2 Function in Architecture   83
5.3 Analysis of External Function and Value   89
5.4 Analysis of Internal Function   94
5.5 Analysis of Functional Interactions and Functional Architecture   98
5.6 Secondary Value-Related External and Internal Functions   108
5.7 Summary   109
References   109

Chapter 6   System Architecture   110
6.1 Introduction   110
6.2 System Architecture: Form and Function   111
6.3 Non-idealities, Supporting Layers, and Interfaces in System Architecture   121
6.4 Operational Behavior   125
6.5 Reasoning about Architecture Using Representations   129
6.6 Summary   136
References   136

Chapter 7   Solution-Neutral Function and Concepts   137
7.1 Introduction   137
7.2 Identifying the Solution-Neutral Function   140
7.3 Concept   142
7.4 Integrated Concepts   152
7.5 Concepts of Operations and Services   157
7.6 Summary   158
References   159

Chapter 8   From Concept to Architecture   160
8.1 Introduction   160
8.2 Developing the Level 1 Architecture   162
8.3 Developing the Level 2 Architecture   166
8.4 Home Data Network Architecture at Level 2   170
8.5 Modularizing the System at Level 1   173
8.6 Summary   175
References   176

PART 3   Creating System Architecture   177

Chapter 9   The Role of the Architect   178
9.1 Introduction   178

9.2 Ambiguity and the Role of the Architect   178
9.3 The Product Development Process   184
9.4 Summary   192
References   196

Chapter 10   Upstream and Downstream Influences on System Architecture   197
10.1 Introduction   197
10.2 Upstream Influence: Corporate Strategy   198
10.3 Upstream Influence: Marketing   201
10.4 Upstream Influence: Regulation and Pseudo-Regulatory Influences   204
10.5 Upstream Influence: Technology Infusion   206
10.6 Downstream Influence: Implementation—Coding, Manufacturing, and Supply Chain Management   207
10.7 Downstream Influence: Operations   210
10.8 Downstream Influence: Design for X   212
10.9 Downstream Influence: Product and System Evolution, and Product Families   214
10.10 The Product Case: Architecture Business Case Decision (ABCD)   217
10.11 Summary   221
References   224

Chapter 11   Translating Needs into Goals   226
11.1 Introduction   226
11.2 Identifying Beneficiaries and Stakeholders   227
11.3 Characterizing Needs   236
11.4 Interpreting Needs as Goals   244
11.5 Prioritizing Goals   250
11.6 Summary   253
References   259

Chapter 12   Applying Creativity to Generating a Concept   262
12.1 Introduction   262
12.2 Applying Creativity to Concept   263
12.3 Develop the Concepts   268
12.4 Expand the Concepts and Develop the Concept Fragments   269
12.5 Evolve and Refine the Integrated Concepts   274
12.6 Select a Few Integrated Concepts for Further Development   277
12.7 Summary   279
References   284

Chapter 13   Decomposition as a Tool for Managing Complexity   286
13.1 Introduction   286

13.2 Understanding Complexity   286
13.3 Managing Complexity   295
13.4 Summary   303
References   308

PART 4 Architecture as Decisions   309

Chapter 14 System Architecture as a Decision-Making Process   311
14.1 Introduction   311
14.2 Formulating the Apollo Architecture Decision Problem   312
14.3 Decisions and Decision Support   317
14.4 Four Main Tasks of Decision Support Systems   319
14.5 Basic Decision Support Tools   320
14.6 Decision Support for System Architecture   326
14.7 Summary   328
References   328

Chapter 15 Reasoning about Architectural Tradespaces   331
15.1 Introduction   331
15.2 Tradespace Basics   332
15.3 The Pareto Frontier   334
15.4 Structure of the Tradespace   341
15.5 Sensitivity Analysis   345
15.6 Organizing Architectural Decisions   350
15.7 Summary   356
References   357

Chapter 16 Formulating and Solving System Architecture Optimization Problems   359
16.1 Introduction   359
16.2 Formulating a System Architecture Optimization Problem   361
16.3 NEOSS Example: An Earth Observing Satellite System for NASA   363
16.4 Patterns in System Architecting Decisions   365
16.5 Formulating a Large-scale System Architecture Problem   389
16.6 Solving System Architecture Optimization Problems   394
16.7 Summary   402
References   402

Appendices   406

Chapter Problems   421

Index   448

Foreword

Norman R. Augustine

A particularly promising trend that has been taking place in healthcare is the marriage of biomedical research with engineering practices. A friend of mine, an engineer, recently described to me a meeting that took place at one of America's most prestigious universities between the faculties of the engineering department and the cardiology department exploring just such opportunities. Having decided to focus on constructing a practicable mechanical human heart, the head of cardiology began his presentation with a description of the properties of the human heart. Almost immediately an engineer interrupted, asking "Does it have to be in your chest? Could it be, say, in your thigh where it would be easier to reach?" No one in the room had ever considered that possibility. Nonetheless, the presentation continued. Soon another interruption occurred; this time it was another engineer asking, "Instead of just one heart could you have three or four small hearts integrated in a distributed system?" No one had thought of that either.

System Architecture, so insightfully presented in this book by three of the field's most highly regarded leaders, is about asking—and—answering just such questions. In my own career I have encountered system architecture questions in fields ranging from engineering to business to government. When established practices of the field of system architecture are applied, far superior outcomes seem to result.

Applying such practices has not always been the case. Early in my career I recall asking various of my colleagues who were working "together" on a guided missile program why they had chosen a particular design approach for their specific element of the product. One replied, "Because it is the lowest weight." Another assured me that his part would have the lowest radar cross-section. Still another answered because her component would be less costly. And yet another had focused on minimizing volume. And so it went.

What was missing? The answer is a *system architect*.

This shortcoming is too often encountered, usually in more subtle ways. Consider the case of the Near-Sonic Transport aircraft that was in the early stages of development a few years ago. A marketing survey had indicated that airline passengers want to get to their destinations faster. To an aerodynamicist (my own early field), if one wishes to avoid the penalties of supersonic flight, that translates into more closely approaching Mach One, creeping up on the drag curve into a regime wherein fuel consumption abruptly increases. This was, in fact, the underlying concept of the Near-Sonic Transport.

But when viewed from a system architecture perspective, the appropriate question is not how to fly faster; rather, it is how to minimize the time to get from one's home, to the airport, check-in, pass through security, board the aircraft, fly, collect baggage and travel to one's final destination. Placed in this context, an even more fundamental question arises: "How much will a passenger pay to save five or ten minutes of flying time?" The answer turns out to be, "not much"—and the Near-Sonic Transport aircraft thus met its early, and deserved, demise. There are clearly better

---

Norman R. Augustine has served in industry as chairman and CEO of Lockheed Martin Corporation, in government as Under Secretary of the Army, in academia as a member of the engineering faculty of Princeton University and as a trustee of MIT, Princeton, and Johns Hopkins and as a regent of the University System of Maryland's 12 institutions.

opportunities in which to invest if one's objective is to help passengers reach their *destinations* more rapidly. The failing in this case was to not recognize that one was dealing with a problem of *system architecture*...not simply a problem of aerodynamics and aircraft design.

My own definition of a "system" evolved over years of experience. It is "two or more elements that interact with one another." The authors of this book wisely add that the resultant functionality must exceed the sum of functionalities of the individual elements. Thus simple in concept, the complexity of most real-world systems is enormous. In fact, the equation describing the number of possible states a system of several elements (that interact in the simplest of all manners) has been aptly named, "The Monster!" And when a system includes humans, as many systems do, the challenge of system architecting becomes all the more immense due to the presence of unpredictability. But these are the kind of systems that one encounters, and are the kind of systems that the authors show how to deconstruct and address.

One such system that I had the occasion to analyze concerned provisioning the (human occupied) U.S. station at the Earth's South Pole. Setting the specific objective of the evaluation in itself required care...as is often the case. Was it to minimize expected cost? Or to minimize worst-case cost in the face of uncertainty, say, due to weather? Or perhaps to minimize "regret"— that is, when supplies are not delivered at all? Or...?

In the case of this particular system there are a number of elements that must interface with one-another: cargo ships, ice breakers, aircraft of various types, ice piers for off-loading, storage facilities, traverse vehicles, communications...and, underlying all decisions, was the ever-present danger of single-point failure modes creeping into the architecture.

In the business world one of the more complex problems faced in my career was whether— and how—all or major parts of seventeen different companies could be combined to create the Lockheed Martin Corporation. Each of the "elements" had its strengths and its weaknesses; each involved large numbers of humans, each with their own goals, capabilities, and limitations; and critical to the decision, the whole had to have significantly greater functionality than the sum of the parts. If the latter were not the case, there would be no reason to pay the financial premium that is implicit in most mergers and acquisitions.

Sadly, in engaging complex questions of this type there is no simple mathematical formula that will reveal the "right" answer. However, the discipline of systems thinking proves to be an invaluable tool in assessing exposure, opportunities, parametric sensitivities, and more. In the above case, most people judge that the answer came out "right"—which, incidentally, contrasts with nearly 80 percent of similar undertakings.

One of the authors of this book and I, along with a group of colleagues, had the occasion to propose to the President of the United States a human spaceflight plan for America for the next few decades. In this instance perhaps the most difficult challenge was to define a useful *mission*, as opposed to the (non-trivial) task of defining an appropriate hardware configuration. Fortunately, such issues are amenable to solution through system thinking.

As the authors point out in the material that follows, the process of establishing the architecture of systems is both a science and an art. But, as is so elegantly portrayed herein, there is a Darwinian phenomenon wherein systems embodying the mistakes of the past do not survive; whereas those that embody sound architectures generally do survive—and even prosper.

That, of course, is what architecting complex systems is all about.

Preface

We wrote this book to capture a powerful idea. The idea of the "architecture of a system" is growing in recognition. It appears in diverse fields including the architecture of a power grid or the architecture of a mobile payment system. It connotes the DNA of the system, and the basis for competitive advantage. There are over 100,000 professionals with the title system architect today, and many more practicing the role of the architect under different titles.

Powerful ideas often have nebulous boundaries. We observed that many of our co-workers, clients, students had a shared recognition of system architecture issues, but used the term in very different scopes. The term is often used to differentiate between existing systems, as in "the architecture of these two mountain bikes is different."

What exactly constitutes the architecture of a system is often a subject of great debate. In some fields, the term is used for a singular decision that differentiates two types of systems at a high level, as in "packet-switched architecture" vs. "circuit-switched architecture." In other fields, the term is used to describe a whole implementation, save for some smaller details, as in "our software as a service architecture."

Our goal was to capture the power of the idea of architecture, and to sharpen the boundaries. Much of the power of idea originates with the potential to trade among several architectures early, to look downstream and identify which constraints and opportunities will be central to value. It isn't possible to trade among early ideas if the architecture encompasses all details, nor is it a meaningful exercise if important drivers of value are missing.

We wrote this book to build on the idea that the architect is a specialist, not a generalist, as proposed by Eberhardt Rechtin. Our intent is to showcase the analysis and methodologies of system architecture, and to develop the 'science' of system architecture. This text is less prescriptive in places than the discipline of product design, as the systems tackled are more complex. Where the product development community has a stronger focus on design, our focus centers more on emergence—the magic of functions coming together to produce a coherent whole.

We've imbued this book with our past experience. We've been fortunate to be involved in the early development of a number of complex systems in communications, transportation, mobile advertising, finance, robotics, and medical devices, ranging in complexity from farm equipment to the International Space Station.

Additionally, we have included case studies from the experience of other system architects, in disciplines ranging from hybrid cars to commercial aircraft. Our intent was that this book can only advance system architecture if it works from challenges faced by system architects today.

We wrote this book for two core audiences—professional architects and engineering students. System architecture as an idea grew out of practitioners' wisdom and attempts to codify the challenges of developing new architecture. One core audience is senior professionals who are faced with architectural decisions. The field encompasses a variety of professionals in senior technical and managerial roles in technical industries—software, electronics, industrial goods, aerospace, automotive, and consumer goods.

This book is also focused on engineering students as a core audience. This text grew out of the graduate course we have taught at MIT for the past 15 years, where we've been fortunate to educate many leaders in the private sector and government. The lens of architecture helps us understand how a system operates today, but moreover, we believe that it is a necessary competency to learn in the management of technical organizations.

Acknowledgments

We'd like to thank the many people that made this book possible. First and foremost, our thanks to Bill Simmons, Vic Tang, Steve Imrich, Carlos Gorbea, and Peter Davison who contributed sections from their expertise, and who all provided comments on early drafts. We're indebted to Norm Augustine, who in addition to contributing the foreword, shaped our thinking on the topic.

Our reviewers Chris Magee, Warren Seering, Eun Suk Suh, Carlos Morales, Michael Yukish, and Ernst Fricke helped us deliver crisp messages and helped identify where we had missed key ideas. We also received a number of anonymous reviews, whose feedback improved the book. Dov Dori has been an invaluable partner as the developer of the OPM.

Pat Hale supported the development of the curriculum at MIT, and provided feedback on an early draft. The 63 students of the MIT System Design and Management Class of 2011 reviewed each chapter in detail and provided mountains of suggestions. In particular, our thanks to Erik Garcia, Marwan Hussein, Allen Donnelly, Greg Wilmer, Matt Strother, David Petrucci, Suzanne Livingstone, Michael Livingstone, and Kevin Somerville. Ellen Finnie Duranceau at MIT Libraries helped us choose a publisher wisely.

Our graduate students over the years have helped shape the book's content – much of their work appears here in one form or another. In addition to those mentioned above, we'd like to thank Morgan Dwyer, Marc Sanchez, Jonathan Battat, Ben Koo, Andreas Hein, and Ryan Boas.

The staff at Pearson made our book a reality—Holly Stark, Rose Kernan, Erin Ault, Scott Disanno, and Bram van Kempen. Thanks for all your hard work.

Finally, to our wives Ana, Tess, and Karen, thanks for your patience as we labored on weekends and during vacations, enduring the risk that this project become a "forever book."

Edward Crawley   Bruce Cameron   Daniel Selva
Cambridge, MA

About the Authors

Edward F. Crawley

Edward Crawley is the President of the Skolkovo Institute of Science and Technology (Skoltech) in Moscow, Russia, and a Professor of Aeronautics and Astronautics and Engineering Systems at MIT. He received an S.B. and an S.M. in Aeronautics and Astronautics and an Sc.D. in aerospace structures, all from MIT.

From 1996 to 2003, he was head of the Department of Aeronautics and Astronautics at MIT. He has served as founding co-director of an international collaboration on the reform of engineering education and was the lead author of *Rethinking Engineering Education: The CDIO Approach*. From 2003 to 2006, he was the Executive Director of the Cambridge-MIT Institute, a joint venture with Cambridge University funded by the British government and industry; the Institute's mission was to understand and generalize how universities can act effectively as engines of innovation and economic growth.

Dr. Crawley has founded a number of companies. ACX, a product development and manufacturing firm; BioScale, a company that develops biomolecular detectors; Dataxu, a company in Internet advertising placement; and Ekotrope, a company that supplies energy portfolio analysis to businesses. From 2003 to 2012, he served on the Board of Directors of Orbital Sciences Corporation (ORB).

Professor Crawley is a Fellow of the AIAA (American Institute of Aeronautics and Astronautics) and Royal Aeronautical Society (UK) and a member of the Royal Swedish Academy of Engineering Science, the Royal Academy of Engineering (UK), the Chinese Academy of Engineering, and the National Academy of Engineering (US).

Bruce G. Cameron

Bruce Cameron is the founder of Technology Strategy Partners (TSP), a consulting firm, and the Director of the System Architecture Lab at MIT. Dr. Cameron received his undergraduate degree from the University of Toronto, and graduate degrees from MIT.

As a Partner at TSP, Dr. Cameron consults on system architecture, product development, technology strategy, and investment evaluation. He has worked with more than 60 Fortune 500 firms in high tech, aerospace, transportation, and consumer goods, including BP, Dell, Nokia, Caterpillar, AMGEN, Verizon, and NASA.

Dr. Cameron teaches system architecture and technology strategy at the Sloan School of Management and in the School of Engineering at MIT. Previously at MIT, Dr. Cameron ran the MIT Commonality Study, which comprised over 30 firms spanning 8 years.

Previously, Dr. Cameron worked in high tech and banking, where he built advanced analytics for managing complex development programs. Earlier in his career, he was a system engineer at MDA Space Systems, and has built hardware currently in orbit. He is a past board member of the University of Toronto.

Daniel Selva

Daniel Selva is an Assistant Professor in Mechanical and Aerospace Engineering at Cornell. He has degrees in electrical engineering and aeronautical engineering from Polytechnic University of Catalonia (UPC), Supaero, and MIT.

Professor Selva's research focuses on applications of system architecture, knowledge engineering, and machine learning tools to early design activities. His work has been applied to the NASA Earth Science Decadal Survey, the Iridium GeoScan Program, and the NASA Tracking and Data Relay Satellite System (TDRSS), where he developed architectural analysis in support of system architects and executives. He is the recipient of Best Paper and Hottest Article awards.

Between 2004 and 2008, he worked for Arianespace in Kourou, French Guiana, as a member of the Ariane 5 Launch team, specializing in the On Board Data Handling, and Guidance, Navigation and Control. He has previously worked for Cambrian Innovation in the development of novel bioelectromechanical systems for use on orbit, and at Hewlett Packard on the monitoring of banking networks. He is a member of the Board of Advisors for NuOrion Partners, a wealth management firm.

Part 1
System Thinking

Part 1: System Thinking focuses on the opportunities presented in system architecture, namely, the opportunity to articulate the key decisions that define a system and to choose an architecture to match complex challenges.

Chapter 1: Introduction to System Architecture presents the idea of architecture with examples, identifies good architecture, and outlines the book. *Chapter 2: System Thinking* assembles the ideas necessary for system analysis. *Chapter 3: Thinking about Complex Systems* identifies the constituent modes of thinking we will use to analyze system architecture.

Chapter 1
Introduction to System Architecture

Architecture of Complex Systems

In June 1962, NASA made the decision to use a dedicated capsule to descend to the surface of the Moon from lunar orbit, rather than to descend to the surface with the Command/Service Module used to bring astronauts to lunar orbit. This decision implied that the dedicated capsule, later named the Lunar Module, would have to rendezvous in lunar orbit with their ride home and support a crew transfer between vehicles.

This decision was made in the first year of the Apollo program, seven years before the maneuver would be executed in lunar orbit. It was made before the majority of program staff was hired and before the design contracts were awarded. Yet the decision was formative; it eliminated many possible designs and gave the design teams a starting point. It guided the work of hundreds of thousands of engineers and an investment that in 1968 exceeded 4% of federal outlays.

We conceive, design, implement, and operate complex and sometimes unprecedented systems. The largest container ship today carries 18,000 containers, up from 480 containers in 1950. [1], [2] Cars built today routinely have 70 processors scattered through the vehicle, connected by as many as five separate buses running at 1 Mbit/s [3]—a far cry from early electronics buses used to communicate fuel injection at a mere 160 bit/s. Oil platforms costing $200 to 800 million [4] are developed and produced almost routinely; 39 were delivered between 2003 and 2009. [5]

These systems are not merely large and complex. They are sometimes configurable for each customer and are often very costly to deliver. Customers of consumer products expect unprecedented levels of customization and configurability. For example, BMW calculated that it offered 1.5 billion potential configurations to its customers in 2004. [6] Some complex systems are very costly to deliver. Norm Augustine points out that the unit cost of a fighter aircraft rose exponentially from 1910 through 1980, predicting that in 2053 the entire U.S. defense budget would procure exactly one aircraft. [7] Interestingly, Augustine's prediction has held up well for 30 years: In 2010 an F-22 raptor cost $160 million, or $350 million if the development costs are included. [8]

The Advantages of Good Architecture

Do these complex systems meet stakeholder needs and deliver value? Do they integrate easily, evolve flexibly, and operate simply and reliably?

Well architected systems do!

FIGURE 1.1 Complex systems: The heavy-lift ship MV *Blue Marlin* transporting the 36,000 metric ton drilling platform SSV Victoria. (Source: Dockwise/Rex Features/Associated Press)

The simplest notion of architecture we will use is that *architecture* is an abstract description of the entities of a system and the relationship between those entities. In systems built by humans, this architecture can be represented as a set of decisions.

The premise of this text is that our systems are more likely to be successful if we are careful about identifying and making the decisions that establish the architecture of a system. This text is an attempt to encode experience and analysis about early system decisions and to recognize that these choices share common themes. Over the past 30 years, analysis and computational effort have opened a broad tradespace of options, and in many areas, that tradespace grew faster than our ability to understand it. The field of system architecture grew out of practitioners' attempts to capture expert wisdom from past designs and to structure a broader understanding of potential future designs.

The market context in which our products and systems compete does not offer any comfort. Consider Boeing's decision to "bet the company" on the development of the 787 aircraft and the associated composite technology. Boeing is half of a global duopoly for large passenger aircraft, yet in its core business, rather than spreading risk across many small programs, the firm turns on a single product's emergent success or failure. The global market for mobile devices is larger and more competitive still. Although it can be argued that the product risk is more diversified (that is, an individual product development investment is a smaller fraction of firm revenues) in the mobile sector, witness the declines of former giants BlackBerry and Ericsson. To capture market share, systems must innovate on the product offering, incorporate novel technologies, and address multiple markets. To compete on tight margins, they must be designed to optimize manufacturing cost, delivered through multi-tiered supply chains. We will argue that good architectural decisions made by firms can create competitive advantage in difficult markets, but bad decisions can hobble large developments from the outset.

Every system built by humans has an architecture. Products such as mobile phone software, cars, and semiconductor capital equipment are defined by a few key decisions that are made early in each program's lifecycle. For example, early decisions in automotive development, such as the mounting of the engine, drive a host of downstream decisions. Choosing to mount an engine transversely in a car has implications for the modularization of the engine, gearbox, and drivetrain, as well as for the suspension and the passenger compartment. The architecture of a system conveys a great deal about how the product is organized.

In the design of complex systems, many of these early architectural decisions are made without full knowledge of the system's eventual scope. These early decisions have enormous impact on the eventual design. They constrain the envelope of performance, they restrict potential manufacturing sites, they make it possible or impossible for suppliers to capture after-market revenue share, and so forth. As an example of gathering downstream information for upstream consumption, the width of John Deere's crop sprayers is constrained to be less than the column separation at the manufacturing site. In this case the width constraint is obvious to the development team and was not uncertain or hidden, but it is one of the main variables in the productivity equation for a crop sprayer.

The central assertion of this text is that these early decisions can be analyzed and treated. Despite uncertainty around scope, even without knowing the detailed design of components, the architecture of the system merits scrutiny. Architecting a system is a soft process, a composite of science and art; we harbor no fantasies that this can or should be a linear process that results in an optimal solution. Rather, we wrote this text to bring together what we've learned about the core ideas and practices that compose system architecture. Our central assertion is that structured creativity is better than unstructured creativity.

This focus on decisions enables system architects to directly trade the choices for each decision, rather than the underlying designs they represent, thus encouraging broader concept evaluation. At the same time, this decision language enables system architects to order decisions according to their leverage on the system performance, in recognition that system architectures are rarely chosen in one fell swoop; rather, they are iteratively defined by a series of choices.

The failed National Polar-Orbiting Environmental Satellite System (NPOESS) is an exemplar of architectural decisions handicapping a system. NPOESS[1] was created in 1994 from the merger of two existing operational weather satellite programs, one civilian (weather prediction) and one military (weather and cloud cover imagery). The rationale for the merger was not ill-founded; these two systems collecting related data presented a $1.3 billion cost consolidation opportunity. [9] Early in the merged program, a decision was made to include the superset of instruments capability from both historical programs. For example, the VIIRS (Visible Infrared Image Radiometer Suite) instrument was expected to combine the capabilities of three historical instruments.

The assumption underlying the program was that the functional complexity of the merged program would scale linearly with the sum of the two historical programs. This might have held, had the program derived needs and concepts from the heritage instruments. However, a second decision to list new functions independent of the system concept trapped the architectural performance in an

---

[1] The prevalence of challenges with government programs cited here reflects a bias: We have more information about government programs than about private programs. Our intent is to learn from the challenges, not to comment on public vs. private.

unreachable corner of its envelope. For example, the VIIRS instrument was to accomplish the tasks of three instruments with less mass and volume than a single historical instrument.

A series of early architectural decisions placed NPOESS on a long and troubled development path, attempting to create detailed designs that ignored fundamental system tensions. Further, a failure to appoint a system architect responsible for managing these trades during the early years of the program foreshadowed challenges to come. The program was canceled in 2010, $8.5 billion *over* the original $6.5 billion estimate. [10]

This text is not a formula or a manual for product development. Success is not assured. Experience suggests that getting the architecture wrong will sink the ship but that getting it "right" merely creates a platform on which the execution of the product can either flourish or flounder.

There are many aspects of this text that are applicable to all systems, whether built by humans, evolved by society, or naturally evolved. The analysis of architecture can be applied to built or evolved systems. For example, brain researchers are trying to unfold the architecture of the brain, urban planners deal with the architecture of cities, and political and other social scientists strive to understand the architecture of government and society. But we will focus predominantly on built systems.

Learning Objectives

This is a text on how to think, not what to think. Our intent is to help the reader develop a way to think about and create system architecture, not to provide a set of procedures. Experience suggests that the best architects have a remarkably common understanding of architecture and its methods, but the content they work with and the context in which they work vary widely.

This text aims to help system architects to *structure and lead* the early, conceptual phases of the system development process, and to *support* the process throughout its development, deployment, operation, and evolution.

To these ends, this text provides guidance to help architects:

- Use system thinking in a product context and a system context
- Analyze and critique the architecture of existing systems
- Identify architectural decisions, and differentiate between architectural and non-architectural decisions
- Create the architecture of new or improved systems, and produce the deliverables of the architect
- Place the architecture in the context of value and competitive advantage for the product and the firm
- Drive the ambiguity from the upstream process by defining the context and boundaries of the system, interpreting needs, setting goals, and defining the externally delivered functions
- Create the concept for the system, consisting of internal function and form, while thinking holistically and out of the box when necessary
- Manage the evolution of system complexity and provide for future uncertainty so that goals are met and functions are delivered, while the system remains comprehensible to all during its design, implementation, operation, and evolution
- Challenge and critically evaluate current modes of architecting

- Identify the value of architecting, analyze the existing product development process of a firm, and locate the role of architecting in the product development process
- Develop the guiding principles for successful architecting

To accomplish these objectives, we present the principles, methods, and tools of system architecture. *Principles* are the underlying and long-enduring fundamentals that are always (or nearly always) valid. *Methods* are the ways of organizing approaches and tasks to achieve a concrete end; they should be solidly grounded on principles, and they are usually or often applicable. *Tools* are the contemporary ways to facilitate process; they are applicable sometimes.

One of our stated goals is for readers to develop their own principles of system architecture as they progress through the text. The architect should base decisions, methods, and tools on these principles.

> *"Principles are general rules and guidelines, intended to be enduring and seldom amended, that inform and support the way in which an organization sets about fulfilling its mission. In their turn, principles may be just one element in a structured set of ideas that collectively define and guide the organization, from values through to actions and results."*
>
> U.S. Air Force in establishing its "Headquarters Air Force Principles for Information Management," June 29, 1998

> *"Principles become modified in practice by facts."*
>
> James Fenimore Cooper, *The American Democrat*, 1838, Ch. 29

We have scattered our own principles throughout this text, but we encourage you to develop your own principles as you reflect on your own experience.

Organization of the Text

This text is organized into four parts.

Part 1: System Thinking (Chapters 1 to 3) introduces the principles of system thinking and then outlines the tools for managing complexity. These principles and tools are echoed through the remainder of the text. The notions are expressed in terms of running examples: an amplifier circuit, the circulatory system, a design team, and the solar system.

Part 2: Analysis of System Architecture (Chapters 4 to 8) is focused on the analysis of architecture. We provide an in-depth exploration of form in an effort to separate it from function, and then we deconstruct function. We introduce the ideas of solution-neutral function and concept, and we analyze the architecture of existing simple systems. Analysis can be applied to any system—both to those intentionally built by humans and to those that evolve, such as organizations, cities, or the brain. In many sections of Part 2, we begin with very simple systems. This is not intended as an insult to the reader's intelligence. Rather, we chose for analysis those systems that can be completely understood in their constituent parts, in order to hone the methods that we later scale up to complex systems. Working with simple systems eliminates the concern that the product cannot be treated as a system because it is impossible to comprehend all of its constituent parts at one time.

Part 3: Creating System Architecture (Chapters 9 to 13) is focused on the creation of architecture through decision making. It traces the forward process of identifying needs through to choosing an architecture. Whereas Part 2 works backwards from architecture to solution-neutral function, Part 3 deals directly with the ambiguity of the upstream process of goal setting, when no legacy architecture is available. Part 3 is organized around three ideas: reducing ambiguity, applying creativity, and managing complexity.

Part 4: Architecture as Decisions (Chapters 14 to 16) explores the potential of a variety of computational methods and tools to help the architect reason through decisions. Parts 1 to 3 are deliberately focused on the architect as a decision maker. We layer analysis and frameworks on top of the domain expertise of the architect, but the architect performs the integration among the layers, weighing priorities and determining salience. Part 4 explores the idea of encoding architectural decisions as parameters in a model that attempts to capture the salient pieces of many layers or attributes. We will show that there are applications for which the complexity of the architecting problem may be usefully condensed in a model, but it is important to remember that no model can replace the architect—accordingly, we emphasize decision support. In our experience, this decision representation serves as a useful mental model for the tasks of architecting.

References

[1] "Economies of Scale Made Steel," *Economist*, 2011. http://ww3.economist.com/node/21538156

[2] http://www.maersk.com/innovation/leadingthroughinnovation/pages/buildingtheworldsbiggestship.aspx

[3] "Comparison of Event-Triggered and Time-Triggered Concepts with Regard to Distributed Control Systems," A. Albert, Robert Bosch GmbH Embedded World, 2004, Nürnberg.

[4] J.E. Bailey and C.W. Sullivan. 2009–2012, *Offshore Drilling Monthly* (Houston, TX: Jefferies and Company).

[5] "US Gulf Oil Profits Lure $16 Billion More Rigs by 2015," Bloomberg, 2013. http://www.bloomberg.com/news/2013-07-16/u-s-gulf-oil-profits-lure-16-billion-more-rigs-by-2015.html)

[6] E. Fricke and A.P. Schulz, "Design for Changeability (DfC): Principles to Enable Changes in Systems throughout Their Entire Lifecycle," *Systems Engineering* 8, no. 4 (2005).

[7] Norman R. Augustine, *Augustine's Laws*, AIAA, 1997.

[8] "The Cost of Weapons—Defense Spending in a Time of Austerity," *Economist*, 2010. http://www.economist.com/node/16886851

[9] D.A. Powner, "Polar-Orbiting Environmental Satellites: Agencies Must Act Quickly to Address Risks That Jeopardize the Continuity of Weather and Climate," Washington, D.C.: United States Government Accountability Office, 2010. Report No.: GAO-10-558.

[10] D.A. Powner, "Environmental Satellites: Polar-Orbiting Satellite Acquisition Faces Delays; Decisions Needed on Whether and How to Ensure Climate Data Continuity," Washington, D.C.: United States Government Accountability Office, 2008. Report No.: GAO-08-518.

Chapter 2
System Thinking

2.1 Introduction

System thinking is, quite simply, thinking about a question, circumstance, or problem explicitly as a system—a set of interrelated entities. System thinking is *not* thinking systematically. The objective of this chapter is to provide an overview and introduction to systems and system thinking.

System thinking can be used in a number of ways: to understand the behavior or performance of an existing system; to imagine what might be if a system were to be changed; to inform decisions or judgments that are of a system nature; and to support the design and synthesis of a system, which we call system architecture.

System thinking sits alongside other modes of reasoning, such as critical reasoning (evaluating the validity of claims), analytic reasoning (conducting an analysis from a set of laws or principles), and creative thinking, among others. Well-prepared thinkers use all of these modes of thought (cognition) and recognize when they are using each one (meta-cognition).

This chapter begins by defining what a system is and exploring the property of emergence that gives systems their power (Section 2.2). Subsequently, we examine four tasks that aid us in system thinking:

1. Identify the system, its form, and its function (Section 2.3)
2. Identify the entities of the system, their form and function, and the system boundary and context (Section 2.4)
3. Identify the relationships among the entities in the system and at the boundary, as well as their form and function (Section 2.5)
4. Identify the emergent properties of the system based on the function of the entities, and their functional interactions (Section 2.6)

These tasks will be explained sequentially, but real reasoning is rarely sequential and more often iterative. As discussed in Chapter 1, *methods* are the ways of organizing such tasks to achieve a concrete end. Methods are usually or often applicable. The *principles* on which the methods of system thinking are based are also presented in this chapter.

2.2 Systems and Emergence

Systems

Because system thinking is reasoning about a question, circumstance, or problem explicitly as a system, our starting point for system thinking should be a discussion of *systems*. Few words in the modern English language are as widely applied or defined as the word "system." The definition that we use in this text is given in Box 2.1.

> ### Box 2.1 Definition: *System*
>
> A system is a set of entities and their relationships, whose functionality is greater than the sum of the individual entities.

The definition has two important parts:
1. A system is made up of entities that interact or are interrelated.
2. When the entities interact, there appears a function that is greater than, or other than, the functions of the individual entities.

At the core of all definitions of the word "system" is the first property listed here: the presence of *entities* and their *relationships*. Entities (also called parts, modules, routines, assemblies, etc.) are simply the chunks that make up the whole. The relationships can exist and be static (as in a connection) or dynamic and interactive (as in an exchange of goods).

Based on this part of the definition, what does *not* qualify as a system? If something is uniform in consistency throughout, it is not a system. For example, a brick (at a macroscopic level) is not a system, because it does not contain entities. However, a brick wall would qualify as a system, because it contains entities (many bricks and much mortar) and relationships (load exchange and geometry). Likewise, if a set of entities have no relationships (say, a person in Ukraine and a bag of rice in Asia), they do not constitute a system.

Notice how hard one must work to define things that are not systems! Someone might argue that at the right scale, a brick is a system: It is made of clay, which itself is a mixture of materials, and the materials have relationships such as sharing load and being in a geometric form (a parallelepiped). Likewise, a person in Ukraine could spend a euro to buy Asian rice, linking these entities into a trading system.

In fact, broadly construed, almost any set of entities can be interpreted as a system, and this is why the word is so commonly used. A closely related concept is the adjective "complex," which (in its original and primary sense) means having many entities and relationships. In some languages, the noun "complex" is used to mean a system, as it sometimes is in technical English (as in "Launch Complex 39A" at the Kennedy Space Center).

Two ideas that are often confused are the concepts system and product. A *product* is something that is, or has the potential to be, exchanged. Thus some products are not systems (rice) and some systems are not products (the solar system), but many of the things we build are both products (exchanged) and systems (many interrelated entities), so the two words have become mixed in common usage.

Another closely related concept is architecture, the subject of this text. In its simplest form, *architecture* can be defined as "an abstract description of the entities of a system and the relationships between those entities." [1] Clearly, the notion of a system (that exists and functions) and architecture (the description of the system) are intimately related.

TABLE 2.1 | Types of emergent functions

	Anticipated Emergence	Unanticipated Emergence
Desirable	Cars transport people Cars keep people warm/cool Cars entertain people	Cars create a sense of personal freedom in people
Undesirable	Cars burn hydrocarbons	Cars can kill people

Emergence

System thinking emphasizes the second property listed in the definition of a system: A system is a set of entities and their relationships, *whose functionality is greater than the sum of the individual entities.*

This emphasized phrase describes what is called *emergence,* and it is the power and the magic of systems. Emergence refers to what appears, materializes, or surfaces when a system operates. Obtaining the desired emergence is why we build systems. Understanding emergence is the goal—and the art—of system thinking.

What emerges when a system comes together? Most obviously and crucially, function emerges. *Function* is what a system does: its actions, outcomes, or outputs. In a designed system, we design so that the anticipated desirable primary function emerges (cars transport people). This primary function is often linked to the benefit produced by the system (we buy cars because they transport people). Anticipated but undesirable outcomes may also emerge (cars burn hydrocarbons). Sometimes, as a system comes together, unanticipated function emerges (cars provide a sense of personal freedom). This is a desirable unanticipated outcome. An undesirable unanticipated function can also emerge (cars can kill people). As suggested by Table 2.1, emergent function can be anticipated or unanticipated, and it can be desirable or undesirable. It is also clear that more than the primary desirable function can emerge from a system (cars can also keep us warm or cool, and cars can entertain people).

The essential aspect of systems is that some new functions emerge. Consider the two elements shown in Figure 2.1: sand and a funnel-shaped glass tube. Sand is a natural material and has no anticipated function. A funnel concentrates or channels a flow. However, when they are put together, a new function emerges: keeping time. How could we have ever expected that sand + funnel would produce a time-keeping device? And how did two mechanical elements, sand and shaped glass, produce an informational system that keeps track of the abstraction called "time"?

In addition to function, *performance* emerges. Performance is how well a system operates or executes its function(s). It is an attribute of the function of the system. How quickly does the car transport people? How accurately does the hourglass keep time? These are issues of performance. Take as an example the human system shown in Figure 2.2, a soccer (or football) team. The function of all soccer teams is the same: the team members must work together to score more goals than the opponent. However, some soccer teams have better performance than others — they win more games. The team portrayed in Figure 2.2 was arguably the highest-performing team in the world in 2014 — the German national team that won the 2014 World Cup.

CHAPTER 2 • SYSTEM THINKING   **11**

FIGURE 2.1 Emergent function from sand and a funnel: Time keeping. (Source: LOOK Die Bildagentur der Fotografen GmbH/Alamy)

The first principle of system architecture deals with emergence (Box 2.2). Principles are long-enduring truths that are always, or nearly always, applicable. The principles we introduce will generally begin with quotations illustrating how great systems thinkers have expressed the principle. These quotations suggest the timelessness and universality of the principle. Each principle also includes a descriptive part and a prescriptive part (which guide our actions), as well as some further discussion.

There are other attributes of operation that emerge from a system, such as reliability, maintainability, operability, safety, and robustness. These are often called the "ilities." In contrast with functional and performance emergence, which tend to create value immediately, the emergent value created by these "ilities" tends to emerge over the lifecycle of the system. How safely does a car transport people? How reliably does the hourglass keep time? How robustly does the German national soccer team win? How robustly or reliably will the software run? When a car

FIGURE 2.2 Emergent performance: The German soccer team in the 2014 World Cup. (Source: wareham.nl (sport)/Alamy)

> **Box 2.2 Principle of Emergence**
>
> *"A system is not the sum of its parts, but the product of the interactions of those parts."*
> Russell Ackoff
>
> *"The whole is more than the sum of the parts."*
> Aristotle, *Metaphysics*
>
> As the entities of a system are brought together, their interaction will cause function, behavior, performance, and other intrinsic properties to emerge. Consider and attempt to predict the anticipated and unanticipated emergent properties of the system.
>
> - The interaction of entities leads to emergence. Emergence refers to what appears, materializes, or surfaces when a system operates. It is this emergence that can give systems added value.
> - As a consequence of emergence, change propagates in unpredictable ways.
> - It is difficult to predict how a change in one entity will influence the emergent properties.
> - System success occurs when the anticipated properties emerge. System failure occurs when the anticipated emergent properties fail to appear or when unanticipated undesirable emergent properties appear.

breaks down at the side of the road, is it a mechanical "ility" problem or an embedded software "ility" problem?

The final class of emergence is so important that it merits a separate discussion: severe unanticipated and undesirable emergence. We usually call this an *emergency* (from the same word root as emergence!). Cars can lose traction and spin or roll. A soccer team could develop conflicts and lose its effectiveness on the day of an important match. Pictured in Figure 2.3 is a

FIGURE 2.3 Emergency as emergence: Hurricane Katrina. (Source: Image courtesy GOES Project Science Office/NASA)

natural example of emergence: Hurricane Katrina as it bore down on New Orleans. The devastation from this system was enormous.

These emergent properties associated with function, performance, the "ilities," and the absence of emergencies are closely related to the *value* that is created by a system. Value is *benefit at cost*. We build systems to deliver the benefit (the worth, importance, or utility as judged by a subjective observer).

In summary:

- A system is a set of entities and their relationships, whose functionality is greater than the sum of the individual entities.
- Almost anything can be considered a system, because almost everything contains entities and relationships.
- Emergence occurs when the functionality of the system is greater than the sum of the functionalities of the individual entities considered separately.
- Understanding emergence is the goal—and the art—of system thinking.
- Function, performance, and the "ilities" emerge as systems operate. These are closely linked to benefit and value, as is the absence of emergencies.

2.3 Task 1: Identify the System, Its Form, and Its Function

Form and Function

Systems simultaneously have the characteristics of *form* and *function*. Form is what the system *is*. Function is what the system *does*. To aid in developing an understanding of form and function in systems and system thinking, we will use four running examples: an amplifier, a design team, the circulatory system, and the solar system. Figures 2.4 through 2.7 show simple illustrations or schematics of these four systems. Note that the examples are chosen to include built and evolved systems, as well as informational, organizational, mechanical, and natural systems.

Each of these systems clearly has a *form*. Form is what a system *is*; it is the physical or informational embodiment that exists or has the potential to exist. Form has shape, configuration, arrangement, or layout. Over some period of time, form is static and perseverant (even

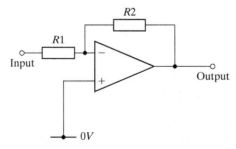

FIGURE 2.4 Amplifier circuit as a system. An operational amplifier and other electronic components that amplify signals.

FIGURE 2.5 Design team (Team X) as a system. Three people whose job it is to come up with a new device design. (Source: Edyta Pawlowska/Fotolia)

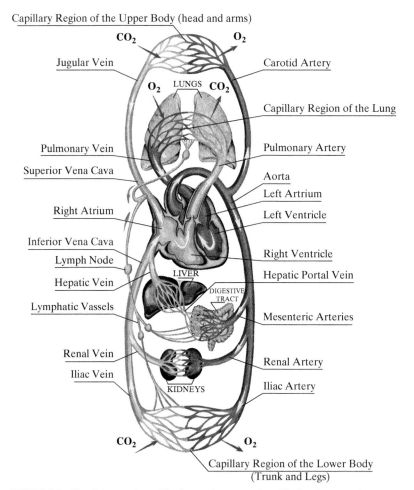

FIGURE 2.6 Circulatory system. The heart, lungs, and capillaries that supply oxygen to tissue and the organs. (Source: Stihii/Shutterstock)

FIGURE 2.7 Solar system. Our sun, and the planets and smaller bodies that orbit it. (Source: JACOPIN/BSIP/Science Source)

though form can be altered, created, or destroyed). Form is the thing that is built; the creator of the system builds, writes, paints, composes, or manufactures it. Form is not function, but form is necessary to deliver function.

Function is what a system *does*; it is the activities, operations, and transformations that cause, create, or contribute to performance. Function is the action for which a thing exists or is employed. Function is not form, but function requires an instrument of form. Emergence occurs in the functional domain. Function, performance, the "ilities," and emergencies are all issues of functionality. Function is more abstract than form, and because it is about transitions, it is more difficult to diagram than form.

Function consists of a *process* and an *operand*. The *process* is the part of function that is pure action or transformation, and thus it is the part that changes the state of the operand. The *operand* is the thing whose state is changed by that process. Function is inherently transient; it involves change in the state of the operand (creation, destruction, or alteration of some aspect of status of the operand). In organizations, function is sometimes referred to as role or responsibilities.

We are now prepared to state Task 1 of System Thinking (Box 2.3).

> Box 2.3 Methods: **Task 1 of System Thinking**
>
> Identify the system, its form, and its function.

Now we can apply this first task to our four running examples and identify their form and function, as summarized in Table 2.2.

For each of the built systems, there is an instrument of form, a process, and a value-related operand, whose change in state is the reason for the existence of the system. For the amplifier circuit, the **output signal** is the value-related operand. There may be more than one operand of

TABLE 2.2 | Form and function in simple systems

System	Form	Function	
		Process	Operand
Amplifier system	The Amplifier Circuit	Amplifies	output signal
Design team (Team X)	The Team	Develops	design
Human circulatory system	The Circulatory System	Supplies	oxygen
Solar system	The Solar System	Maintains	constant solar flux

a process (for example, the input voltage), but creating the amplified **output signal** is why we build this sort of device, and producing this amplified output signal is the primary function of the amplifier.

The design team (Team X) is a type of built system, in that someone assembled the team. The form is the collective of the individuals, and its primary function is to `develop` a **design**. In addition to its primary function, a built system may also deliver secondary functions. For example, Team X may also `present` the **design**. Primary and secondary functions are discussed in more detail in Chapter 5.

For natural systems, such as the solar system and the circulatory system, it is a bit more difficult to identify function. For the circulatory system, the form is clearly the collective of the heart, lungs, veins, arteries, and capillaries. The function could be expressed as `supply` **oxygen** to cells, as in Table 2.2 (supply is the process, and oxygen is the operand). But one could also say that the function is to `absorb` **CO_2** from cells, or, more generally, to `keep in balance` the **gas chemistry of the cells**. In systems that have evolved (rather than having been designed), identifying a crisply defined function is a bit more difficult, because there is no statement of design intent from a designer. (In principle, we could talk with the designer of the amplifier circuit and Team X and ask, "What was the function you were trying to produce in the system?" or "What functional and performance emergence did you anticipate?").

The solar system is even more of a challenge. The elements are unquestionable: the Sun, planets, and so on. But what is the function? An Earth-centric view might be that the solar system `maintains` **Earth's temperature** to make life as we know it possible. This is certainly a function, but there are dozens of other formulations of the function of the solar system that are equally valid. The function listed in Table 2.2 is a more general statement: The solar system `maintains` an **approximately constant solar energy flux to the planets**. This is truly emergence, because it requires both a constant solar output and a roughly constant planetary orbital radius. The problem is not that the solar system does not have a function, but that it has so many! And it is very difficult to question the designer about design intent.

The distinction between form and function also appears in what business calls goods and services. *Goods* are products that are tangible (what we would call form). Services are products that are less tangible and more process-oriented (what we would call function). In fact, every system can be sold as form, which delivers value by performing the function. Or the system can be sold as function (service) that implicitly requires the instrument of form in order to be executed.

Instrument – Process – Operand: Canonical Patterns in Human Thought?

In each of the four systems of Table 2.2, one can always identify the canonical characteristics of the system: the instrument of form (something that *is*) and the function (what it *does*), which in turn is composed of a process (the transformation, shown in a `special font`) and the operand (what is transformed, shown in **bold**). All systems have these three characteristics.

When Noam Chomsky developed transformational grammar, proposing the deep structure that underlies all human natural language, he found that the underlying structure of language has three parts: a noun that is the instrument of the action (what we call form), a verb that describes the action (what we call process), and a noun that is the object of the action (what we call the operand). The basic unit of all human language is the sentence, which has two nouns (the instrument and the operand) and one verb. Thus, this noun–verb–noun or instrument–process–operand model is either fundamental to all systems or fundamental to the way the human brain understands all systems. In either case, it is very useful!

In summary:

- All systems have form (what the system is) and function (what the system does). The form is the instrument of the function.
- Function further breaks down into process (the transformation) and the operand (the object that is transformed or whose state is changed).
- The primary function of most human-built systems is usually clear.
- The primary function of evolved systems is more difficult to discern and is often subject to interpretation.
- The proposed representation of a system (instrument form–process–operand) very closely resembles the deep structure of natural language (noun–verb–object).

2.4 Task 2: Identify Entities of a System, Their Form, and Their Function

We have established that a system itself can be considered as a single entity that has form and function. Now we will examine how a system decomposes into entities, each of which also has form and function. These entities are usefully represented by abstractions. The selection of the entities is guided by holistic thinking and by focus. The system is surrounded by a boundary that separates the system from its context. The following discussions examine Task 2 of System Thinking, which is stated in Box 2.4.

> **Box 2.4 Methods: Task 2 of System Thinking**
>
> Identify the entities of the system, their form and function, and the system boundary and context.

Entities with Form and Function

As a matter of definition, systems are composed of *a set of entities*. These entities are the constituents of the system.

TABLE 2.3 | System entities and their form and function

System function	Entity function	Entity Form	System Form
Amplifies signal	Sets gain	Resistor 1	Amplifier Circuit
	Sets gain	Resistor 2	
	Amplifies voltage	Operational amplifier	
Develops design	Interprets requirements	Amy	Team X
	Develops concepts	John	
	Evaluates, approves design	Sue	
Supplies oxygen to organs	Pumps blood	Heart	The Circulatory System
	Exchange gasses with atmosphere	Lungs	
	Exchange gasses with organs	Capillaries	
Zooming > < Emergence		Aggregation > < Decomposition	

In general, each of the entities of the system will also have form and function. We will sometimes use the word "element" (whose synonyms include "part" and piece") when we want to emphasize some aspect of the form of the system. The word "entity" (whose synonyms include "unit" and "thing") tends to more generically evoke both form and function.

The two columns on the right side of Table 2.3 show the decomposition of the form of our example systems into its constituent elements of form. The amplifier circuit decomposes into the two resistors and the operational amplifier; Team X decomposes into its members, and so on. Reading right to left, breaking a system into smaller pieces of form is called *decomposition*. Reading left to right, collecting pieces into the form of the system is called *aggregation*.

The middle two columns of Table 2.3 indicate the *mapping* of the form of the entities onto the function of the entities. Resistors 1 and 2 set the gain. The heart pumps blood. Each of the entities of the system also has form and function. Note that the function of each entity is written as a process and operand.

The two columns on the left side of Table 2.3 show the way in which the function of the system and the function of the entities are related. Reading left to right, breaking the function into constituents is called *zooming*. Reading right to left, when the functions of the entities combine to produce the function of the system, we find the sought-after property of *emergence*.

Function is a quasi-static view of the process acting on an operand. When a number of functions act in a sequence of operations, a more dynamic *behavior* emerges, as will be discussed in Chapter 6.

Sometimes in system thinking, it is useful to think just about function and zooming. For example, to think about the amplifier circuit in terms of amplification and gain setting. This kind of functional thinking is often used early in analysis and design. On the other hand, sometimes it is enough to reason about form and decomposition, such as when you are developing a "parts list" (amplifier, resistor 1, resistor 2). Reasoning about form or function separately is a convenience: it does not imply that both are not ultimately present or that they are not linked.

Our initial reference for "the system" is arbitrary. We could have started higher up with the human body as "the system," which could then be decomposed to find the circulatory system, the digestive system, and so on. Or we could have started lower down, defining the heart as "the system," which is composed of chambers, valves, and so on. This leads us to a generalization: *All systems are composed of entities that are also systems, and all systems are entities of larger systems.*

Because the initial choice of "the system" in these hierarchies is arbitrary, all systems must be made up of systems, which are made up of smaller systems, and so on. There are limits: the cosmos on one end (but is there really only one cosmos?) and the quarks on the other end (but is there really nothing smaller than quarks?). What matters is that we choose a system boundary that is useful, so that we train our system-thinking lens on the most important part of the problem.

In practice, definition of the entities and boundary is important and challenging. There are five issues the systems thinker faces:

- Defining the initial decomposition into entities
- Identifying the potential entities using holistic thinking
- Winnowing down to the consequential entities using focus
- Creating abstractions for the entities
- Defining the boundary of the system, and separating the system from context

The remainder of Section 2.4 addresses these five issues.

Define the Initial Decomposition into Entities

The level of difficulty encountered in defining the entities, and therefore the internal boundaries, of the system depends on whether the system is made up of *distinct* elements, is *modular*, or is *integral*. Sometimes the system is made up of clearly *distinct* entities, and the decomposition is obvious. Team X is made up of three people. Any other decomposition of the team would not make sense. Likewise, the solar system is clearly made up of the sun, the planets, and the smaller bodies (of which there are a multitude). Unambiguous decomposition into entities is a trait of systems that really are made up of discrete entities brought together and defined as a system (a fleet of ships, a herd of horses, a forest of trees, a library of books, and the like).

For systems that are fundamentally *modular*, the decomposition is more challenging but still relatively clear. Modules are relatively independent, especially in function. Internal relationships are dense within a module, and relationships between modules are weaker or less dense. For the amplifier circuit system, there are inputs, resistors, amplifiers, internal nodes, and connections. There may be some fuzziness about where the resistor ends and the connector begins, but it is largely clear.

Integral systems are the most difficult to decompose. Integral systems cannot be easily divided with their function intact. They are often highly interconnected systems, such as the components in the steering mechanisms of a car (tires, wheels, suspension, steering gears, column), some of which are simultaneously also components of other systems (ride quality, drive). Truly integral mechanical elements (such as complex forgings and machined parts) and integrated circuits are examples of integral elements. Many information systems are highly integral.

Identify the Potential Entities of the System—Holistic Thinking

Holism insists on the intimate interconnection of things—on the idea of the whole—and to think holistically is to think deliberately about the whole. Holistic thinking seeks to identify all of the entities (and other issues) that might be important to the system (see the Principle of Holism in Box 2.5). We think holistically in order to bring into view all aspects of the system at hand, taking into account the influences and consequences of anything that might interact with the system. We use holism to expand our thinking about the problem or issue at hand.

> **Box 2.5   Principle of Holism**
>
> *"Always design a thing by considering it in its next larger context—a chair in a room, a room in a house, a house in an environment, an environment in a city plan."*
>
> Eliel Saarinen
>
> *"No man is an island, entire of itself; every man is a piece of the continent, a part of the main."*
>
> John Donne
>
> Every system operates as a part of one large system or several larger systems, and each is itself composed of smaller systems. Think holistically about all of these relationships, and develop architectures that are in harmony with the larger, smaller, and peer systems.
>
> - Holism holds that all things exist and act as wholes, not just as the sum of their parts. Its sense is the opposite of that of reductionism, which suggests that things can be understood by carefully explaining their parts.
> - To think holistically is to encompass all aspects of the system at hand, taking into account the influences and consequences of anything that might interact with the system.
> - In more simple terms, to think holistically is to think about all the things (entities, relationships, and so on) that may be important to the question, circumstance, or problem at hand.
> - Methods to stimulate holistic thinking include structured and unstructured brainstorming, frameworks, thinking from different perspectives, and thinking about context.

By thinking as widely as is feasible about what might be important to the system, we increase the chances that we will move something into consideration that will ultimately be important. Holistic thinking gets issues onto the "radar screen."

There is a distinction between the *known-unknowns*, and the *unknown-unknowns*. A *known-unknown* is something that you know is there but don't know much about. Its presence is known, but its features are unknown. However, you know that you should know more about it. An *unknown-unknown* is something that you don't even know is there, so you have no way to evaluate its importance. Holistic thinking works to identify as many potential unknown-unknowns as possible so that their potential importance can be considered.

There are various methods to help stimulate holistic thinking, including: structured and unstructured brainstorming (Chapter 11); the development of frameworks to ensure that relevant issues have been considered (Chapters 4 through 8); thinking from various perspectives (Chapter 10); and thinking explicitly about context (Chapter 4).

TABLE 2.4 | The Evolution of System Thinking about Team X

Initial Thinking about Entities	After Holistic Thinking	After Focus	After Creating Abstractions	After Defining System Boundary
John develops concept	John develops concept	John develops concept	John develops concept	John develops concept
Sue evaluates, approves design	Sue evaluates, approves design	Sue evaluates, approves design	Sue evaluates, approves design	Sue evaluates, approves design
	Amy interprets requirements	Amy interprets requirements	Amy interprets requirements	Amy interprets requirements
	Heather determines customer needs	Heather determines customer needs	Marketing does market analysis	Marketing does market analysis
	Chris does competitive analysis	Chris does competitive analysis		
	Karen plans manufacturing	Karen plans manufacturing	Operation plans manufacturing and supply chain operations	Operation plans manufacturing and supply chain operations
	James plans supply chain	James plans supply chain		
	Nicole interprets regulation			
	Meagan coaches team			
	John models project finance			

As a running example in this section, we will apply the five issues facing a systems thinker to Team X. In a team with discrete individuals, identifying the entities at first looks easy. As indicated in the first column of Table 2.4, let's assume that we first looked narrowly at design as only concept generation (John) and design approval (Sue). The initial consideration was then expanded by holistic thinking into a longer list of potential team members, including requirements analysts, finance analysts, team coaches, and experts in marketing, manufacturing, and supply chains, as indicated in the second column.

The desired outcome of holistic thinking is a longer list of all the potentially important entities to consider in defining the system and its context. This list is then narrowed by focus.

Include the Important Entities of the System—Focus

The next issue that the system thinker faces is *focus*—that is, to identify what is important to the question at hand (see the Principle of Focus in Box 2.6). This means separating the wheat from the chaff. It means cutting down the list of everything generated in holistic thinking to a shorter list of things that are truly consequential.

> **Box 2.6 Principle of Focus**
>
> *"I see no more than you, but I have trained myself to notice what I see."*
>
> Sherlock Holmes in "The Adventure of the
Blanched Soldier" by Sir Arthur Conan Doyle
>
> *"The question is not what you look at but what you see."*
>
> Henry David Thoreau
>
> The number of identifiable issues that will influence a system at any point is beyond one's ability to understand. One must identify the most critical and consequential issues, and focus on them.
>
> - At any given time, there are tens or even hundreds of issues identified by holistic thinking that could impact the system under consideration. This is too many for any individual or small team to simultaneously understand.
> - To sustain close consideration of the important issues at any moment, one must be prepared to leave others behind.
> - Failure rarely occurs in aspects on which you focus.
> - Process or filter this larger set of issues to identify those that are important to that day or activity. Focus on the hard issues, and avoid the temptation to address the easier ones first.

The pivotal step in focusing is defining the question, circumstance, or problem at hand and articulating what is important about it. More specifically, what is important to you and your stakeholders? What outcomes are important? Is it the emergent behavior of the system? Is it satisfaction of some specific set of criteria?

Then you can begin to reason through the entities in the whole and ask a simple question that is very difficult to answer: Is this entity important in determining the outcome and the emergence that I am interested in? We could make a very long list of things, but the human brain can only reason about a finite number of things simultaneously, while remaining able to understand their interaction. This manageable number is conventionally thought of as seven +/− two. [2]

What we are doing by focusing is being aware of a longer list of things that are potentially important, and then "swapping in" up to seven of them at any time to really focus on. When the circumstances change, we will swap in another set of issues to reason about.

Returning to the Team X example in Table 2.4, we might reason that the main outcome of the team is a good design, the inputs are the requirements, and that the supporting entities include an understanding of the supply chain and manufacturing. Therefore, we would keep under consideration the three members of the team (Sue, John, and Amy), as well as those who determine customer needs, those who analyze the competitive environment, and the experts on manufacturing and supply chain, as shown in the third column of the table. Based on our focus analysis, the experts on finance, team dynamics, and regulation would therefore be omitted from today's system thinking.

This leads to the final part of the focus issue: performing a sanity check to make sure that the entities still under consideration are broad enough to cover the important question, circumstance, or problem, but small enough so that they can be carefully examined with the resources at hand.

Create or Recognize Abstractions for the Entities

Once you have a sense of what is important to the question, circumstance, or problem (the outcome of the holistic thinking and focus), the next issue is to define or recognize the appropriate abstractions to represent the entities in the system. An *abstraction* is defined as "expression of quality apart from the object" or as a representation "having only the intrinsic nature rather than the detail." Many problems come with predefined abstractions (people, layers, control volumes), which can either enable or disable your reasoning. Creating useful abstractions means bringing to the surface important details about the entity and hiding, within the abstractions, any details and complexity that you do not need to consider.

Let's look at some abstractions in our four running examples. In the amplifier circuit, we abstracted the operational amplifier into a device that has an inverting input, a non-inverting input, and an output and that amplifies the difference between the inputs; see Figure 2.4. The actual circuit to do this is shown in Figure 2.8. The abstraction hides all of this detail and allows us to reason about the function on the "surface"—amplification. In Team X, we abstracted a physiologically and psychologically complex person into a "team member" who could create concepts. For the circulatory system, we abstracted the complex organ called the heart into a simple pump. In the solar system, we abstracted the entire solid mass, ecosystem, and population of the planet Earth into a sphere.

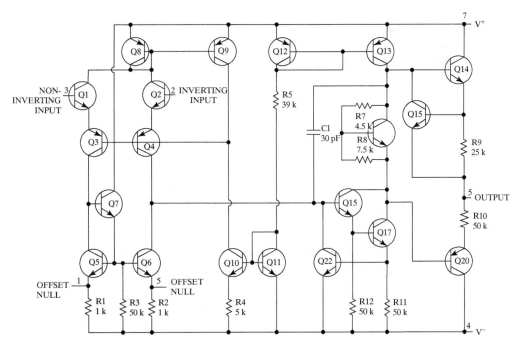

FIGURE 2.8 Hidden details in the operational amplifier abstraction.

From these examples, we can generalize to the following guidelines on creating abstractions:

- Create abstractions of form and function with the important information represented on the surface, and with less important details concealed.
- Create abstractions that allow for representation of appropriate relationships (see Section 2.5).
- Create abstractions at the right level of decomposition or aggregation.
- Create the minimum number of abstractions that will effectively represent the aspects of the system at hand.

It is possible to create less-than-useful abstractions. For example, we would violate the first guideline above if we abstracted an operational amplifier as a heat source. Doing so would be technically correct, but it would not bring to the surface the entity's important role in amplification. We could violate the third guideline by representing too much detail on the components of an operational amplifier. Again, this might be true, but so much detail may not be necessary to understand the operational amplifier's role in a circuit.

When creating abstractions, you would naturally loop back many times to the focus issue to ensure that you were creating abstractions that captured the important issues. You might even loop back to the holism issue if you were reminded of something that was missing from the holistic view.

Note that abstractions are not unique, and there may be other abstractions of the same entities that are also completely valid. Which abstraction is the right one to choose depends on the nature of the question, circumstance, or problem at hand. You usually cannot make universal abstractions.

To return to the Team X example of Table 2.4, we have maintained as abstractions the three individuals John, Sue, and Amy, but we have abstracted Heather and Chris into "Marketing" and likewise their functions into "market analysis." Similarly, we have abstracted Karen and James and their functions into "planning manufacturing and supply chain operations." Although this reduction from seven entities to five seems trivial, we will see in Section 2.5 that the relationships among the entities scale like N^2 (N-Squared). By defining this smaller set of abstractions, we have probably reduced the possible relationships from 49 to 25, a significant improvement!

The outcome is a set of abstractions that are *important to the system* but have not yet been defined to be *in the system*. In other words, we have not yet drawn the system boundary.

Define the Boundary of the System, and Separate It from Context

In defining the entities of "the system," it will often be necessary to define a *boundary* of the system. The boundary makes it clear what is in "the system" and what is outside it. All systems, perhaps short of the cosmos, have boundaries. When we examine systems, we always define them to be of limited extent, either because we are simply not able to consider a more extensive set of entities (a human capability limitation) or because we believe it is not useful to do so (a human judgment).

In defining the boundary of the system, we separate the system from its context. *Context* is what surrounds the system. It is the entities that are "just on the outside of the system" but are relevant to it.

Between the system and the context sits the *system boundary*. In drawing the system boundary, we might consider

- Including the entities to be analyzed (if the goal is understanding)
- Including what is necessary to create the design (if the goal is design)
- Including what we are responsible for implementing and operating (if the goal is delivery of value)
- Formal boundaries, established by law, contract, or other legal regime
- Traditions or conventions that distinguish the system from context
- Interface definitions or standards that we must respect, including supplier relationships

When a relationship crosses a boundary, it defines an *external interface* between the system and the context. These external interfaces are critical for the system and will be discussed in Section 2.5.

Table 2.4 represents the outcome of Task 2 of System Thinking for Team X. If we think that the job is to produce a design, then a logical place to put the system boundary is with John, Susan, and Amy within the system and to place marketing and operations outside it. The system boundary will be consistently shown in this text as a dashed line.

As we conclude the discussion of Task 2 of System Thinking (identifying the entities of the system, their form, their function, and the system boundary and context), we are left with the information shown in Figure 2.9. The entities are shown in the boxes, with the form and function described in the text. The boundary, indicated by the dashed line, separates the system from the context.

In summary:

- All systems are composed of entities, which have form and function and are themselves likely to be systems.
- Defining the composition of the system as entities is easy for a system made up of distinct entities, is moderately difficult for a modular system, and is quite difficult for integral systems.

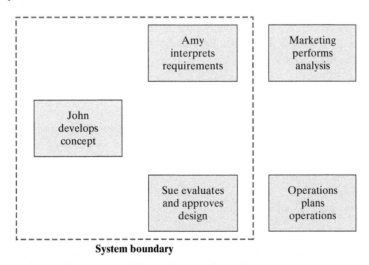

FIGURE 2.9 Team X entities and system boundary.

- Thinking holistically helps to identify all of the entities that might be important to represent in the system, but it often yields too many entities to usefully consider.
- Focus helps to reduce the entities to those that are consequential to consider at that moment, but the set may change with time and activity.
- Creating abstractions helps bring to the surface the essential details of an entity, while hiding the rest of the complexity.
- Defining a boundary separates the system from the context.

2.5 Task 3: Identify the Relationships among the Entities

Form and Function of Relationships

As a matter of definition, systems are composed of a set of entities *and their relationships.* By now, the reader should have come to expect that these relationships can have two characters: they can be functional relationships or formal relationships.

Functional relationships are relationships between entities that do something; they involve operations, transfers, or exchanges of something between the entities. We sometimes call functional relationships *interactions* to emphasize their dynamic nature. During interactions, operands are exchanged by the entities or acted on jointly. Thus a heart exchanges blood with a lung, and a team member shares results with a colleague. A more thorough presentation of functional interactions can be found in Chapter 5.

Formal relationships are relationships among the entities that exist or could exist stably for some period of time. The adjective "formal," derived from "form," is not just a way to describe a dinner party! Formal relations often include a connection or a geometric relationship. For example, when a lung is connected to the heart, or when a person joins a team, a formal relationship exists. We sometimes call formal relationships *structure* to emphasize their static nature. We discuss formal relationships in more detail in Chapter 4.

In general, a functional relationship usually requires a formal relationship. The formal relationship is the instrument of the functional relationship. The heart cannot exchange blood with the lung without a connection. Team members cannot share results without either being in proximity or having an information link.

These formal and functional relationships can be represented in either a relationship diagram or an N-Squared table. Figure 2.10 shows a relationship diagram of a system with two entities within the system and one outside in the context. The formal interactions are represented by the thin double-headed arrows, and the functional interactions are represented by broader arrows, which can be single- or double-headed, depending on the nature of the interactions. Some relationships are within the system, and some cross the system boundary, which is labeled as such and indicated by the dashed line. Best practice for such diagrams is to label the entities with their form and function, and also to label the relationships, but this often produces a very cluttered diagram.

We can now use our running examples to illustrate Task 3 of System Thinking, as expressed in Box 2.7. The relationship diagram for the amplifier circuit is shown in Figure 2.11. The structural relations indicate that the input circuit voltage is connected to Resistor 1, which is connected at its other port to Resistor 2 and the Op Amp. The output of Resistor 2 is connected to the Op Amp at a different port, and to the output circuit. This link provides a type of feedback

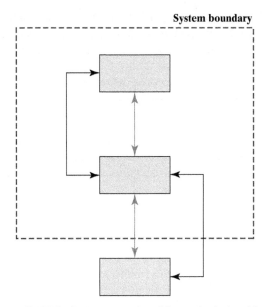

FIGURE 2.10 A system of entities and relationships.

around the Op Amp. For a circuit, these structural connections or formal relationships are electrical connections. The arrows indicate the functional interactions, which for a circuit are flows of current. In contrast, the standard electrical diagram for this circuit (Figure 2.4) would have only one relationship shown between elements, which would imply to an electrical engineer both a connection and a flow of current.

An alternative representation of relationships uses two N-Squared tables, as shown in Table 2.5. Each N-Squared table lists the N entities on each side. The first shows the formal

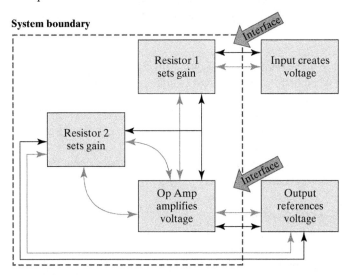

FIGURE 2.11 Formal structure and functional interaction for the amplifier circuit.

> **Box 2.7 Methods: Task 3 of System Thinking**
>
> Identify the relationships among the entities in the system and at the boundary, as well as their form and function.

relationships, the second the functional relationships. The entities within the system are above and to the left of the dashed lines. Off diagonal terms indicate an internal connection or interface. There is value in both the relationship diagram, which is easier to visualize, and the N-Squared table, which has more detail and does not become visually congested as nodes and connections proliferate.

Formal relationships tend to be more concrete, and they are a good starting point for thinking about relationships in general. As you examine each formal relationship, however, try to understand the functional relationship that it enables. The formal relationships are important primarily because they are instruments of the functional relationships. Because emergence occurs in the functional domain, it is really the functional interactions that are of primary importance, as will be discussed further in Section 2.6.

External Interfaces

Formal and functional relationships can exist across the boundary of the system, between entities in the system and its context. These are called the *external interfaces* of the system. In the N-Squared table, Table 2.5, these external interfaces are indicated as any relationship outside the part of the table reserved for the system (such as "connected at input"). In the relationship diagram, Figure 2.11, these external interfaces show up as arrows crossing the system boundary and are marked. Similar external interfaces can be found between Team X members and supporters, and between the circulatory system and the air. In fact, it is nearly impossible to define a system that is not connected somehow to entities beyond its boundary by external interfaces.

In summary:

- Fundamental to the definition of a system is the existence of relationships among the entities, which can be formal (they exist or could exist—structure) or functional (they do something—interactions).
- In general, some of the entities of a system will also have both formal and functional relationships with context entities outside the system; such relationships occur across external interfaces.
- Formal and functional interactions can be effectively represented by either relationship diagrams or N-Squared tables.

2.6 Task 4 Emergence

The Importance of Emergence

Emergence is the magic of a system. As the entities of a system are brought together, a new function emerges as a result of combination of the function of the entities and the functional interactions among the entities. Recall that a system is *a set of entities and their relationships, whose*

TABLE 2.5 | The N-Squared tables that represent the formal and functional relationships of the amplifier circuit

Formal Relationships	Resistor 1	Resistor 2	Operational amplifier	Input	Output
Resistor 1		Connected at V−	Connected at V−	Connected at input	
Resistor 2	Connected at V−		Connected at V− and connected at output		Connected at output
Operational amplifier	Connected at V−	Connected at V− and connected at output			Connected at output
Input	Connected at input				
Output		Connected at output	Connected at output		

Functional Relationships	Resistor 1	Resistor 2	Operational amplifier	Input	Output
Resistor 1		Exchanges current at V−	Exchanges current at V−	Exchanges current at input	
Resistor 2	Exchanges current at V−		Exchanges current at V− and also at output		Exchanges current at output
Operational amplifier	Exchanges current at V−	Exchanges current at V− and also at output			Exchanges current at output
Input	Exchanges current at input				
Output		Exchanges current at output	Exchanges current at output		

functionality is greater than the sum of the individual entities. This second phrase focuses purely on emergence. "Greater functionality" is delivered through emergence.

Nothing "emerges" in the domain of the form of the system. When the form of chunk A and chunk B are brought together, nothing more happens in the form domain than the sum A plus B. Properties of the aggregation of the form are relatively easy to calculate. The mass of the system is just the mass of chunk A and chunk B. Form is "linear."

However, in the functional domain, the sum of A and B is much more interesting, is much more complicated, and has nothing to do with linearity. When the function of chunk A interacts with the function of chunk B, almost anything can happen! It is exactly this property of emergence that gives systems their power.

Striving to understand and predict emergence, and the power it brings to systems, is the primary goal of system thinking.

System Failure

Another way to understand the importance of emergence is to think what might happen if the anticipated emergence does not occur. This can happen in two ways: The anticipated desirable emergence can fail to occur, or undesirable unanticipated emergence can occur; see Table 2.1. Both are bad situations.

Consider some potential system failures in our examples. In the human circulatory system, a connection from the heart may partially clog, raising blood pressure in other parts of the system; thus an unanticipated undesirable emergence occurs, impacting the entire system. In Team X, a team member may develop good requirements but not communicate them effectively, and the team consequently makes a design that causes the company to lose a sale; thus an anticipated desirable emergence fails to occur.

Gridlock on a four-lane urban expressway is a classic example of system failure. Every car is doing exactly what its function calls for, transporting people. The roadway is performing its function, supporting car travel. The drivers are staying safely behind each other and within their lanes. Yet the anticipated desirable performance of cars traveling rapidly down the road fails to materialize.

A well-documented case of unanticipated undesirable emergence is shown in Figure 2.12. An Airbus A320 tried to land in a crosswind with the upwind wing held low. On the slippery runway, the wheel brakes were not effective, so after touch-down the pilot tried to apply the engine-based thrust reverser, but it did not operate. What happened? As a safety measure, the software system was designed not to deploy the thrust reverser until the plane had "landed,"

FIGURE 2.12 Failures are often system emergence: A320 crash in Warsaw, Poland. (Source: STR News/Reuters)

which was signaled by weight compressing *both* landing gear assemblies. But because one wing was held low, the landing gear on one side was not compressed. Everything worked exactly as planned that day, but the result was a system failure—an emergency.

Trying to understand and anticipate such system failures is also a goal of system thinking.

Predicting Emergence

As suggested by Box 2.8, the last task in system thinking is to predict emergence. It is hard to predict *a priori* what will emerge from the combinations of the functions of the various entities of a system. The anticipated desirable function may emerge (system success) or may fail to emerge, or something unanticipated and undesirable may emerge (system failure).

> **Box 2.8 Methods: Task 4 of System Thinking**
>
> Based on the function of the entities, and their functional interactions, identify the emergent properties of the system.

We have seen examples of anticipated function emerging. In the amplifier circuit, amplifying voltage plus setting gain created amplification. In Team X, the members working effectively and communicating created a good design. But if the systems thinkers did not already know about this emergence, how would they predict it?

There are three ways to predict emergence. One is to have done it in the past. This is prediction based on *precedent*. We look for identical or very similar solutions in our experience and implement them with at most small changes. We build the pendulum mechanism in grandfather clocks because our grandfathers built it that way. When Team X was put together, there was experience that suggested that this group of people would form an effective team.

Another way to predict emergence is to do *experiments*. We simply try putting together the entities with the proposed relations to see what emerges. This can range from tinkering to very highly structured prototyping. You could explore the output of the Op Amp by building one, applying an input voltage, and monitoring the output. Spiral development is a form of experiment in which some of the system is first built to check emergence before the rest of the system is built (in later spirals).

The third way to predict emergence is *modeling*. If the function of the entities and the functional interaction can be modeled, then it may be possible to predict emergence from a model. One example of spectacular success in modeling occurred in the development of integrated circuits. IC's with billions of gates are now routinely manufactured and produce the correct emergent properties. How? The fundamental element is a transistor, which can be modeled simply, and from there up it is all mathematics. Our amplifier circuit can be modeled with a few lines of algebra if one knows the constitutive relations for the Op Amp and resistors, and Kirchhoff's voltage and current laws.

What do you do if you need to predict emergence for systems that are without precedent, cannot be experimented on, and cannot be reliably modeled? Welcome to the question at the crux of system thinking! Such issues arise routinely in many domains, including new product development. In these situations, we are left to reason about what will emerge. This reasoning may be informed partially by precedent (observing results in similar but not identical systems) and partially by experiments and incomplete modeling, but the projection about emergence ultimately depends on human judgment.

Emergence Depends on Entities and Relationships

Notice that the emergence of function from a system depends on the function of the entities and their functional interactions. The form enables the function of the entities, and the formal relationships are instrumental in functional interactions. This implies that both form and formal relationships (structure) are important to consider in predicting emergence. To see this more clearly, let's look at the simple systems shown in Figure 2.13:

- An electrical low pass filter, consisting of a single resistor and a capacitor. Place the resistor between the input voltage and output, and the capacitor between the output and ground, and high frequencies are attenuated. Change the *pattern of connectivity* by switching the two components, and low frequencies are attenuated.
- A mechanical lever, consisting of a bar and a fulcrum. Put the fulcrum near the end farther from the operator, and you get the desired emergence of "magnify force." Change the *location,* putting the fulcrum closer to the operator, and this desired emergence disappears.
- A simple software segment, with a conditional statement and a simple computation. Put the IF statement first, and the "a = 100" statement is executed only if the condition is true. Change the *sequence,* placing the a = 100 statement first, and it is always executed.

The formal relationships are critical to the emergence: the *pattern of connection* of the electrical components, the *location* of the fulcrum, and the *sequence* of the software instructions. The formal relationships are important in guiding a certain specific functional interaction that leads to a specific system-level emergence.

In summary:

- Emergence occurs when the function of the entities and their functional interaction combine to produce a new functionality, which is more than the "sum of the parts."
- System success and system failure often hinge on emergence.

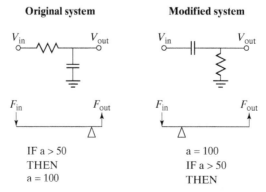

FIGURE 2.13 Emergence depends on structure.

- Emergence can be predicted *a priori* by relying on precedent, experimentation, and modeling. For unprecedented systems for which experimentation and modeling are not easy, humans must reason about emergence based on available information.
- Emergence depends on the function of the entities enabled by form, and on the functional relationships enabled by the formal relationships.
- It is the property of emergence that gives systems their power and also creates the challenges in understanding and predicting them.

2.7 Summary

Because system thinking is simply thinking about a question, circumstance, or problem explicitly as a system, it makes sense that there is a strong parallel between the essential features of a system and the tasks of system thinking. The essential features of a system are listed in the left column of Table 2.6, and the tasks that constitute system thinking, which were introduced in Section 2.1 and developed in Sections 2.2 through 2.6, are shown in the right column of Table 2.6.

At the highest level, the goal of system thinking is to enable us to reason about systems. We are surrounded by systems, and they are becoming increasingly complex. System thinking should help us to *make these complex systems appear less complicated.*

There are a variety of outcomes that can be achieved with system thinking. An entry-level goal of system thinking is *understanding what is,* or examining a system and trying to make sense of it. A more advanced goal of system thinking is *predicting what might be if something changes.* This assertion should be greeted with caution, because predicting what will happen

TABLE 2.6 | The essential features of systems and the tasks of system thinking

Essential Features of Systems	Tasks of System Thinking
Systems have form and function, and form is the instrument of function.	Identify the system, its form, and its function.
Systems are composed of entities, each of which also has form and function. These entities are in general also systems, and the system itself may be an entity of larger systems.	Identify the entities of the system, their form and function, and the system boundary and context.
The entities of the system are linked through relationships that have a formal and functional character. Some of the entities are linked through relationships with entities outside the system, which are said to be in the context of the system.	Identify the relationships among the entities in the system and at the boundary, their form, and their function.
Function and other characteristics of the system emerge as the functions of the entities interact, guided by the form of the relationships among the entities. It is emergence that gives systems their power: that they possess functionality greater than the sum of the functionalities of their parts.	Based on the function of the entities, and on their functional interactions, identify the emergent properties of the system.

if something changes requires the ability to predict emergence—a task that is fraught with difficulty!

Working upward, the next role of system thinking is in informing *judgment and balance in decision making.* Decision making requires the identification, analysis, and weighing of options. A systems thinker might ask: What tensions are apparent in the system? What alternatives or solutions balance the various factors and resolve tensions? How will the solution react to future change? At the heart of informed decision making is ensuring that all the important implications of the decision are identified and considered, which is an application of system thinking.

At the pinnacle of system thinking is *"synthesizing a system,"* which is the domain of system architecting and the subject of Part 3 of this text.

References

[1] Edward Crawley et al., "The Influence of Architecture in Engineering Systems," *Engineering Systems Monograph* (2004).

[2] George Miller, "The Magical Number Seven, Plus or Minus Two: Some Limits on our Capacity for Information Processing," *Psychological Review* 63, no. 2 (1956): 81–97.

Chapter 3
Thinking about Complex Systems

3.1 Introduction

In Chapter 2, we explored system thinking and how to think about things as systems. A number of key ideas were introduced in Chapter 2, including the power of systems, form and function, entities and relationships, abstractions and emergence, and boundaries and context.

The approaches described in Chapter 2 were deliberately demonstrated on simple abstractions of systems with two or three important entities and relationships. This was done to focus the reader's attention on the system issues rather than on the high level of complexity of systems we normally encounter. However, most of the systems that we deal with professionally are complex. These include the artifacts we design, the systems used to build and operate them, the stakeholder systems involved, and the organizations in which we work. Hence, in this chapter we will extend the approaches of system thinking to more complex systems. In other words, we will start developing the methods of system architecting.

Complexity is inherent in system architecture, and before proceeding with a rigorous development of the analysis of architecture (Part 2) and the synthesis of architecture (Part 3), we should spend a short time dealing with the issues of complexity in systems and the approaches that have been developed to enable us to comprehend complex systems.

This chapter starts with a brief discussion of what makes systems complex. We then summarize the approaches that enable people to better understand and work with complex systems. These include decomposition and hierarchy, different types of relationships among entities (such as class/instance relationships and specialization/generalization), and specific tools to reason through complex systems, such as zigzagging, views, and projections. The chapter ends with an introduction to SysML and OPM: two tools that are often used to represent complex systems.

3.2 Complexity in Systems

Complexity

A system is *complex* if it has many interrelated, interconnected, or interwoven entities and relationships (Box 3.1). The fuzziness of the word "many" in this definition is deliberate for now.

Complexity is driven into systems by "asking more" of them: more function, more performance, more robustness, and more flexibility. It is also driven into systems by asking systems to

> **Box 3.1 Definition: *Complex System***
>
> A complex system has many elements or entities that are highly interrelated, interconnected, or interwoven.

work together and interconnect—your car with the traffic control system, your house with the Internet, and so on.

Complex systems require a great deal of information to specify and describe. Therefore, some measures of complexity are based on the information content of the description of the system. Other measures of complexity take the approach of trying to categorize what the system does; this is a function-based approach to complexity.

An idea closely related to complex is *complicated*. The definition of complicated takes into consideration the finite ability of the human to perceive and understand complexity. The systems of Chapter 2 are not complicated. We can understand them reasonably well after a few minutes of study. Complicated things have high *apparent complexity*. They are complicated because they stretch or overwhelm our ability to understand them.

Dealing with complexity is not new. In fact, by the time of classical Rome, complex systems were routinely built (such as large water delivery networks). However, over the last century, systems and the context in which they operate have become more complex (compare the telephone of 1900 and 2000). This increasing complexity has strained our ability to comprehend systems.

Our job as architects is to train our minds to understand complex systems so that they do not appear to be complicated to us. We should seek to build architectures that do not appear complicated to all the others who have to work on the system—designers, builders, operators, and so on.

The task of good architecture design can be summarized as follows: *Build systems of the necessary level of complexity that are not complicated!*

Introducing Team XT

In this chapter, we will analyze the case of Team XT, which is an extended (XT) version of the Team X system introduced in Chapter 2. The team's role is still to produce a design. All of the entities and relations of Team X are still valid, but more detail is considered. We have chosen to continue with the example of a more complex organization because many readers can easily relate to the example, and because the lessons of system thinking are readily applicable to organizations.

Imagine that we walk into a room at a company and are introduced to Team XT. We begin to apply the methods of Chapter 2—first by asking, "What is the system?" In this case, the system is still the group of people and process responsible for developing the design. Next, what is the form? In this case, the form is the sum of the people. The function is what the system does, its activity or transformation. Here it is to develop a design. We have completed Task 1 of System Thinking (Box 2.3).

If the entities of form of Team XT are the people, then determining the function for each member of the team requires careful observation, extensive experience, or interviews with the individuals. The resulting function of each of the members is listed in the third column of Table 3.1.

TABLE 3.1 | The full Team XT, an extended team whose role is to develop an engineering design

Team XT	Form	Function	Function interaction	From/To	Location	Connectivity	Role
	Sue	Evaluates and approves design	Gets finalized concept/requirements	John, Amy	Cambridge	Share design tool	Team XT manager
	Amy	Finalizes requirements	Gets draft requirements document	Jose, Vladimir	Cambridge	Share requirement tool	Requirements group leader
	Jose	Develops requirement documents	Gets strategy and regulatory input	Mats, Ivan	Cambridge	Share requirement tool	Group member
	Vladimir	Develops requirement documents	Gets needs and competitive analysis	Heather	Cambridge	Share requirement tool	Group member
	Ivan	Interprets regulations	N/A		Cambridge	Share requirement tool	Group member
	Mats	Interprets corporate strategy	N/A		Cambridge	Share requirement tool	Group member
	John	Develops finalized concepts	Gets evaluated options	Mark	Moscow	Share design tool	Concept group leader
	Natasha	Develops conceptual options	Gets finalized requirements	Amy	Moscow	Share requirement tool	Group member
	Analyst 1	Analyzes designs	Gets conceptual options	Natasha	Moscow	Share design tool	Group member
	Analyst 2	Analyzes designs	Gets conceptual options	Natasha	Moscow	Share design tool	Group member
	Mark	Evaluates design options	Gets analysis	Analyst 1,2	Moscow	Share design tool	Group member
	Phil	Leads supporters	N/A		Cambridge		Support group leader
	Nicole	Provides IT support	Supports design tool	Design tool users	Cambridge		Group member
	Meagan	Coaches team	Supports group leaders	John, Amy	Cambridge		Group member
	Alex	Models project finances	Supports design analysts	Analyst 2	Cambridge	Share design tool	Group member
	Dimitri	Provides administrative support	Supports manager	Sue	Cambridge		Group member
System Boundary					Cambridge		
Marketing	Heather	Determines needs	Gets competitive analysis	Chris	Cambridge		
	Chris	Does competitive analysis	N/A		Cambridge		
Operations					Cambridge		
	Karen	Plans manufacturing	Gets final design	Sue	Cambridge		
	James	Plans supply chain	Gets final design	Sue	Cambridge		

38 PART 1 • SYSTEM THINKING

Each of these functions has an operand and a process. As discussed in Chapter 2, we think holistically (Box 2.5) to identify all of the entities that might be important, and then we draw the boundary (Box 2.6) to focus on the designers and design processes. This makes the Marketing and Operations functions part of the context external to Team XT, as indicated by the dashed line in Table 3.1. Now we have completed Task 2 of System Thinking (Box 2.4).

Task 3 of System Thinking is to identify the relationships among the entities in the system and at the boundaries. In Table 3.1 we have identified the principal functional interactions among the Team XT participants, both within the team itself and across the boundaries to Marketing and to Operations. The fourth column shows the interaction, and the fifth column shows the person at the other end of the interaction. These are also shown in Figure 3.1.

The formal structure is shown in the last three columns of Table 3.1. Geographically, some of the team members are located in Cambridge, while the rest are in Moscow. There are several software tools and associated databases that help connect the team members. These are

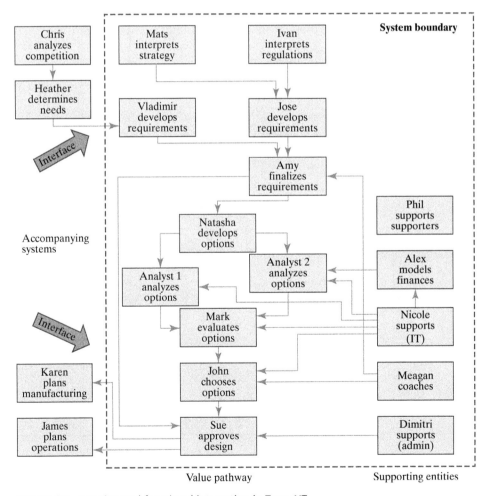

FIGURE 3.1 Function and functional interaction in Team XT.

connection relationships. Finally, there is a reporting structure; these are human relationships that do not by themselves imply any interaction. The various kinds of formal relationships are discussed in more detail in Chapter 4.

Task 4 of System Thinking, identifying the emergent properties, will become more evident as we develop and apply the tools for thinking about more complex systems in the sections that follow.

3.3 Decomposition of Systems

Decomposition

Decomposition is the dividing of an entity into smaller pieces or constituents. It is one of the most powerful tools in our toolset for dealing with complexity. "Divide and conquer" is a fundamental strategy. Break a problem down into smaller problems until each is tractable. It is no coincidence that Julius Caesar's account of the Gallic Wars opens with the declaration "All Gaul is divided into three parts." By the time of classical Rome, this approach was understood and widely applied.

As discussed in Section 2.4, sometimes defining decomposition is easy—for example, when the system is made up of distinct elements. Sometimes the system is modular, which suggests decomposition, and sometimes it is integral, where the decomposition can be somewhat arbitrary.

The difficulty with breaking things apart is not in the breaking apart, but in the process of bringing together the decomposed entities to build the whole. This process is often called integration. In the formal domain, we say that form aggregates, and we have to worry about the physical/logical fit between the elements as they are brought together. In the functional domain, we decompose functions into more basic functions, and then, in recombining the entities of function, we encounter emergence: the real challenge.

Let's begin the analysis of Team XT. Above, we jumped to the conclusion that the right way to decompose the system was based on the distinct elements of form—the members of the team. If the form is not so distinct, or is not yet defined, we are more likely to consider decomposing based on entities of function. This might produce entities tagged with functions (steering mechanism, compressor, sorting routine). Examining Table 3.1, we see that there are several places where focusing on function would yield a different decomposition. For example, the function "Develops requirement documents" is common to both Jose and Vladimir, so an alternative decomposition might have included Jose and Vladimir in one entity. Such a function-based decomposition might have better supported the understanding of emergence.

Hierarchy

Hierarchy is another powerful approach that can be used to understand and reason about complexity in systems. *Hierarchy* is defined as a system in which entities belong to layers or grades, and the layers are ranked one above the other. Hierarchy is very commonly found in social systems. For example, in the military there are grades: generals, colonels, majors, captains, and so on. In large companies there are presidents, executive vice presidents, senior vice presidents, and so on.

What causes some grades to be higher in the hierarchy than others? Generally:

- They have more scope: Governors rank above mayors because they have more scope (a state is larger than a city).

Julius Caesar, *The Gallic Wars,* Translated by W. A. McDevitte and W. S. Bohn.

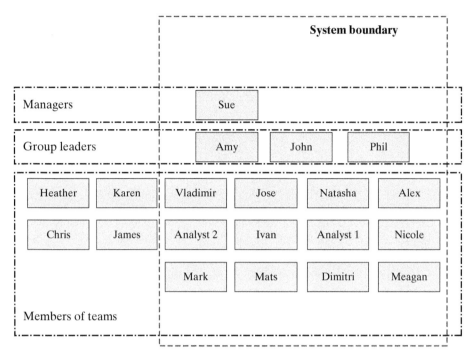

FIGURE 3.2 Hierarchy in the Team XT members.

- They have more importance or performance: Black belts rank above brown belts because they perform to a higher standard and have greater skills.
- Their functions entail more responsibility: Presidents rank above vice presidents because they have more responsibility.

Hierarchy is not always obvious. Even close examination of Figure 3.1 and Table 3.1 does not reveal the actual hierarchy in Team XT. This is because the group leaders and managers actually work in the delivery of value, rather than simply reviewing the work of others. We need to examine Figure 3.2, which explicitly shows hierarchy, to see that Sue is ranked above Amy, John, and Phil, who themselves are ranked above all others on the extended team. We extracted this information from the last column of structural information in Table 3.1. The hierarchic view presents quite a different impression than the egalitarian view of Figure 3.1; now it is obvious that some team members are somehow more important than others. Note that hierarchy does not imply that anyone on a particular level reports to any specific person on the next level. That occurs only in hierarchic decomposition.

Hierarchic Decomposition

Often, decomposition and hierarchy are combined into a multilevel or hierarchic decomposition: a decomposition with more than two levels, as shown in Figure 3.3. Figure 3.3 is cognitively much more satisfying than the undifferentiated second level of Figure 3.2. It appeals to our desire to group things in sets of seven +/− two. [1] Figure 3.3 shows three group leaders reporting to Sue and about four people reporting to each of the group leaders.

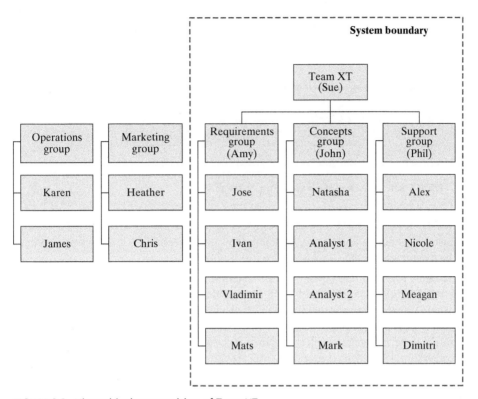

FIGURE 3.3 Hierarchic decomposition of Team XT.

The "group" is a useful abstraction suggested by Table 3.1 and Figure 3.2, but it is not a unique way to decompose the team. We could have clustered the team members in several other ways at the level below Sue, including by location, connectivity, or functional relationships. In fact, Figure 3.2 does not imply anything about who is close to or connected to whom (formal relationships) or who exchanges information with whom (functional relationships).

Simple Systems, Medium-Complexity Systems, and Complex Systems

We will adopt a scheme of classifying systems as simple, of medium complexity, or complex complexity, or complex. We call a system that can be completely described by a one-level decomposition diagram of the type shown in Figure 3.4 a *simple system*. All four of the systems we examined in Chapter 2 were represented as simple system: even the solar system had only one level of decomposition, the planets and smaller bodies. In a simple system there are generally no more than seven +/– two elements of form at Level 1. When you get to these elements, they are more or less *atomic parts* (see the discussion below).

If there are no more than (seven +/– two)^2 parts—that is, no more than about 81 entities—a system can be represented by a two-level decomposition diagram of the type shown in Figure 3.3. We call these *medium-complexity systems*.

A *complex system* has the same representational diagram as a medium-complexity system (Figure 3.3), but at the second level (down) there are still only abstractions of things below.

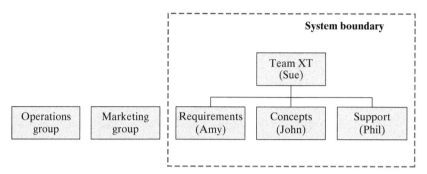

FIGURE 3.4 One-level decomposition of Team XT.

These are significantly harder to analyze, and more common. For example, if Natasha were really just the leader of another organization that is responsible for developing concepts, this would be a complex system.

One rarely sees a single decomposition diagram more than two levels deep. There are two reasons why such drawings are not used. First, a three-level diagram could have as many as (seven +/− two)^3 elements at the third level, or as many as 729, which is far more than humans can routinely process. Second, when we look down into an organization or system, we find that we know the Level 1 elements well (the "direct reports" in the organization), and we more or less comprehend things on the next level (the direct reports of your direct reports), but beyond that it is more or less a haze.

We will simply refer to Level 0, Level 1 (down), and Level 2 (down) when discussing the decomposition of systems. As we learned in Chapter 2, virtually everything is a system, so the designation of Level 0 as "the system" is somewhat arbitrary and depends on the viewpoint of the architect. There are dozens of words to designate the layers below "the system"; they may be referred to as modules, assemblies, sub-assemblies, functions, racks, online replacement units (ORUs), routines, committees, task forces, units, components, sub-components, parts, segments, sections, chapters, and so on. Unfortunately, there is little consensus about how these terms are applied. One person's assembly is another person's component. There are relatively fewer words for the levels above "the system"; people speak of systems of systems, complexes, and collections. If needed, we will call these Level 1 (up) and Level 2 (up).

Atomic Parts

Our discussion hinges somewhat on how we define the term "atomic part," which does not have an exact definition. Its meaning derives from the Greek word ἄτομος (*atomos*), which means indivisible. In mechanical systems, we will stipulate that the atomic parts are those that cannot easily be "taken apart." A person, a screw, and a processor chip all pass this test. Of course, a processor chip has many internal details that might be architecturally important. How many of what type of transistor or gate are included? How are they connected? Even a simple screw has architecturally important details. Is it a straight head or Phillips head? Yes, this definition is a bit fuzzy, but here is a simple rule: "Call it a part when you can't take it apart."

In information systems, the definition of the term "atomic part" is even fuzzier. A useful test is to call something an atomic part when it possesses a semantic meaning (as does a word or an instruction) or when it is a unit of data or information. These words, instructions, or data units would include details, of course. Because all information is an abstraction (as we will discuss in Chapter 4), defining meaningful abstractions of abstractions is necessarily more fuzzy, but here is another simple rule: "Call it a part if it loses meaning when you take it apart."

3.4 Special Logical Relationships

The Class/Instance Relationship

Another tool that helps us understand and manage complexity is the class/instance relationship. The term "class/instance relationship" is more commonly used in software than in hardware. A *class* is a construct that describes the general features of something, and an *instance* is a specific occurrence of the class. The instance is often referred to as the instantiation of the class.

The idea of class/instance is quite common. For example, if we referred to a model of a car (say, a Ford Explorer), that is the class, and an instance would be a specific Vehicle Identification Number (VIN). A part number refers to a class, and a serial number refers to an instance. In many computer languages, the class/instance relationship is explicitly supported. For example, when programming the Web interface of an online service application, we may use different instances of the class "Button" to represent different buttons on the interface that do different things (such as refresh the screen or submit information to the server) when we click on them.

We have already subtly made use of the class/instance relationship. A careful examination of Table 3.1 reveals that there are two analysts. We could have created a class called "Analyst," listed it on the team roster, and then identified two instances of that class. The class/instance relationship is a useful way to manage complexity when an entity recurs many times and when classes can easily be instantiated. The 11 players on a soccer team and the 88 keys of a piano are instances of the classes "Player" and "Key," respectively.

The Specialization Relationship

A related idea is the specialization/generalization relationship. This relationship describes the connection between a general object and a set of more specific objects.

As we will see in Part 2, the specialization operation is used extensively in design. In shopping for a house, choosing which house to buy for a given set of occupants, budget, and setting is a process of specialization. The general object is "House," and specific objects could be different house styles, such as a colonial house, a Victorian house, and a modern house. In Team XT, we created three specializations of the generalization "Group": a requirements development group, a design group, and a support group.

Specialization is akin to the concept of inheritance in object-oriented programming, in which a class can be created starting from a more general class, from which we inherit some attributes and functionality and to which we can add some new attributes and functionality.

Recursion

Recursion occurs when the process or object uses itself within the whole. Said another way, it is the use of entities or relationships in a self-similar way. If a routine or function uses itself, or an approach used at one level of a system is used again at other levels within the system, this is recursion.

In the example of Team XT, we used a mild form of recursion. The entire sum of the members is a team. Then, within the team, there are three groups, each of which is a mini-team, as shown in Figure 3.4.

Recurrence is commonly and very explicitly used in software engineering. For example, in creating a network with nodes and edges, one possible implementation is to have a class "Node" that has an attribute "Neighbors" that is an array of Nodes. Thus, the class "Node" is defined recursively, in terms of itself.

3.5 Reasoning through Complex Systems

When they are presented with a complex system, in either analysis or synthesis, system thinkers use several techniques to reason their way through the system.

Top-down/Bottom-up Reasoning

Top-down/bottom-up refers to the direction in which you approach a system. Most of the development in this text is based on the *top-down* approach. You start from the goals of a system and proceed to concept and the high-level architecture. Then you develop the architecture in increasing detail until you reach the smallest entities of interest to you. This method follows the "left-hand side" of the systems engineering V model. [2]

An alternative is the *bottom-up* approach. Here, you think about the artifacts, capabilities, or services that are available in the lowest-level entities and build upward from them, predicting emergence. A variant is to start at both top and bottom, and work toward the middle, which might be called an *outer-in* approach.

In the case of Team XT, we primarily used a top-down approach, defining the system and its function, and then decomposing and identifying internal processes. Of course, the CEO of Team XT's company could have practiced bottom-up thinking when the team was formed by putting the 16 most appropriate employees in a room and inviting them to self-organize to develop a design most effectively.

Because there is really no top or bottom to truly complex systems, in reality we always apply a *middle-out* approach; we start at some arbitrary point in the system hierarchy and try to reason one or two levels up or down. Good architects should be able to practice all of these approaches.

Zigzagging

Zigzagging is a term coined by Nam Suh in his work on axiomatic design. [3] Suh observed that when you reason about a system, you alternate between reasoning in the form domain and reasoning in the function domain. You tend to start in one domain, work as long as is practical, and then switch to the other.

When we first started reasoning about Team XT, we identified the function as "design developing." Then we switched to the form domain and identified the concept of form as small groups. We decomposed the team into small groups, and then we switched back to the function side to identify what function each of them performed, and how developing a design emerged. This alternating pattern of thinking about form and then about function would continue through the levels of the system.

3.6 Architecture Representation Tools: SysML and OPM

Views and Projections

A description of the architecture of a complex system contains an enormous amount of information—far more than any single human can readily comprehend. How should we present this information? There are essentially two choices. We can maintain an integrated model and form projections when they are needed, or we can maintain multiple views in the model.

These two approaches can be seen in traditional civil architecture. Before there were 3D computer rendering tools, an architect would draw several *views* (a floor plan of a building, façades, and several sections), and these views would be the documents that guided construction. But there would be no guarantee that the views would be consistent and that they would produce a building with all parts connected.

Since the advent of 3D rendering, civil architects have been able to construct an *integrated* 3D model. If the civil architect needs a specific view, the software *projects* the model onto the 2D plane that shows us what we want to emphasize: a floor plan, façade, or section. The projections are certain to be consistent, because they came from the same 3D model.

The same options exist for system architectures. We can construct a larger integrated model and, when necessary, project it to get views. Or we can start by constructing the views, with only limited confidence that they will yield a consistent whole. Both of these approaches are in common practice.

The integrated-model school is represented in contemporary tools by the Object Process Methodology (OPM). [4] These approaches try to incorporate information about form, function, entities, and relationships into one model. An alternative approach is to use a representation that develops individual views. Both the Systems Modeling Language (SysML) [5] and the Department of Defense Architecture Framework (DoDAF) [6] use this approach.

SysML and OPM are discussed below. Both tools appeared at approximately the same time in the early 2000s, and both are widely used. Note that this introduction provides only an overview of the tools, without going into the details of the different views and diagrams, which are best obtained by reviewing the references. Further details for both approaches are described in Part 2.

SysML

SysML was developed in 2003 as a joint effort of the Object Management Group (OMG) and the International Council on Systems Engineering (INCOSE) to adapt the software engineering Unified Modeling Language (UML) [7] to the needs of systems engineers. SysML uses a subset of UML and adds several features oriented toward modeling system requirements and performance.

The original UML consists of thirteen diagrams or views. Six of them are used to describe software structure (class diagram, package diagram, object diagram, component diagram, composite structure diagram, and deployment diagram). The other seven UML diagrams are used to describe software behavior (state machine diagram, activity diagram, use case diagram, sequence diagram, communications diagram, timing diagram, and interaction overview diagram).

As shown in Figure 3.5, the most recent version of SysML (2009) consists of nine views. Seven of these views are directly taken from UML: the class diagram (renamed the block definition diagram), the package diagram, the composite structure diagram (renamed the internal block diagram), the activity diagram, the state machine diagram, the use case diagram, and the sequence diagram. Two new views were added: a parametric diagram and a requirement diagram.

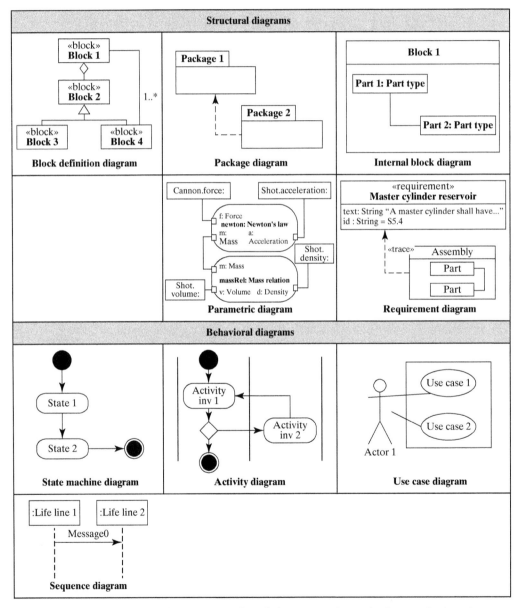

FIGURE 3.5 SysML diagrams. (Source: Jon Holt and Simon Perry, SysML for Systems Engineering, Vol. 7, IET, 2008)

The reader may wish to refer to *SysML for Systems Engineering* (Holt and Perry 2008) for a more comprehensive discussion. [8] In Figure 3.5, the block definition diagram and the internal block diagram represent the system form: The block definition diagram represents the elements, and the internal block diagram represents their structure. We develop these ideas further in Chapter 4.

The requirements diagram captures the text for and relationships among requirements. (This is analogous to our discussion of stakeholders and goals in Chapter 11.) The parametric diagram represents information on property values and constraints; it contains more detail than is generally encountered in architectural analysis. The package diagram is used to bookkeep the model elements.

The four behavioral diagrams represent the functional domain and associated behavior. The use case diagram portrays what in Chapter 5 we call externally delivered function and value. The remaining three diagrams represent various aspects of function and time-dependent behavior discussed in Chapters 5 and 6.

OPM

OPM was developed by Professor Dov Dori at Technion with the goal of unifying the object- and process-oriented paradigms for describing systems in a single methodology.

In OPM, objects (which appear in SysML Structural Diagrams) are represented by boxes and are discussed in Chapter 4. Processes (which appear in SysML Behavioral Diagrams) are represented by ovals and are discussed in Chapter 5. In OPM, however, objects and processes appear in one diagram and are connected through different types of relationships: The object can be an instrument of a process, or the object can be changed by the process (Figure 3.6). We discuss these relationships in Chapter 6.

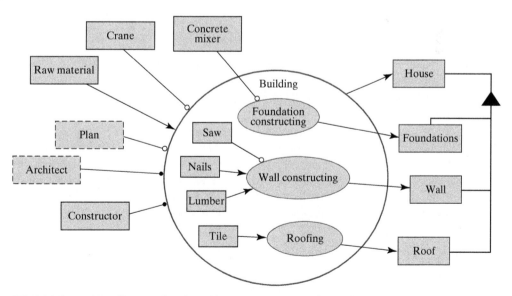

FIGURE 3.6 An OPM diagram showing objects, processes, and their relationships in a single view. System Architect: Dov Dori, Inventor of OPM.   (Source: Dori, Dov, www.gollner.ca/2008/03/object-process.html)

TABLE 3.2 | SysML and OPM representations of hierarchy and the logical relationships among entities

Relationship	SysML	OPM
Decomposition/ aggregation	A with filled diamond to B, C	A with filled triangle to B, C
Specialization/ generalization	A with open triangle to B, C	A with open triangle to B, C
Class/instance	A; B:A, C:A	A with double triangle to B, C

An important feature of OPM is that it does not have different views or types of diagrams for a system but, rather, a single integrated model. The information in many of the SysML diagrams can be represented with a single diagram in OPM, using objects, processes, and relationships.

Both SysML and OPM are expressive enough to represent system architecture, and they contain many of the same features. For example, Table 3.2 provides equivalent representations of the decomposition and logical relationships between entities introduced in this chapter. Translating entire representations is a more complicated task. See, for example, Grobshtein and Dori (2009) on how to generate SysML views from an OPM model. [9] We will use OPM as the primary tool for architecture representation in this text, but we will refer to relevant SysML diagrams as needed.

3.7 Summary

This chapter has presented different approaches to thinking about complex systems. These approaches are summarized in Table 3.3.

TABLE 3.3 | The features of complex systems and approaches to thinking about complex systems

Features of Complex Systems	Approaches to Thinking about Complex Systems
Systems can be decomposed to isolate increasingly smaller and more specialized entities (Section 3.3)	Decompose the system into entities
	Identify hierarchy if present among the entities
	Organize the decomposition into hierarchy
	Continue decomposition until two levels of abstraction or atomic parts are reached
Certain logical relationships can exist among the entities of a system (Section 3.4)	Identify classes and instances of the classes among the entities
	Identify specializations and generalization among the entities
	Identify when entities are used or appear recursively
There are a number of patterns of relationships that exist in a complex system that can be exploited when reasoning through systems (Section 3.5)	Reason about systems from the top down, bottom up, and middle out
	Zigzag between form and functional domains
	Identify the system, the accompanying system, and the use context
	Within a level, identify layers of value pathway and supporting entities
	Function-Goal reasoning: the function at one level becomes the goal at at the next lower level
	Work vertically through levels by reasoning 2 down and then 1 up to identify organization and clustering
There are ways to view or project the architecture to enhance understanding (Section 3.6)	Construct consistent views of the system (for example, SysML)
	Project from a higher dimensional model to create consistent views (for example, OPM)

We saw in Section 3.3 that systems can be decomposed into smaller and more specialized entities, and that a hierarchy that goes two levels below the reference level is a useful way of representing this decomposition.

In Section 3.4, we introduced two types of relationships between entities, namely the class/instance relationship, and the specialization relationship. We also saw how systems are sometimes most naturally represented recursively—that is, by repeating a certain structure within itself.

In Section 3.5, we discussed different ways in which we can reason through a complex system, such as top-down, bottom-up, outer-in, and middle-out. We showed that this usually requires zigzagging between the function and form domains. In subsequent chapters, four additional techniques are introduced. The distinctions among the system, those entities that accompany the system to deliver value, and the overall use context are described in Chapter 4. The identification of the value pathway along which value is delivered, along with other layers in the architecture, are presented in Chapters 5 and 6. In Chapter 8 we discuss an approach to thinking about multilayer architectures: function–goal reasoning, in which the function at one level becomes the goal at the next lower level. Finally, in the discussion of complexity in Chapter 13, we present 2 Down 1 Up reasoning, which is used to identify the clustering at Level 1 based on information at Level 2.

In Section 3.6 we introduced the idea of representing systems with views and projections. This leads us to the two mainstream frameworks used for representing complex systems, SysML and OPM. SysML uses multiple views or diagrams to represent different aspects of a system (such as structure and behavior), whereas OPM maintains a master model that represents many aspects of the system.

In the remainder of this text, we will make substantial use of these approaches to reasoning about complex systems.

References

[1] George A. Miller, "The Magical Number Seven, Plus or Minus Two: Some Limits on Our Capacity for Processing Information," *Psychological Review* 63, no. 2 (1956): 81.

[2] Cecilia Haskins, Kevin Forsberg, Michael Krueger, D. Walden, and D. Hamelin, *INCOSE Systems Engineering Handbook,* 2006.

[3] Nam P. Suh, "Axiomatic Design Theory for Systems," *Research in Engineering Design* 10, no. 4 (1998): 189–209.

[4] D. Dori, *Object-Process Methodology: A Holistic Paradigm.* Berlin, Heidelberg: Springer (2002), pp. 1–453.

[5] Tim Weilkiens, *Systems Engineering with SysML/UML: Modeling, Analysis, Design.* Morgan Kaufmann, 2011.

[6] DoD Architecture Framework Working Group, "DoD Architecture Framework Version 1.0," *Department of Defense* (2003).

[7] http://www.omg.org/spec/UML/

[8] Jon Holt and Simon Perry, *SysML for Systems Engineering,* Vol. 7, IET, 2008.

[9] Variv Grobshtein and Dov Dori, "Creating SysML views from an OPM model," *Model-Based Systems Engineering, 2009. MBSE'09. International Conference on* 2–5 March 2009, IEEE, 2009.

Part 2
Analysis of System Architecture

We now begin the analysis of architecture in depth. In Part 1, we built up a framework for analyzing systems, and we introduced some of the tools that will be useful in that analysis. Here in *Part 2: Analysis of System Architecture*, we will work to develop a deep understanding of form and function and will show that these are the building blocks of system architecture.

What is architecture? Two possible definitions are:

The arrangement of the functional elements into physical blocks.

Ulrich and Eppinger [1]

The whole consists of parts; the parts have relationships to each other; when put together, the whole has a designed purpose and fills a need.

Reekie and McAdam [2]

What these two definitions have in common is that they describe the key elements of a system: form and function, parts and whole, relationships, and emergence.

Our experience is that there are many different definitions of architecture, which vary in their emphasis on form or function, and which vary enormously in the level of detail required. Consider two other definitions:

The fundamental organization of a system, embodied in its components, their relationships to each other and to the environment, and the principles governing its design and evolution.

ISO/IEC/IEEE Standard 42010 [3]

A formal description of a system, or a detailed plan of the system at component level to guide its implementation.

The Open Group Architecture Framework (TOGAF) [4]

Notice that one definition connotes a high-level description ("fundamental organization," "principles"), while the other emphasizes detail and implementation. How much detail is required to capture architecture?

The architecture of the system should ensure that the function and performance that are linked to benefit emerge. *Benefit* is the worth, importance, or utility of a system as judged by a subjective observer. *Value* will be defined here as the delivery of benefit at cost. A system can deliver high value by producing benefit at low cost, or high benefit at modest cost. A deeper discussion of value identification and delivery will be conducted in Chapters 5 and 11. We will develop a more detailed definition of architecture in Chapter 6, as we build up the ideas necessary for the analysis of architecture.

Our approach in Part 2 is to start with very simple systems, even though this may seem counterintuitive for a text on complex systems. Our intent is to showcase the ideas of architecture in a context where the system can be fully represented—all the pieces are present. This alleviates concern that we have somehow left out the kernel of architecture by not representing enough detail, and we will show what information we believe is sufficient to capture the architecture of the system.

As we proceed through Part 2, we will work on progressively larger systems, but our intent remains to showcase methods and diagrams that can directly represent architecture. Then, in Part 3, we will turn to complex systems, where we will make use of the constituent ideas of architecture (such as form and function), but we will focus on architecture in a management context where specific diagram languages like OPM and SysML are not always appropriate.

In order to allow the reader to reason from the concrete to the abstract, we will begin Part 2 with a discussion of the form of a system and then work toward function in Chapter 5 and the mapping of form to function to define architecture in Chapter 6. This is the direction of reasoning in reverse engineering. We just assume that the architecture exists and proceed to analyze it. Learning analysis first creates the conditions for successful synthesis. Chapter 7 adopts a forward engineering approach, examining how systems derive from a solution-neutral statement of function and arrive at a concept. Finally, in Chapter 8, we demonstrate how this concept flows to an architecture.

References

[1] K.T. Ulrich and S.D. Eppinger, *Product Design and Development* (Boston: Irwin McGraw-Hill, 2000).

[2] John Reekie and Rohan McAdam, *A Software Architecture Primer,* Software Architecture Primer, 2006.

[3] "Systems and Software Engineering: Architecture Description," ISO/IEC/IEEE 42010: 2011.

[4] http://en.wikipedia.org/wiki/TOGAF

Chapter 4
Form

4.1 Introduction

We begin Part 2 with Form. This chapter lays the groundwork for the analysis of architecture by first defining its most concrete aspect, form. We will discuss form as an idea, the organization of form by decomposition and hierarchy, and the representation of form. Establishing a clear understanding of form is the foundation for the analysis of architecture.

We begin Chapter 4 with a rigorous discussion of form in Sections 4.2 and 4.3 and then address formal relationships in Section 4.4. The context that surrounds the form of the system is described in Section 4.5. Section 4.6 reviews the main ideas of form again via an example of a software system, the bubblesort algorithm.

In each of the chapters in Part 2, we will begin with a list of questions that guide us in the analysis of form and suggest the outline of the chapter. Table 4.1 also contains the outcomes produced by answering the questions.

4.2 Form in Architecture

Form

It is surprisingly difficult to separate form from function. In common speech, we refer to form with function words. Try describing a paper coffee cup, a pencil, or a spiral notebook without any reference to function. If you used the words "handle," "eraser," and "binding," you were using words rooted in a function. To stay entirely in the form domain, we might use "flat cardboard half-circle," "rubber cylinder," and "metal spiral." One objective of this chapter is to clearly distinguish form from function.

Form is what has been or is eventually *implemented*. Form is eventually built, written, composed, manufactured, or assembled. Form is about existence. As indicated by the definition in Box 4.1, the first test for form is that it exists. Form is what the system *is*. It is the concrete and often visible manifestation of the system. The clause "for some period of time" is included in the definition to enable us to discuss the form of things that have existed in the past (analysis of historical systems) or will exist in the future (analysis of system designs).

The second test for form is that it is instrumental in the execution of function. For this function to occur, there usually needs to be some instrument to facilitate the function, or to "carry" the function. Form is what is eventually *operated* to deliver function and value.

TABLE 4.1 | **Questions for defining form**

Question	Produces
4a. What is the system?	An object that defines the abstraction of form for the system
4b. What are the principal elements of form?	A set of objects that represent the first and potentially second level downward abstractions of the decomposed system
4c. What is the formal structure?	A set of spatial and connectivity relationships among the objects at any level of decomposition
4d. What are the accompanying systems? What is the whole product system?	A set of objects in the whole product system that are essential for the delivery of value—and the relationships to those accompanying systems
4e. What are the system boundaries? What are the interfaces?	A clear definition of the boundary between system and context and a definition of the interfaces
4f. What is the use context?	A set of objects that are not essential to value delivery, but that establish place, inform function, and influence design

Box 4.1   Definition: *Form*

Form is the physical or informational embodiment of a system that exists or has the potential for stable, unconditional existence, for some period of time, and is instrumental in the execution of function. Form includes the entities of form and the formal relationships among the entities. Form exists prior to the execution of function.

Form is a product/system attribute.

Consider the beach house shown in Figure 4.1. What is the form? The form can be interpreted as the spaces in the house—the grand room, the kitchen, and so on. The kitchen space exists and facilitates food preparation and conversation (function). In the emergency instructions shown in Figure 4.2, the structure is that some of the individual pictures are in sequence (reinforced by the numbering), and then, at a higher level, some of the sequenced pictures are prefaced by other, smaller pictures or letters.

Form includes the sum of the entities, which are the elements or chunks of form. These are elements of the whole, and this is the whole–part relationship for form. Form also includes the relationship among the entities of form, which is often called the structure. Form is the entities of form plus the structure.

In the case of the beach house, the formal relationships are easily represented by the floor plan, which shows the spatial relationships among the spaces. In the emergency instructions, the structure is that some of the individual pictures are in sequence (reinforced by the numbering), and then, at a higher level, some of the sequenced pictures are prefaced by other, smaller pictures and are in blocks indicated by horizontal rules.

Form is a *system attribute*; it is an independent way in which to view or characterize the system. Form is also part of the solution that the architect is proposing.

CHAPTER 4 • FORM   55

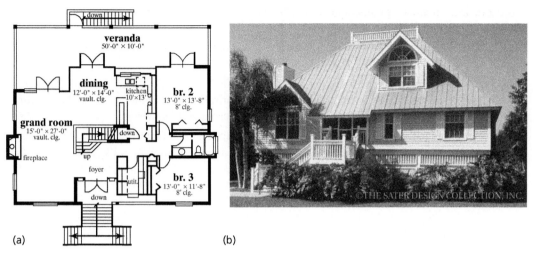

FIGURE 4.1 (a) A beach house as an example of civil architecture (b) Spy glass house.   (Source: (a) and (b) © The Sater Design Collection, Inc.

Each discipline has developed its own shorthand for representing form. Form was first represented by words, such as the way national constitutions describe the form of government. Next, representational maps and drawings were developed: illustrations, schematics, drawings, three-dimensional views, and perspective diagrams. These attempt to represent some sub-scale yet realistic image of the system. Later, symbolic representations were developed: equations,

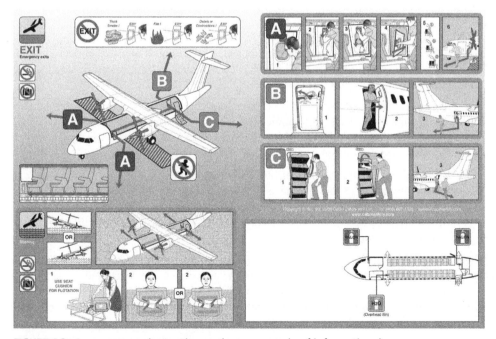

FIGURE 4.2 An emergency instruction card as an example of informational architecture.   (Source: Copyright © Cabin Safety International Ltd.)

pseudocode, circuit diagrams, and the like. The symbols used in these diagrams have only abstract meaning.

Analytical Representation of Form; Objects

We seek a way to represent the form of any system—natural, evolved, built, mechanical, electrical, biological, or informational—with the same set of semantically exact representations. These facilitate analysis, communication across disciplines, and the capture of knowledge.

The representation we adopt in this text closely follows Dori's Object Process Methodology (Chapter 3). Dori determined that systems can be represented with only two classes of abstractions: objects and processes (hence the name). "Together, objects and processes faithfully describe the system's structure, function and behavior in a single, coherent model in virtually any domain." [1]

The definition of an object in Box 4.2 was deliberately phrased in such a way that it closely reflects the first part of the definition of form, so that objects could be used to represent form. However, the definition of form (Box 4.1) contains the additional criteria that the form must be instrumental of function and must exist before function. Objects that do not meet these additional criteria are operands, as discussed in Chapter 5. Generally, the names of objects are nouns.

Box 4.2 **Definition: *Object***

An object is that which has the potential for stable, unconditional existence for some period of time.

Figure 4.1 shows a beach house that is an object, which is both physical and tangible. The rooms of the house are also objects. In Figure 4.2, the instructional drawings are objects, as are the smaller drawings to the upper left of the boxes.

Objects can also be informational. Dori defines an informational object as anything that can be comprehended intellectually, which implies that it exists. Informational objects include ideas, thoughts, arguments, instructions, conditions, and data. Domains such as control systems, software, mathematics, and policy make extensive use of informational objects.

One of the defining characteristics of objects is that they have characteristics or *attributes* that describe the object. An object can have physical, electrical, or logical attributes. Some of these attributes can be considered as *states*, in which an object can exist for some duration of time. A process can change an object's state. The combination of all states describes the possible configuration of state of the object.

When our beach house is being built, its state of construction goes from "not built" to "built." When people arrive, its state of occupancy goes from "unoccupied" to "occupied." When the furnace runs, its state of temperature goes from "cooler" to "warmer." In this example, the attributes and corresponding states are construction (unbuilt/built), occupancy (unoccupied/occupied), and temperature (cool/warm).

The graphical representations for an object are adopted from Object Process Methodology (OPM), where objects are denoted with rectangles (Figure 4.3). The representation on the left

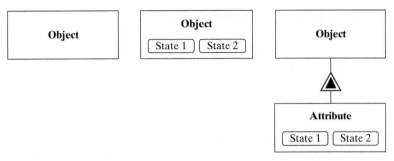

FIGURE 4.3 Three OPM representations of objects: A simple object, an object with states illustrated, and an object with an attribute and states explicitly illustrated through the "is characterized by" relationship.

simply defines an object, and the one in the center shows possible states of the object. On the right, the "double triangle" on the line from the object to the attribute is read as the relationship "is characterized by." The lower box lists the attribute and possible states.

Note that objects in SysML have different representations, depending on whether they are physical objects, human operators or users, or something else.

Decomposition of Form

Decomposing entities into appropriate abstractions was discussed in Section 3.3. Decomposing form comes naturally to architects. Form is concrete, it decomposes easily, and the aggregation of form traces the decomposition in a simple way.

Decomposition is represented in OPM by a graphical tree diagram, as shown in Figure 4.4. The black triangle is used to indicate decomposition. This diagram shows that System 0 (the object at Level 0) *decomposes* into the objects at Level 1, which correspondingly *aggregate* into System 0.

In summary:
- Form is a system attribute that includes the elements that exist and are instrumental in the execution of function. Form is implemented and eventually operated.
- Form will be modeled as objects of form, plus the formal structure among them.
- Objects are static entities that exist and have attributes; among these are states that can be changed by a process.
- Form can be decomposed into smaller entities of form, which in turn aggregate into larger entities of form.

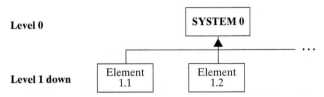

FIGURE 4.4 OPM representation of decomposition.

4.3 Analysis of Form in Architecture

With this background, let's begin the analysis of the form of a real engineering system, the centrifugal pump of Figure 4.5. Box 4.3 contains an explanation of the principles of operation of this pump. This example was chosen as the reference case because it is "modular"; the parts are not

Box 4.3 Insight: Centrifugal Pump—How It Works

A centrifugal pump works by increasing the energy of the fluid that passes through it. The turning impeller does the work on the flow. Depending on the design of the pump, the increase in energy can be delivered either as an increase in velocity of the outlet flow or as an increase in static pressure.

The flow enters the pump on the axis of rotation of the impeller through the hole in the center of the cover shown in Figure 4.5. The impeller turns the flow radially and does work on the flow, increasing its velocity internal to the pump. The housing contains a passage that slows the velocity of the water, or diffuses the flow. The newly gained energy is converted from kinetic energy into higher pressure. Mass is conserved, so the same amount of fluid exits the pump as enters it. Energy is conserved, so the electrical power going into the electrical motor is reflected in the increased pressure of the outlet mass flow.

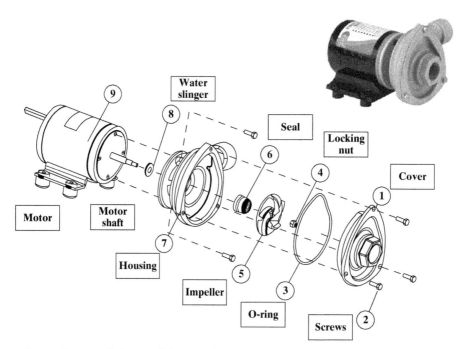

FIGURE 4.5 Expanded view of the centrifugal pump. (Source: PumpBiz.com)

absolutely discrete like the members of a team, nor are they integral like the parts of the heart. In addition, the pump has only nine parts. It is a "simple system," judging on the basis of the guideline presented in Section 3.3 that we find atomic parts at Level 1.

We acknowledge that the pump is a simple, almost trivial example, and that applying the techniques of system architecture to such a simple system is like taking a sledge hammer to a thumb tack. However, we can learn a lot from a simple system before tackling a truly complex one.

Defining the System

We will use the questions in Table 4.1 to organize our discussion of form. The first question asks us to identify the system and its form: "What is the system?"

The *procedure* to answer Question 4a is to examine the system and create an abstraction of form that conveys the important information and implies a boundary of the system that is consistent with the detailed boundary that will be drawn in answering Question 4e. Abstractions were discussed in Section 2.4.

For Question 4a in this case, we create the abstraction of form "pump," which not only brings to the surface the idea of something that moves fluid but also hides all details of motors, impeller, and the like. We implicitly draw the system boundary around the parts listed in Figure 4.5, which separates the system from the context. Outside the system, we presumably have the hoses that connect to the pump input or output, the mechanical support for the pump, and the controller and power supply for the motor. The outcome of this task is simply an object that represents the abstraction for the entire system, the object labeled "Pump" in Figure 4.6.

We could have chosen "centrifugal pump" as a more specific abstraction than "pump" for our system. Other types of pumps include axial flow pumps and positive displacement pumps.

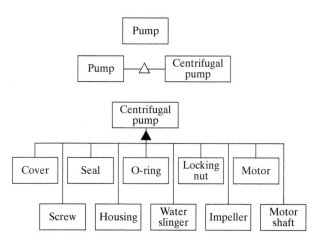

FIGURE 4.6 Specialization and decomposition of a centrifugal pump in OPM.

Figure 4.6 shows the objects called pump and centrifugal pump connected by the OPM symbol for the specialization relationship (an unfilled triangle). Specialization was introduced in Section 3.4 and is used extensively in design.

Identifying the Entities of Form

Continuing to Question 4b of Table 4.1, we now seek to identify the entities in the form domain called elements of form. The procedure to answer Question 4b is explained below by example. In brief, we will start with a reference parts list as the initial set of abstractions, decompose some elements further if necessary, combine or eliminate elements where possible, and use hierarchy to identify the most important elements of form.

The "parts list" for the pump is indicated by the numbered parts in Figure 4.5 and is a useful place to start creating abstractions. For this simple pump system, we will use the parts list as the set of abstractions defining elements of form, with a few exceptions. The first is the motor, which actually has a non-rotating element and a rotating motor shaft. Some elements connect to the motor and some connect to the shaft, and this difference is important. Calling all of this "the motor" would hide too much information. Anticipating the internal function of the elements of form, one can guess that the shaft also is involved with different functions than the motor. Therefore, we create two elements to describe the motor and the motor shaft, producing the full list of entities in Figure 4.6. To illustrate that this decomposition can be shown as a graph or as a list, we have included a list-like representation in Table 4.2. In general, information about a system can be contained in a graph or list, and we will use the two interchangeably.

A careful examination of Figure 4.5 reveals that there are five screws. Yet in the decomposition of Figure 4.6, we identified only one abstraction, called "screw." Implicitly, we created a class called "Screw" and then identified five instances of the class. The class/instance relationship was discussed in Section 3.4, and the OPM symbol was introduced in Table 3.2.

TABLE 4.2 | Parts list and elements of form for a centrifugal pump

Parts List	Abstractions Used to Designate Elements of Form
Cover	Cover
Screws	Screws (class of 5)
O-ring	O-ring
Locking nut	Locking nut
Impeller	Impeller
Seal	Seal
Housing	Housing
Water slinger	Water slinger
Motor	Motor
	Motor shaft

TABLE 4.3 | Hierarchy in the elements of form of the centrifugal pump

Centrifugal Pump
Cover, Impeller, Housing, Motor shaft, Motor
O-ring, Seal, Water slinger
Screws (class), Locking nut

It is not uncommon, in decomposing systems, to find entities that simply can be combined for convenience, because they are integrally part of delivering a single function. For example, in the pump list of elements in Table 4.2, we could have combined the locking nut with the impeller, since the only function of the locking nut is to secure the impeller to the shaft.

The Pump as a Medium-Complexity System

If we briefly think of the pump as a medium-complexity system (in which we find atomic parts at Level 2), we can apply hierarchic reasoning, following the approach of Section 3.3. Remember that *hierarchy* is defined as a system in which grades are ranked one above the other because they have more scope, importance, performance, responsibility, or function.

Hierarchy can be applied to built systems. A simple table showing hierarchy for the pump is shown in Table 4.3. We identify five key parts: The cover and housing contain the fluid and provide interfaces; the impeller does the work on the fluid; the motor drives the shaft and supports the housing; and the motor shaft drives the impeller. We can also identify the fasteners as the least important elements, leaving the remaining three parts in the middle rank. This table gives us a quite different impression than the graphical decomposition of Figure 4.6 and the list in Table 4.2. Based on the information in Table 4.3, we might first try to reason about the five highest-grade objects, to understand them thoroughly before including the lower-priority elements.

Figure 4.7 shows an OPM representation of a hierarchic decomposition that extends two levels down. There is no formal structure (other than decomposition) and no functional interaction implied by the diagram!

Even though the pump has only ten entities of form, it is possible to convert its representation from a one-level decomposition (Figure 4.6) to a two-level decomposition, as shown in Figure 4.8.

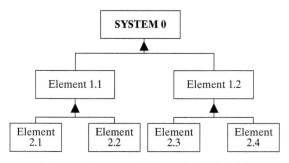

FIGURE 4.7 OPM representations of multilevel decomposition.

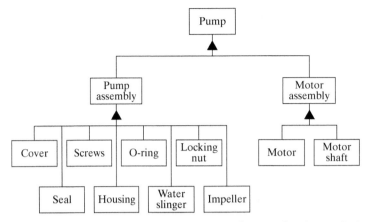

FIGURE 4.8 Two-level decompositional view of a centrifugal pump in OPM.

The elements at Level 1 in Figure 4.8 are now not real parts but simply abstractions. In this case, the abstractions "pump assembly" and "motor assembly" were selected. These could have been chosen because the motor is highly integral and the other elements highly interconnected. Or perhaps it was because the pump assembly has a distinct function (increase the pressure of water) and the motor has another distinct function (drive the pump).

The creation of Level 1 abstractions is *not unique,* as discussed in Section 3.3. For example, the Level 1 abstractions could have been "rotating components" (which would include the lock nut, impeller, water slinger, and motor shaft) and "non-rotating components" (the rest of the elements).

To summarize the approaches used to identify the entities of form of a medium-complexity system, consider the OPM diagram shown in Figure 4.9.

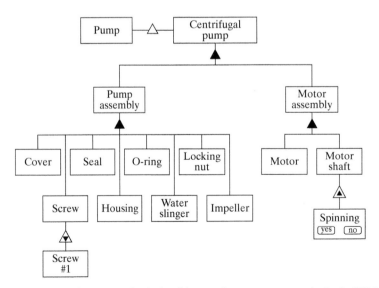

FIGURE 4.9 Summary of relationships used to manage complexity in OPM.

Starting at the top left, the more general idea of *pump* specializes (Section 3.4) to the more specific idea of *centrifugal pump*. Centrifugal pump decomposes through two levels of hierarchy (Section 3.3). At the first level, we have created the intermediate abstractions called *pump assembly* and *motor assembly*. These then decompose to ten elements of form. One of these is in fact not a single object, but a class called "Screw," of which there are actually several instances (Section 3.4). One of the other objects, the motor shaft, exhibits the attribute of state called spinning, which can have the states of yes (the shaft is turning) and no (it is not turning).

In summary:

- Analyzing form requires creating an abstraction of form that conveys the important information, but not too much information, and implies a boundary of the system.
- The elements of form can be represented as a hierarchic decomposition—a set of objects that represent the first- and potentially second-level abstractions of the decomposed system, which can be represented in either a graphical or a list format.

4.4 Analysis of Formal Relationships in Architecture

Formal Relationships

In this section, we will focus on the relationships of form, which are sometimes called *formal relationships* or (more commonly) *structure*. Question 4c in Table 4.1 suggests that this is the next task in analysis.

As defined in Box 4.4, *structure* is the set of formal relationships among the elements of form of a system. Structure is not conveyed by Figure 4.9! Structure is additional information. If you are merely given a random set of mechanical parts or lines of code (the set of objects of form of the system), you do not know how to "assemble them." What is missing is information about the structure. Structure shows where the elements of form are located and how they are connected. After elements of form, formal relationships are the next most tangible aspects of a system.

> **Box 4.4**   Definition: *Formal Relationships* or *Structure*
>
> Formal relationships, or structure, are the relationships between objects of form that have the potential for stable, unconditional existence for some duration of time and may be instrumental in the execution of functional interactions.

Structure is often instrumental in functional interactions. If there are functional relationships among the elements of form, in most cases there must also be a carrier of these functional interactions. Formal relationships often carry the functional interactions. If A supplies power to B (a functional interaction), there may be a connecting wire (the structure). If A provides an array to B (a functional interaction), there may be a link or shared address space (the structure). Since functional interactions are crucial to emergence, and since functional interaction is often enabled by formal structure, understanding structure is essential to understanding the system.

Structure is the set of relationships that exist and do not require change, exchange, or anything that happens along a timeline. At any instant of time, they just *are*. Like all of form, formal relationships are about what exists, not what occurs. In our pump, the motor is connected to the housing—a connection that simply exists.

Formal relationships can be changed. Elements can be connected, unconnected, or moved. Formal relationships often implicitly represent things that have happened in the past. The relationship "A is next to B" implies that at some time in the past, A was placed next to B. The relationship "A is connected to B" implies that at some earlier time, there occurred a joining process in which A was connected to B.

The two main types of structural relationships are *spatial/topological relationships* and *connections*. In a circuit design, for example, the designer produces the layout drawing (Figure 4.10) that shows where components are located—the spatial/topological view.

The designer also produces the circuit diagram (Figure 4.11) that shows how components are connected. Both are needed; from Figure 4.10 alone it is not possible to figure out what is connected, and from Figure 4.11 alone it is not possible to determine where things are located. As will be discussed, there are also other types of structural relationships that are sometimes important to a system.

Spatial/Topological Formal Relationships

Spatial/topological relationships explain *where things are*: the location or placement of the objects of form. These relationships imply only location and placement, *not* the ability to transmit anything between the objects. In physical systems, spatial and topological relationships are largely issues of geometry, but in information systems, one must also consider the more abstract issues of space and topology, as we shall see in Section 4.6.

Spatial relationships capture absolute or relative location or orientation. Above and below, ahead and behind, left and right, aligned and concentric, near to and far from—all of these relationships convey relative location. Absolute location may be given with some reference, such as an Earth geodesic reference or a local reference system.

Topology is about placement and suggests relationships such as within, contained in, surrounded by, overlapping, adjacent, touching, outside of, or encircling. In practice, the differences between spatial and topological relationships are very subtle, and both can be considered part of one set of spatial/topological relationships.

Spatial/topological relationships are often very important. For example, it may be important to know that the lungs are not only connected to the heart but also near it. At compilation, it is important to know where a certain code is located. In a factory, it may be important to know the relative locations of two workstations that exchange parts. Optical elements have to be aligned to work properly.

The procedure for identifying the spatial/topological relationships (Question 4c of Table 4.1) is to examine each pairwise set of objects of form and ask, "Is there an important spatial/topological relationship between these two objects?" Thinking holistically, we recognize that there is a spatial relationship between every pair of objects. We must apply focus, and because we are trying to use structure to eventually understand functional interaction and emergence, the key question is the following:

Is this spatial or topological relationship key to some important functional interaction or to the successful emergence of function and performance?

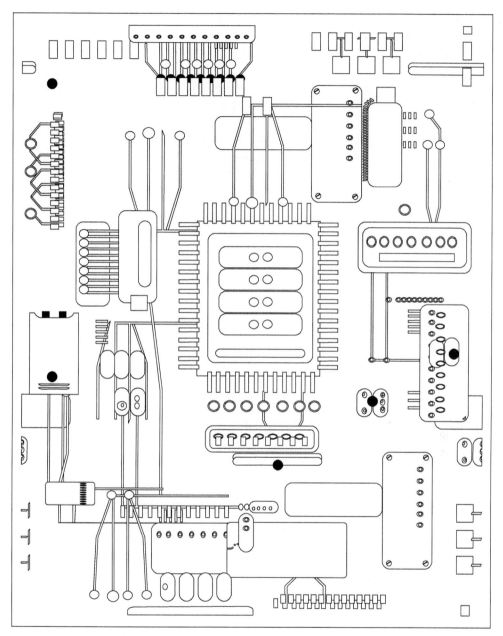

FIGURE 4.10 Spatial/topological representation of a circuit: Location of components. (Source: Vectorass/Fotolia)

A typical mechanical engineering representation of spatial/topological information for an assembled system is shown in Figure 4.12 for the pump.

From this three-view drawing, we can see approximately where elements are located, what touches what, and so on. A more illustrative drawing is called the expanded view, an example

66 PART 2 • ANALYSIS OF SYSTEM ARCHITECTURE

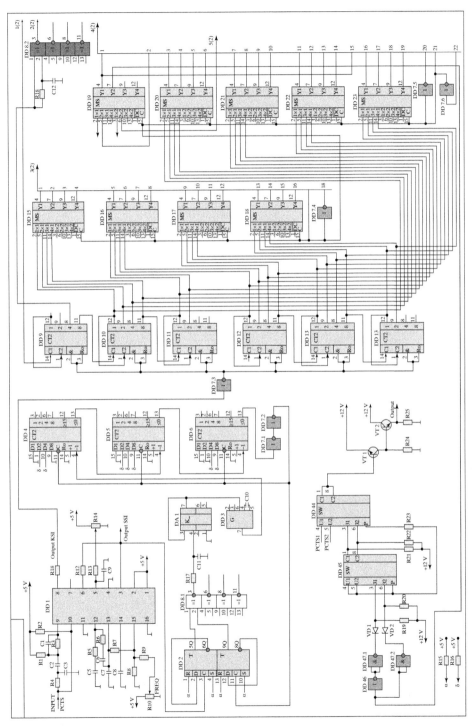

FIGURE 4.11 Connectivity relationships representation of a circuit: How components connect. (Source: ID1974/Fotolia)

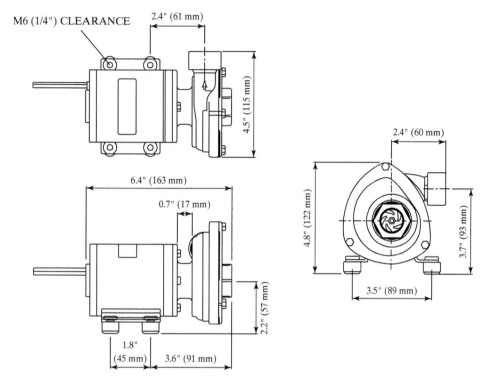

FIGURE 4.12 Spatial/topological representation of a pump: Where components are located.

of which is shown in Figure 4.5. From this and several other types of drawings (such as section views), all of the important spatial/topological information can be extracted. Figure 4.11 shows a typical spatial/topological view of a circuit.

Representing Formal Relationships with Graphs and Diagrams; OPM

We seek a generalized way to represent structural information for an architecture. As was suggested in Chapter 2, there are essentially two choices: We can represent structure with graphs and diagrams or with tables and matrix-like views. Since both have their advantages and disadvantages, we will develop both approaches.

Graph representations generally represent the objects of form, with structure indicated by various types of lines and arrows connecting the objects. In some notational schemes, the arrows are semantic and express meaning. In others they are more informal.

In this text, we will consistently use the OPM graphical representation of structure shown in Figure 4.13, which is a binary link generally shown as a single-headed arrow that runs from one object to another, with an accompanying label.

There is a "forward" version of the link—see the link from object A to B—with a label that describes the nature of the relationship in simple language. For example, examining Figure 4.5, we see that the housing surrounds the impeller but does not touch it. In Figure 4.13, this is shown by an arrow from housing to impeller with the label "Surrounds" so that one reads, "the housing surrounds the impeller."

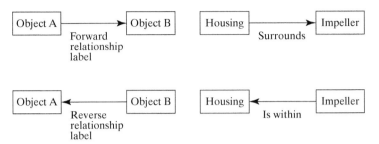

FIGURE 4.13 OPM graphical representation of structure.

The direction of the arrow from A to B is arbitrary and does not imply any exchange, interaction, or causality. It is simply a relationship that exists. Therefore, we could just as accurately have drawn the reverse, or "backwards," link from B to A and labeled it "Is within" so that one would read "the impeller is within the housing." Even though different words are used to label the arrow, the relationship has the same meaning. Generally, we show the arrow in only one direction, and the other is implicit. Sometimes the same words can be used in both directions. "The cover is aligned with the motor shaft" would reverse to "the motor shaft is aligned with the cover."

We are now ready to apply the procedure for identifying and diagramming important spatial/topological relationships of the pump, where we have restricted ourselves to the five important elements from the hierarchy of Table 4.3.

To begin, consider the cover object. It touches the housing and is aligned with the motor shaft (they are co-axial). What may not be obvious is that the cover object must also be close to the impeller; the pump would not work if there were a large gap between the impeller and the cover. In Figure 4.14, these relationships are represented by an arrow that runs from the cover to the housing and is labeled "Touches"; another arrow that runs to impeller and is labeled "Is close to"; and a third arrow that runs to the motor shaft and is labeled "Is aligned with." We then proceed to the other elements of form, identifying important relationships between pairs, until Figure 4.14 is complete.

A similar diagram showing the structure of the pump could be drawn in SysML, in particular using the block definition diagram and the internal block diagram. This is explained in more detail in Box 4.5.

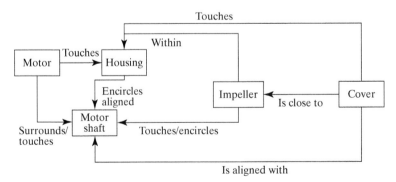

FIGURE 4.14 Simplified spatial/topological structure of a centrifugal pump: Graphical OPM.

Box 4.5 Methods: Representing Form (Elements and Structure) in SysML

What we call form (elements and structural relationships) is represented in SysML using blocks for the entities and several of the SysML relationships between the blocks. However, the issue is that these same blocks and some of the relationships can sometimes represent function and functional interaction as well. Therefore, in representing what we call form (elements and structure) in SysML, a subset of the features of SysML are used selectively.

There are several types of relationships in SysML that are used to represent what we call structure (Figure 4.15). Aggregation and composition are represented by a solid line with a diamond at the beginning of the line (the pump is composed of the motor, the impeller, and so on). Specialization/generalization is represented by a solid line with a triangle at the beginning of the line (for example, the centrifugal pump is a specialization of the generalization "pump"). General dependencies, represented by dashed lines with an arrow at the end of the line, and association, represented by a line with a label, can both be used to represent what

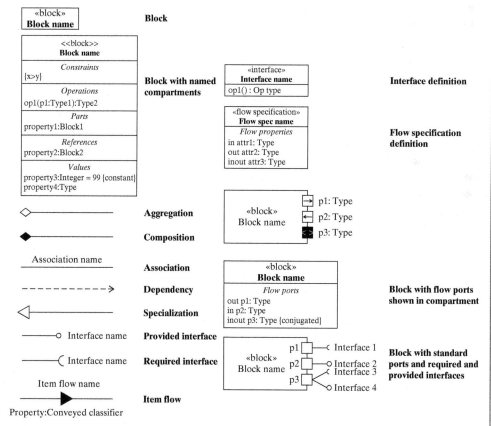

FIGURE 4.15 Legend of graphical elements used in SysML structural diagrams. [2] (Source: Jon Holt and Simon Perry, SysML for Systems Engineering, Vol. 7, IET, 2008)

(continued)

we call structural relationships. Care must be taken, because dependencies and associations can also be used to express functional relationships. Interface links and item flow are explicitly used to represent functional interactions.

SysML has two diagrams primarily used to represent what we call structure.

- *Block definition diagrams* show the main elements of the system and the formal relationships among them, which may be aggregation, composition, association, dependency, or specialization relationships. Blocks can represent elements of the system and attributes of those elements. In block definition diagrams, blocks may also have ports (inputs and outputs of a block) that indicate *functional interaction*—the type of flow that goes through them, and their interfaces.
- *Internal block diagrams* are used to show the "internal" structure of blocks, focusing in particular on instances of the system. This includes setting block properties (e.g., set the dimensions of the impeller) and describing flow paths (e.g., illustrate how water passes through the pump). Note that the block properties can include what we call attributes of form, whereas the flow paths are part of functional interaction.

Representing Formal Relationships with Table and Matrix-like Views; DSM

An alternative representation of structure can be captured with table and matrix-like views, of which the Design Structure Matrix (DSM) is an example (Section 2.5 and Box 4.6). One reads the DSM in Table 4.4 as "Object A has relationship Forward 1 with Object B, and Object B has relationship Backwards 1 with Object A."

Box 4.6   Definition: *DSM*

DSM is an acronym that generally stands for Design Structure Matrix but sometimes stands for Decision Structure Matrix or Dependency Structure Matrix. A DSM is an N-Squared matrix that is used to map the connections between one element of a system and others. It's called N-Squared because it is constructed with N rows and N columns, where N is the number of elements. The normal convention, used in this text, is that one reads down the column to the relationship to the row heading, and that things "flow" from column heading to row heading.

The types of relationships shown in a DSM can vary. In this chapter we show spatial/topological and connectivity relationships in DSMs, but elsewhere we use them to represent a variety of types of information.

Table 4.5 shows the DSM representation for the simplified pump, which contains the same information as Figure 4.14. Again, the cover touches the housing, is aligned with the motor shaft,

TABLE 4.4 | DSM representation of structure

Parts List	Object A	Object B	Object C
Object A	X	Backwards 1	Backwards 2
Object B	Forward 1	X	Backwards 3
Object C	Forward 2	Forward 3	X

TABLE 4.5 | Simplified spatial/topological structure of a centrifugal pump: Design Structure Matrix (DSM) representation

Parts List	Cover	Impeller	Housing	Motor	Motor Shaft
Cover	X	Close to	Touch		Aligned with
Impeller	Close to	X	Surrounds		Touch/ is encircled by
Housing	Touch	Within	X	Touch	Is encircled by/aligned
Motor			Touch	X	Within/touches
Motor Shaft	Aligned with	Touch/ encircles	Encircles/ aligned	Surrounds/ touches	X

and is close to the impeller. The forward and backwards relationships are present; the housing surrounds the impeller, and the impeller is within the housing. Just like the directions of the arrows in the OPM diagram of Figure 4.14, the entries in this DSM do not imply direction or causality. In particular, for structure the forward and backwards relationships must have the *same semantic meaning*.

We will use both OPM and DSM representations. Each has advantages and disadvantages. The graphical representation is normally easier to develop and visualize, but as the system gets more complex, the graphical view becomes cluttered. The matrix representation allows more information to be captured without clutter and can be operated on in computation, but it is harder to visualize.

Connectivity Formal Relationships

A second type of formal relationship captures the idea of connectivity and answers the question "What is connected, linked, or joined to what?" In network systems such as trusses, circuits, and chemical processing plants, connectivity is the most important formal relationships to understand, because it explicitly creates the ability to transfer or exchange something between the connected objects of form. Connectivity relationships are often instruments of functional interaction, so they directly support the emergence of function and performance, as is examined in Chapter 6.

Connectivity and special/topological relationships are quite different in principle. Two elements can be adjacent but not connected (two houses in a neighborhood), and two elements can have unspecified location but be connected (a computer and a server). In practice, the difference between spatial/topological structure and connectivity structure is usually evident if the connectivity refers to electrical, chemical, thermal, or biological connections. In mechanical systems, because things must be adjacent to be connected, the spatial/topological structure and connectivity structure tend to be less distinct.

Connectivity relationships actually encode information about how the objects were implemented: assembled, manufactured, coded, or written. If two things are connected now, there was a process of connecting them in the past. But the connection currently exists, and therefore it is a relationship of form.

In software systems, compilation of one object with another is a type of connection due to implementation structure. Generally in software, connectivity is more subtle. If data or commands are passed, there must exist somewhere a connection point to pass them. These connections are commonly more transient and hidden deep within the compiled version of code and the operating system, which assigns common addresses (sometimes called pointers) to information that is shared. In object-oriented programming, two objects may contain a reference to a common third object, defining a structural relationship. Structure in software systems is discussed in more depth in Section 4.5.

There are a few classes of functional interactions that do not require an explicit formal connection. The most common of these is interaction through electrical forces and the electromagnetic spectrum, which includes visible and radio frequencies, as well as heat transfer through black body radiation. Of course, gravitational interactions also do not require an explicit connection. There is a class of interactions associated with the "ballistic" flow of particles and continua, such as a fluid jet or a thrown baseball. Here mass, momentum, and energy are exchanged without any explicit interconnect other than "space."

The procedure used to answer Question 4c of Table 4.1 and to determine connectivity relationships is similar to that used to identify spatial/topological relationships. We examine each pairwise set of objects of form, and ask,

Is this connectivity relationship key to some important functional interaction or to the successful emergence of function and performance?

Connectivity is usually more evident and less subject to interpretation than special/topological structure.

In identifying connections, one has to consider whether the connector itself must be modeled as an element of form or, conversely, can be abstracted away. In the example of the circulatory system in Chapter 2, we successfully thought about the heart, lungs, and capillaries without explicitly mentioning the veins and arteries that connect them. In general, one first attempts to abstract away the connectors, simply to minimize the number of objects that are being considered. An exception to this rule is the connector at a system interface. Such interfaces are sufficiently important to the system, and such a common source of problems, that modeling the interface as a separate object is often appropriate, as is discussed in Chapter 6.

A typical mechanical engineering representation of connectivity is the expanded view of the pump in Figure 4.5. The connectivity structure for the simplified pump is shown in Figure 4.16. Now the cover "presses" against the housing, which is a kind of mechanical connection. The devices that cause the pressing are the screws, which have been abstracted out of this simplified model. The impeller has a similar kind of press fit, due to its being slid onto a shaft.

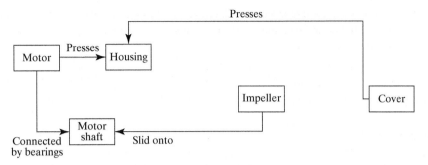

FIGURE 4.16 Simplified connectivity structure of a pump: Graphical OPM.

The relationship between the motor shaft and the motor is quite complex; the shaft is contained within the motor and is supported on bearings that allow rotation. Here it makes most sense to label the connectivity relationship with the type of the connector: bearings.

Comparison of Figure 4.16 with Figure 4.14 shows fewer connections than spatial/topologic relationships. The spatial/topological relationships in Figure 4.14 that are associated with *touching* reappear in Figure 4.16 as a connectivity relationship. However, those associated with pure topology (such as alignment, within, and encircles) are not reflected in Figure 4.16. In mechanical systems, the distinction between the spatial relationship "touching" and the connectivity "pressing" is subtle. One implies only location, and the other also conveys the ability to transmit load. If we had chosen a mechanical example in which the two objects were glued together, then the difference between touching (spatial) and glued (connectivity) would be more evident.

The DSM representation of the connectivity structure of the simplified pump is shown in Table 4.6. Once again, it contains the same information as the graphical representation in Figure 4.16. There is no implication that anything is exchanged; these connections just exist. Therefore, the matrix is symmetric, except for the phrases "Slid onto" and "Had slid onto," which are necessary so that the DSM will make sense when read from the column heading to the row heading: "The impeller slid onto the shaft," whereas "The shaft had slid onto the impeller."

As a practical matter, the spatial/topological and connectivity relationships are often combined into one representation of structure. Table 4.7 shows a summary of the structure of the

TABLE 4.6 | Simplified connectivity structure of a centrifugal pump: DSM representation

Parts List	Cover	Impeller	Housing	Motor	Motor Shaft
Cover	X		Presses		
Impeller		X			Had slid onto
Housing	Presses		X	Presses	
Motor			Presses	X	Connected by bearings
Motor Shaft		Slid onto		Connected by bearings	X

TABLE 4.7 | Simplified combined spatial/topological and connectivity structure of a centrifugal pump: DSM representation. S signifies a spatial relationship and C a connectivity relationship.

Object List	Cover	Impeller	Housing	Motor	Motor Shaft
Cover	X	S	SC		S
Impeller	S	X	S		SC
Housing	SC	S	X	SC	S
Motor			SC	X	SC
Motor Shaft	S	SC	S	SC	X

system, where the details have been abstracted to just the existence of a spatial (S) or a connectivity relationship (C) among the elements.

Other Formal Relationships

In addition to spatial/topological relationships and connectivity, there are other types of formal relationships that simply *exist*. The procedure for identifying these is the same: Examine pairwise sets of objects, and determine whether there is a relationship of these types that has an important role in enabling functional interaction and emergence.

ADDRESS The address of something *is* where it can be found. It is an encoded form of the spatial location, but it is common enough to be considered a separate type of formal relationship, because address can also be virtual. In software, in particular, one makes extensive use of the address of registers and the like. A shared address is a connection through which data can be exchanged.

SEQUENCE The static sequence of entities is a type of formal relationship. If we know that A *is* always after B, then this establishes a formal relationship. For example, in imperative software languages, the writing of a statement 2 after statement 1 implies that 2 is executed after 1; this is not necessarily true in declarative languages. This is a functional statement. However, the fact that 2 follows 1 in the coding is true in both cases, which is a formal statement.

MEMBERSHIP Being a member of some group or class can be considered a type of formal relationship: A *is* a member of group 1. Membership is an abstraction that signifies being in a group. It may lead to interaction and responsibility in a group, but it may not.

OWNERSHIP If you own something, it *is* yours. Like membership, ownership is an abstraction both useful and common. What really separates my land from your land, or my money from your money? Only that it was sometime in the past purchased by me or granted to me. But once I own it, ownership is a static formal relationship between the owner and the owned item.

HUMAN RELATIONSHIPS Another type of relationship exists among humans: the bonds between people and the attitudes they hold toward one another. Knowledge of another person's abilities, trust in another person, and liking of her or him—these are bonds among people. Somewhat uniquely, human relationships are not necessarily reciprocal. For example, you may trust a political figure who has no explicit knowledge of you at all.

All formal relationships are static at any given time, but they can change. Spatial location can be changed by moving things. Connection can be made and broken, as when a lamp is plugged into an outlet. Addresses can be changed by reassignment. Sequence can be changed by reordering. Membership in a club exists, but the members can quit. Ownership is changed by gift giving or sales. And the most fickle of all formal relationships are human bonds, which can deteriorate in a second.

In summary:

- Form consists of objects and *structure*—the elements of form and their *formal relationship*. Both are present in a system, and both must be considered in analysis.
- There are three broad types of formal relationships:
  - Connection (relationships that create the formal connection over which functional interaction can take place)
  - Location and placement (including spatial/topological relations, address, and sequence)
  - Intangible (membership, ownership, human relationship, and so on)
- Formal relationships inform and influence the nature of the functional interaction and the emergence of function and performance.
- Formal relationships are static in that they *exist* (although they can be changed).

4.5 Formal Context

The Accompanying Systems, Whole Product System, and System Boundary

Another way to analyze and understand systems is by taking increasingly broader or more holistic views of the system and its context. This is an application of the Principle of Holism (see Box 2.5). We will see that it is useful to take two increasingly holistic views of a system, which we will call the whole product system and the use context.

Expanding outward from the product/system, the first level of context we encounter includes the other objects that are not part of the product/system but are essential for the system to deliver value. We call these the *accompanying systems*. The sum of the product/system and the accompanying systems is called the *whole product system*. The product/system is separated from the accompanying systems by the *system boundary* (Section 2.4).

What are the accompanying systems and therefore the whole product/system (Question 4d of Table 4.1)? It is vital for the architect to understand and model the accompanying systems, because they may play an important role in emergence. They must be present, connected, and working for the product/system to deliver value. There must be well-defined interfaces to these systems at the system boundary, whose existence is described by the answer to Question 4e.

The procedure for addressing Questions 4d and 4e of Table 4.1 is to look holistically at the objects of form that are near the system or are connected to it and ask, "Are these essential to the delivery of product/system value?" If so, then the abstractions of form are created for the elements of the accompanying systems, and the formal structure between the system and the accompanying systems is identified.

Applying this procedure to the pump in Figure 4.5, we can reason that at a minimum, there must be elements that transport fluid to and away from the pump: the inflow and outflow hoses. The entire assembly must be structurally supported by a pump support, and the motor must be

76 PART 2 • ANALYSIS OF SYSTEM ARCHITECTURE

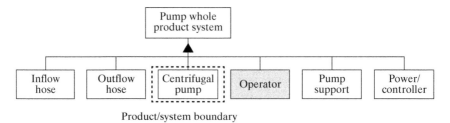

FIGURE 4.17 Decompositional view of a pump whole product system: Graphical OPM.

powered and controlled. The pump whole product system, including these four accompanying systems, is shown in Figure 4.17. Most often there is an operator involved in delivering value, and it is good practice to include the operator in diagrams such as Figure 4.17 to remind the architect of the importance of considering human interactions with the system.

The dashed line is the boundary that surrounds the product/system. It is good practice to draw the product/system boundary clearly, to reduce ambiguity and explicitly identify interfaces.

No structural information is shown in the decomposition view of Figure 4.17. This is best presented in diagrams like the connectivity structure in Figure 4.18. Now one can clearly see the connection between the principal objects of form of the pump and the accompanying systems, and the interfaces at the system boundary.

The Use Context

Expanding one more step outward, we find the next level of context, the *use context*. The whole product system fits within this use context, which includes the other objects that are normally present when the whole product system operates but are *not necessary for it to deliver value.*

The use context is important because it informs the function of the product/system. It gives place to the whole product system, and it gives us information on the environment in which the system operates and informs design. In analysis, the architect may have to speculate on the use

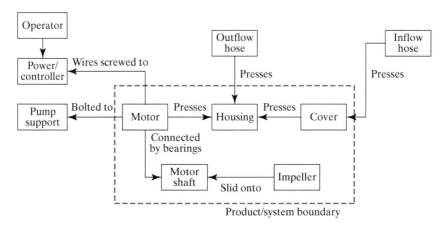

FIGURE 4.18 Connectivity structure of the simplified pump whole product system.

context, because it is rarely documented in system descriptions. The procedure for answering Question 4f of Table 4.1 is to ask, "What elements of context that are not already in the whole product system form our understanding of use and design?"

Applying this idea to the centrifugal pump, we will have to speculate, because the drawings we have available do not contain any hint about use context. We can guess that the pump might be an industrial sump pump for salt water removal or part of a home dishwasher. The system might have different requirements based on these use contexts.

The need for a clear understanding of the whole product system, informed by the use context, is driven by the accountability and responsibility of the architect. Even though the architect is *responsible* only for the product/system, it is inevitable that the architect will be held *accountable* for the function of the whole product system. If it does not work, the architect will likely be accountable regardless. Just as it is important for an architect to understand about two levels down in decomposition, it is important for him or her to understand about two levels out in context—the whole product system and the use context.

Imagine that the inflow hose for our pump is supplied by a third party. If the hose does not connect well to the expensive new pump, it is much more likely that the pump supplier will be held accountable for the bad interface than the hose supplier.

In summary:
- The whole product system consists of the product/system and the accompanying systems that are necessary to deliver value (Question 4d of Table 4.1).
- The system boundary runs between the system and the accompanying systems, and each crossing of the boundary indicates an interface that should be controlled (Question 4e of Table 4.1).
- The use context includes the objects that are normally present when the whole product system operates but are not essential for the delivery of value (Question 4f of Table 4.1). Use context establishes place, informs function, and influences design.

4.6 Form in Software Systems

The Software System: Informational Form and Duality

In this section, we will review the procedure of analyzing the form of a simple system by looking at a software system, bubblesort. Like the pump, this is a far more simple system than the tools of architecture are intended for, but it will bring out some important observations.

Figure 4.19 shows pseudocode for the standard bubblesort algorithm that sorts an array from its smallest to its largest entry by successively switching the location of adjacent entries if they are not in ascending order.

We will analyze the architecture guided by the six questions presented in Table 4.1. Question 4a is "What is the system?" In this case, it is obvious that the system is the pseudocode listed in Figure 4.19. But what is the *object of form* that represents this system? This requires us to think about what a *software object* is. Note that this is *not* the "software object" found in object-oriented programming, even though they share the same nomenclature.

Box 4.1 defined form as "the physical or informational embodiment of a system that exists...and is instrumental in the execution of function...[and that] exists prior to the

```
1   Procedure bubblesort (List array, number length_of_array)
2       for i=1 to length_of_array - 1;
3           for j=1 to length_of_array – I;
4               if array [j] > array [j+1] then
5                   temporary = array [j+1]
6                   array[j+1] = array [j]
7                   array[j] = temporary
8               end if
9           end of j loop
10      end of i loop
11  return array
12  End of procedure
```

FIGURE 4.19 Pseudocode for the bubblesort algorithm.

execution of function." Therefore, in software, the code (or pseudocode) is the form: It exists, it exists prior to the execution of function, and it is instrumental of function. It is implemented (written). When it is operated, this form is interpreted as an instruction that, when executed, leads to function. The passenger emergency instruction card (Figure 4.2) is a metaphor for software code. It had a set of objects (pictures, etc.) that exist, that have been implemented, and that, when operated, are interpreted by humans as instructions and lead to function.

There is a subtle difference between defining abstractions of software and information systems and defining those of a physical system, in that information itself is an abstraction. Informational form must always be stored or encoded in some physical form, and the two are a *duality*. As suggested in Box 4.7 on the Principle of Dualism, informational form must always be represented by physical form. Poems are in print; thoughts are encoded in neural patterns; DVD movies are optical markings encoded on a surface; and images are composed of pixels. In

---

Box 4.7   Principle of Dualism

"Dualism in philosophy, mind/body, free will/determinism, idealism/materialism appear as contradictory only because of underdeveloped formulation of the concepts involved."

Hegel's dialectic,
Science of Logic (1812–1816)

All built systems inherently and simultaneously exist in the physical domain and the informational domain. It is sometimes useful to explicitly consider both the physical and the informational views of a system.

- Dualism holds that things can be usefully considered in two domains simultaneously—in this case, physical and informational domains.
- Systems that we usefully think of as information systems are just abstractions of physical objects that store and process the information.
- Systems that we think of as physical store all the information about their form, but not necessarily about their function.

all cases, information is encoded in some physical form. When we speak about information, it is patterns in physical form that can be encoded and decoded by the user.

Formal Entities and Relationships in Software

Continuing in Table 4.1, Question 4b is "What are the principal elements of form?" Now that we understand that the code is form, the likely elements are obvious: the lines of pseudocode. Alternatively, we could abstract together the "procedure/end procedure," the "if/end if," the "for j/end of j loop," and the "for i/end of i loop" and identify eight important elements.

Decomposing the lines of code further does not provide additional useful information about the system. There are certainly details that are important, such as command semantics, order of variables, and naming conventions. This information is best treated as a detail of an object, much as the details of the impeller of the pump were important to its performance, but not to its architecture.

Question 4c of Table 4.2 is "What is the formal structure?" The procedure for determining formal relationships that is recommended in Section 4.3 is to examine each pairwise set of objects and determine whether there is a spatial/topological or a connectivity relationship between them.

For software, we need to consider the idea of spatial/topological relationships more generally. The structure of objects in software code is implicit and is suggested by (optional) indentations, code conventions, and language syntax. The most important element of spatial/topological structure in imperative languages is *sequence*—what lines of code are executed before what other lines of code. For example in Figure 4.19 it is apparent that lines 5–7 are in a sequence. Also, they are *within* the conditional "if statement," which is *within* the j loop, and so forth. Figure 4.20 shows the structural relationships for lines 4–8 explicitly.

The "precedes" or "follows" sequential spatial/topological relationship informs the transfer of control during execution. Likewise, the "contains" or "within" relationship informs the

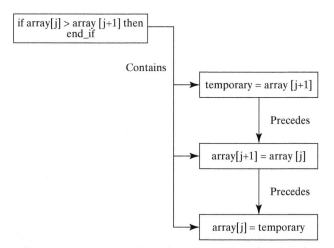

FIGURE 4.20 Spatial/topological structural view of a part of bubblesort procedure: Graphical OPM.

TABLE 4.8 | Spatial/topological structure of bubblesort: Design Structure Matrix (DSM) representation. F means follows, P means precedes, W means within, C means contains.

Object List	1	2/10	3/9	4/8	5	6	7	11	12	Calling routine	Compiler
1	X	F									
2/10	P	X	FW	W	W	W	W				
3/9		PC	X	FW	W	W	W				
4/8		C	PC	X	FW	W	W				
5		C	C	PC	X	F					
6		C	C	C	P	X	F				
7		C	C	C		P	X	F			
11							P	X	F		
12								P	X		
Calling routine											
Compiler											

execution of all the lines that are "contained" within the section of code, if the conditional test is met. In such imperative pseudocode, the spatial/topological structure informs the eventual flow of control, or at least the part that is strictly associated with structure.

The DSM tabular view of the spatial/topological structure of the bubblesort code is shown in Table 4.8. The grey section indicates the lines that were shown in Figure 4.20. This matrix makes explicit what lines are preceded by what other lines of code, as well as lines that are contained within the paired lines such as "if/end if." The DSM indicates that there are no spatial/topological relationships with the two objects outside of the procedure—that is, the compiler and calling routine.

The *connectivity* structural relationships for software are associated with the connections that will carry the functional interaction: the exchange of data or variables. As an example, the connectivity DSM for bubblesort is shown in Table 4.9. Here the entries are used to indicate the variable shared: the array, length of the array, temporary, and the indices for the i and j loops.

Comparing Tables 4.8 and 4.9 for the spatial/topological and connectivity structural relationships of the bubblesort procedure, we see less similarity between the two than in the corresponding arrays for the pump in Tables 4.5 and 4.6. In general, it is possible to make spatial/topological and connectivity structure more distinct in information systems than in physical systems.

Software Whole Product Systems, Boundaries, and Use Context

Table 4.1 next suggests Question 4d: "What are the accompanying systems?" and "What is the whole product system?" Recall that the whole product system contains the accompanying systems

TABLE 4.9 | Connectivity structure of bubblesort: DSM representation. Letters indicate the connection associated with the variable. A means array, L means length, T means temporary, I and J stand for indices, and C stands for sharing the entire procedure as an instruction with the compiler.

Object List	1/12	2/10	3/9	4/8	5	6	7	11	Calling routine	Compiler
1/12	X	L	L	A	A	A	A	A	AL	C
2/10	L	X	LI							C
3/9	L	LI	X	J	J	J	J			C
4/8	A		J	X	AJ	AJ	AJ	A		C
5	A		J	AJ	X	AJ	AJT	A		C
6	A		J	AJ	AJ	X	AJ	A		C
7	A		J	AJ	AJT	AJ	X	A		C
11	A			A	A	A	A	X	A	C
Calling routine	AL							A		
Compiler	C	C	C	C	C	C	C	C		

that are essential to deliver the product/system value. In the case of the bubblesort, value is delivered as a sorted array.

The additional accompanying systems in the whole product system therefore are at least the calling routine and the compiler that compiles the source code into executable instructions. Following on from there, the accompanying systems would be the processor that executes the instructions, and the input and output devices to enter the arrays. Ultimately, there are also the power supply, network, operator, and so forth. For simplicity, we will limit our discussion of the accompanying systems to the compiler and calling routine, as shown by the objects listed beyond the dashed line (system boundary, Question 4e) in Tables 4.8 and 4.9.

Examining these tables reveals the interfaces in the off-diagonal blocks. Although there are no important spatial/topological interfaces, there are important connectivity interfaces. There must be a connection that allows the array and array length to be shared between the calling routine and bubblesort. In addition, there must be a connection that enables the compiler to view the code and create the appropriate machine-level instructions.

Finally, Question 4f of Table 4.1 is "What is the use context?" From the information we have, this would be difficult to say. It could be an instructional module for students, a library application, or a piece of embedded system-critical software. The requirements for quality assurance, reliability, and maintainability would be different in these cases.

In summary:

- The objects of form for a software system are the code, which (when operated) will be interpreted as instructions.

- The software form can be decomposed into modules, procedures, and eventually (if desired) lines of code.
- The structure of software consists of spatial/topological structural relationships that inform control flow during execution, and connectivity structure, which is the instrumental connection that allows data and variable flow during execution.
- The whole product system for software includes the software code, the accompanying system (compiler, calling routine, processor, system software, input and output, and so on) and the use context systems that inform requirements and operations.

4.7 Summary

Form is a system attribute that is the physical/informational embodiment of a system that exists or has the potential to exist and is instrumental of function. Form can be decomposed into objects of form, and those objects have formal relationships or structure. Form is eventually implemented and operated. The form of the system combines with other accompanying systems (with which it interfaces at its boundary) to create the whole product system that generates value. A complete description of form includes the formal objects of the system, their structure, the accompanying systems, the interfaces, and the use context.

In this chapter, we took the approach of reverse engineering. We started the analysis with the more concrete system attribute of form and deferred the less concrete system attribute of function to the next chapter. In Table 4.1, we defined six questions to guide the analysis of form, along with procedures that might be useful in addressing the questions. We applied these procedures to two simple systems, a mechanical pump and bubblesort, and found that they were helpful in guiding the analysis of these systems.

References

[1] D. Dori, *Object-Process Methodology: A Holistic Paradigm* (Berlin, Heidelberg: Springer, 2002), pp. 1–453.

[2] Figure credit: Holt and Perry (2008), *SysML for Systems Engineering*.

Chapter 5
Function

5.1 Introduction

The analysis of function and form are interwoven. This chapter is the complement of Chapter 4 and establishes the groundwork for the analysis of function. When employing zigzagging (Section 3.5), the architect works in one domain for a period and then switches to the other. Therefore, discussing form first does not imply any strict sequence in analysis, but rather is a topical organization of material.

Function is less tangible than form, so we begin by defining and describing function in Section 5.2. In Section 5.3, we discuss the externally delivered function that is linked to the primary purpose of the system. This external function emerges from internal functions, as described in Section 5.4. The internal operands create functional interaction and yield the functional architecture (Section 5.5). A system often delivers secondary externally delivered functions as well, and these are discussed in Section 5.6. A short discussion of representing function in SysML concludes the chapter.

In this analysis of function, we will again be guided by a set of questions, given in Table 5.1.

5.2 Function in Architecture

Function

Function is what a system does. Function is about activity, in contrast with form, which is about existence. As indicated by Box 5.1, function involves operations, transformations, or actions. Performance (how well a system performs) is an attribute of function.

Box 5.1   Definition: *Function*

Function is the activity, operation, or transformation that causes or contributes to performance. In designed systems, function is the actions for which a system exists, which ultimately lead to the delivery of value. Function is executed by form, which is instrumental in function. Function emerges from functional interaction between entities.

Function is a product/system attribute.

TABLE 5.1 | Questions for defining function and architecture

Question	Produces
5a. What is the primary externally delivered value-related function? The value-related operand, its value-related states, and the process of changing the states? What is the abstraction of the instrumental form?	An operand-process-form construct that defines the abstraction of the system
5b. What are the principal internal functions? The internal operands and processes?	A set of processes and operands that represent the first-level and potentially second-level downward abstractions of the decomposed system
5c. What is the functional architecture? How do these internal functions connect to form the value pathway? How does the principal external function emerge?	A set of interactions among the processes at any level of decomposition
5d. What are the other important secondary value related external functions? How do they emerge from internal functions and pathways?	Other processes, operands, and their functional architecture that deliver value in addition to the principal externally delivered function

Just as form has entities of form (objects) and formal relationships among the entities (structure), function has entities of function and functional relationships among the entities (interactions). The function we see on the surface of a system is a result of the emergence that has taken place among these entities of function within the system, and in the whole product system.

Function is vitally important to systems and system architecture. All of the magic of systems and their emergence, and nearly all of the challenge of designing them, is found in the functional domain. Function is therefore a system attribute.

All systems have function, but in built systems, function must be conceived so that the goals of the system are achieved. Because function is less tangible than form, it is expressed in many different ways. In organizations, function is sometimes called tasks, roles, or responsibilities. In the Constitution of the United States, it is called powers. The Constitution lists the functions of the legislature, executive, and judiciary.

Function as Process Plus Operand

In Chapter 2, we found that a function is made up of a process and an operand, leading to the rule *function = process + operand*. For the house and emergency briefing card examples, we have *function = housing + resident* and *function = instructing + passenger*. In order to better understand function, we first need to better understand operands and processes.

OPERANDS Operands are a type of object (Box 4.2). Some objects in a system are elements of form, and some are operands. In human language, all objects are usually named by nouns.

All objects, including operand objects and objects of form, have the potential for stable unconditional existence for some period of time. However, form is a type of object that must exist before the function, is instrumental in the function, and is designed by the architect and supplied with the system.

On the other hand, an operand is the part of function that represents what is changed by the function—the *resident* is housed or the *passenger* is instructed. The more formal definition of the term "operand" is given in Box 5.2. We see that operands are acted upon by the process part of the function. Operands may exist before the function is executed, or they may appear as the function is executed. In general, they will be created, transformed, or consumed by the process.

Box 5.2   Definition: *Operand*

An operand is an object and therefore has the potential for stable, unconditional existence for some period of time. Operands are objects that need not exist prior to the execution of function and are in some way acted upon by the function. Operands may be created, modified, or consumed by the process part of function.

Interestingly, the operands are not usually supplied by the architect or builder of the system. Instead, they appear at the time of operation, usually from other sources. The resident is not supplied with the house, nor is the passenger supplied with the instruction card. Often, as architects, we do not have much control over the operand. The residents may have unique tastes in buildings, and the passenger may not speak any of the languages used on the instruction card. The operand is *almost never shown* on a conventional representation of a system.

PROCESSES   In human language, processes are associated with verbs, which are the aspect of our human language that captures actions or transformations. A process is defined in OPM as a pattern of transformation applied to one or more objects (Box 5.3). Processes generally involve the creation or destruction of an operand or a change in an operand. In our examples, we identify *housing* of the resident, where housing is a process. This process changes a resident from "un-housed" to "housed." For the aircraft emergency instruction card, the process is the *instructing* of the passenger. Instructing changes the knowledge of the passenger.

Box 5.3   Definition: *Process*

Process is a pattern of transformation undergone by an object. Processes generally involve creation of, destruction of, or a change in an operand.

In contrast with objects, a process is *transient* and *dynamic* and takes place along a timeline. You cannot take a photograph of process; you need to take a video. No single frame of the video reveals process; it is only when the entire sequence of frames is viewed that process can be comprehended. Imagine a video of a sprinter crossing a finish line. If we look at one frame alone, or at the international symbol for running (Figure 5.1), we can only infer dynamics. Cartoonists have developed a series of tricks to hint at motion within a single frame. Is the person on the left taking the hubcap off or putting it on?

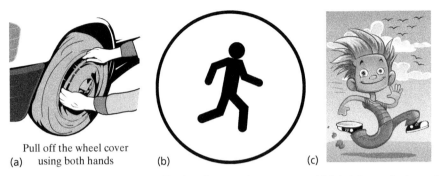

(a) Pull off the wheel cover using both hands (b) (c)

FIGURE 5.1 Examples of the difficulty of expressing process, which is inherently dynamic, with a static image: Illustration showing hubcap removal, a symbol for running, and a cartoon for running.   (Source: (b) Miguel Angel Salinas Salinas/Shutterstock (c) Screwy/Shutterstock.)

Analytical Representation of Function

In order to proceed further with the analysis of function, we need to develop an explicit analytical representation of process and operands. In the imprecise style of the figures of Chapter 2, we simply drew a round-cornered rectangle and wrote the statement of form, the process, and the operand inside.

OPM has a set of symbols used to represent objects and processes and their interactions (see Figure 5.2). The process is shown by the oval, with a label indicating the name of the process. The operand is shown as a rectangle, because it is an object, with the name of the operand written inside. The function is made up of these two OPM elements, the process and the operand, but does not have a separate OPM symbol.

In OPM, there are three ways of representing the relationships between an operand and a process, as shown in Figure 5.2. A single-headed arrow running from the process to the operand implies that the process creates the operand, as a factory would create a car, or as a team develops a design. The car or design did not exist before the process, and after the process, it did.

A single-headed arrow running from the operand to the process implies that the process consumes the operand, as when a factory consumes parts (to make a car), or the lungs consume

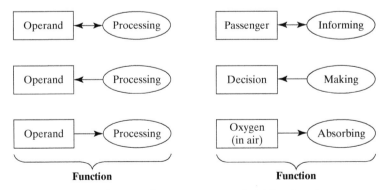

FIGURE 5.2 OPM diagram of process + operand yielding function. From top to bottom, these represent the process affecting the operand, consuming the operand, and producing the operand.

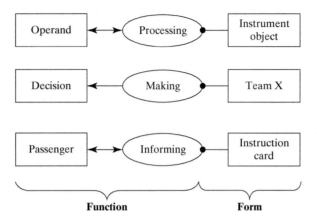

FIGURE 5.3 OPM representation of the canonical system architecture: The function as a process and an operand that the process affects, and the form as an instrument object.

oxygen from the air in the lungs (in order to transport it to the tissue and organs). Consumption is relatively rare. It implies that the abstraction of the operand no longer exists in the original nature and place after the process executes. Note that in the car factory example, the parts still exist after they leave the factory, but they have become part of another abstraction, called the car.

A double-headed arrow implies that the process affects the operand but does not consume or produce it. The residents were people both before and after moving into the house. The only difference was that their attribute of being housed was changed. The existence of the operand did not change: rather, some attribute of the operand was changed by the process.

OPM reserves the round-headed arrow as the symbol that links the instrument object with the process that it enables. The process then connects, with the appropriate pointed arrow (create, destroy, or affect), to the operand.

We are now in a position to analytically diagram form and function, as illustrated in the canonical model of a system shown in Figure 5.3.

A more explicit representation of the canonical system is illustrated in Figure 5.4, in which the states of the operand are explicitly shown. This view emphasizes that the operand changes its state through the action of the process enabled by the instrument objects.

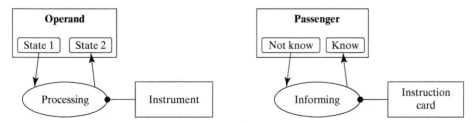

FIGURE 5.4 More explicit OPM representation of the canonical system architecture, showing the value-related states that are changed by the process.

This representation that emphasizes the states of the operand is closer to the way in which SysML expresses function using the state machine diagram. SysML state machine diagrams are made of states (rounded boxes), transitions (arrows), and events/signals (rectangles with an inward or outward triangle). Transitions triggered by events or signals capture the conditions under which the system goes from one state to another.

An example of a typical SysML state machine diagram (from a robot controller) is given in Figure 5.5. Recall that one of the differences between SysML and OPM is that SysML separates description of form and structure from description of function and behavior. Thus the SysML state machine diagram contains only functional information and does not show any relevant objects; note that neither the instruction card nor the air traveler appears on the state machine diagram of Figure 5.6.

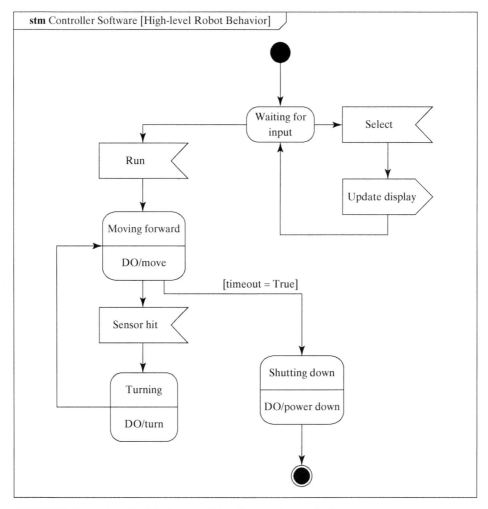

FIGURE 5.5 Example of SysML state machine diagram for a robot.

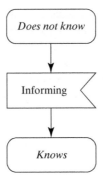

FIGURE 5.6 Simple SysML state machine diagram for the instruction card. Note the absence of objects on the diagram (no instruction card, no traveler).

In summary:
- Function is the activity, operation, or transformation that causes or contributes to performance.
- Function is composed of process and operand. Process is the pure transformation, while the operand is an object acted upon by the process. Processes are generally creation, transformation, and destruction.
- The canonical model of a system consists of the operand, the process designed to change it in such a way as to create value, and the instrument object that enables the process.

5.3 Analysis of External Function and Value

Primary Externally Delivered Function

We now turn our attention to Question 5a in Table 5.1: "What is the primary externally delivered value-related function?" We can parse this statement into two parts. First, the function must be externally delivered; it must cross the boundary of the system and influence something in the context. Second, there is a *primary* function—the function for which the system was built. The primary externally delivered functions for several example systems are given in Table 5.2.

There are systems that apparently are built only for their form. These are often associated with aesthetics, accomplishment, or collecting—art, trophies, and coins, for example. These elements of form actually have a function: pleasing + the viewer, impressing + others, and satisfying + a need to collect, respectively. If you were not pleased by art, you would not buy it or travel to see it. The function of these kinds of specialized forms is tied to the human reaction to the forms. The form, however, still delivers functions.

Function delivers value when it acts externally to the system. It is important that the function is *externally delivered*, that it is not only internal to the system. This implies an important rule of systems: *Function and value are always delivered at an interface in the system boundary.* For example, referring to some of the examples listed in Table 5.2, value is created only when the amplified signals are delivered at the output; only when the team delivers a design; and only when the oxygen is delivered to an organ or tissue.

TABLE 5.2 | Primary value-related externally delivered functions of a system: Function (operand with state and process of transformation) and the associated form

Function			Form
Value-related operand	Value-related attribute and state	Value-related process	System form
Output signal	Magnitude (higher)	Amplifying	Operational amplifier
Design	Completeness (complete)	Developing	Team X
Oxygen	Location (at organs)	Supplying	Circulatory system
Water	Pressure (high)	Pressurizing	Centrifugal pump
Array	Sorted-ness (sorted)	Sorting	Bubblesort code
Bread	Sliced-ness (sliced)	Making	Kitchen

Built systems have a *primary* externally delivered function. This is the function for which the system is built, or at least originally was built. If the built system does not deliver this function, it will be a failure. Operational amplifiers are built to amplify; they may also filter high frequencies, but if they do not amplify, they are a failure. Team X is created to deliver a design: if it does not do so, it is a failure.

The reason we focus on primary function is that in modern practice, engineers have an enormous list of requirements and potential features to deliver. Many products are delivered that fail to provide the primary function, providing instead a host of secondary features. Witness the challenges encountered by Apple when the iPhone case interfered with the radio antenna, disabling the primary function. The architect should try to reason about the primary function, before being distracted by other features.

The test to identify the primary value-related function is therefore:

- What is the function that the system was originally built to deliver, before all of the other value-related functions were added? Or,
- Which function, if it failed to materialize, would cause the operator to discard or replace the system?

We will call all the other externally delivered functions secondary functions and will discuss them in Section 5.6.

Value-Related Operands

Systems deliver *value* when they execute their primary externally delivered function. As indicated in Box 5.4, value is defined as benefit, worth, importance, or utility for some associated

> **Box 5.4** Definition: *Value*
>
> Value is benefit at cost. Benefit is synonymous with the worth, importance, or utility created by a system. An observer judges benefit subjectively. Cost is a measure of the contribution that must be made in exchange for the benefit.

cost. Good value can therefore be created by high benefit at modest cost, or by modest benefit at low cost. Note that others sometimes define the term "value" as a synonym of "benefit," but we will consistently equate value to benefit at cost. A similar definition is found in Lean Enterprise Value [1]: "Value is how various stakeholders find particular worth, utility, benefit, or reward in exchange for their respective contributions to the enterprise."

If the system is working properly, value will be delivered when the function emerges (Box 5.5), but this is far from certain. Value is subjective and is judged by an observer who is not the architect but, rather, is the customer, beneficiary, or user. (The procedure for trying to ensure that systems are designed so that they actually deliver value is the subject of Chapter 11.)

> **Box 5.5** Principle of Benefit Delivery
>
> *"Design must reflect the practical and aesthetic in business, but above all … good design must primarily serve people."*
>
> Thomas J. Watson, Chairman of IBM
>
> *"A cynic is a man who knows the price of everything and the value [i.e., the benefit] of nothing."*
>
> Oscar Wilde
>
> Good architectures deliver benefit, first and foremost, built on the primary externally delivered function of the systems by focusing on the emergence of functions, and their delivery across the system boundary at an interface.
>
> - The benefit delivered by systems is provided by their externally delivered function.
> - There is usually a single externally delivered function for which the system was originally defined, the absence of which will cause failure of the system.
> - The benefit is delivered at the boundary of the system, most likely at an interface.

It is critical to identify the *value-related operand* from among all of the operands on which the system acts. You can identify the value-related operand by asking, "What operand does the system exist to influence: to create, destroy, or affect?" For example, the operational amplifier clearly exists to create the high-amplitude output signal.

We should also consider the *attribute* of the operand whose change is associated with value, in this case the amplitude (Table 5.2). The system delivers value when the process "amplifying"

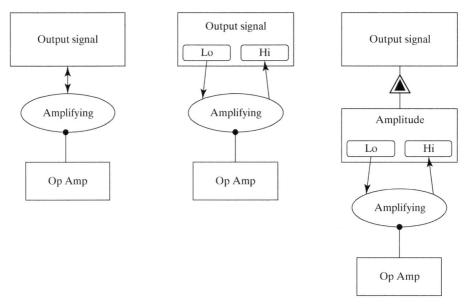

FIGURE 5.7 Three OPM diagrams that represent the primary externally delivered function with increasing detail.

acts on the operand "signal" in such a way as to change its value-related attribute "amplitude" to the state of "high." Figure 5.7 shows several ways in which this external function delivery can be diagrammed in OPM.

For Team X, the final design is the value-related operand. The team will undoubtedly produce other outputs: discarded options, analysis, time cards, and the like. But the reason why Team X exists is to produce the final design. When the boss asks, "Is the design complete yet?" it is an indication of the value-related function.

We can now finish answering Question 5a of Table 5.1. Focus on the primary externally delivered function and ask:

- What is the most specific way to describe the operand on which the system acts on the way to delivering value?
- What is the attribute of this operand whose change is associated with the delivery of value, and what is the final state associated with value?
- What is the specific process executed by the system to change the value-related state of this specific operand?

For our pump example, the answers to these questions are as follows: The operand is water, the value-related attribute is pressure (which should be high), and the process is pressurizing. How do we know that the value-related operand is the high-pressure water? The pump can be thought of as consuming low-pressure water and discharging high-pressure water, as shown in Figure 5.8.

We don't build pumps to consume low-pressure water, but we do build them to produce high-pressure water. Therefore, the high-pressure water is the value-related operand. An alternative

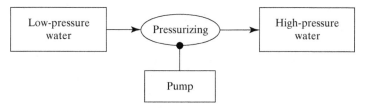

FIGURE 5.8 OPM diagram of the pump, showing the value-related operand.

representation in OPM is illustrated in Figure 5.9, which shows water as the operand, and the process changing the state associated with the attribute of pressure.

In bubblesort, the sorted array is the value-related operand. Sorted-ness is the attribute, and sorted is the state associated with value. The sorting process can be thought of as consuming the unsorted array and producing the value-related operand: the sorted array. Or the process can be thought of as changing the attribute of sorted-ness for the array from unsorted to sorted.

Often, there are operand objects present that are not directly related to the delivery of value, which we will simply call *other operands*. For example, in bubblesort, the array length variable that is passed to the procedure is an operand, but it is not the value-related operand. It is simply another operand used in sorting. Low-pressure water and electrical power are other operands of the pump.

Before we leave this section, we will introduce one more simple system, a kitchen that makes sliced bread. The last line of Table 5.2 summarizes the system: The externally delivered value-related operand is sliced bread. Thus bread is the operand, sliced is the value-related state, and making is the process.

In summary:
- The primary externally delivered function of a system emerges when the process acts across a boundary of the system at an interface.
- The benefit of a system is linked to the externally delivered function.
- Benefit materializes as a consequence of a process acting on the value-related operand. Analyze the operand, the value-related attribute and its change in state, and the process.

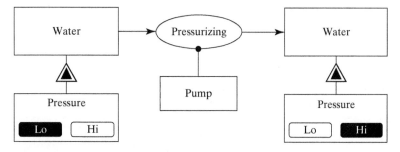

FIGURE 5.9 Alternative OPM diagram of the pump, showing the state of the value-related operand.

5.4 Analysis of Internal Function

Internal Function

Next we analyze the internal function (Question 5b of Table 5.1). Within a system, there are internal functions and relationships among these internal functions. Together these define the *functional architecture* of the system. The externally delivered function, performance, and "ilities" emerge from these internal functions and relationships.

First, we identify the important internal functions or the entities of function (discussed in this section). Then we construct the functional architecture (discussed in Section 5.5) when we connect up the entities of function. These two steps are similar to the procedure used in systems engineering of functional decomposition. [2]

The key task is to identify the *entities of internal function*. We have already encountered internal functions in the simple examples of Chapter 2. Notice that each of these internal functions has an operand part and a process part, as shown in Table 5.3. For the operational amplifier, gain is an operand, and setting is a process. The output signal (with voltage as a state) is an operand, and increasing is the process.

Identifying Internal Function

How did we identify the internal functions? There are several approaches, including reverse engineering of the form, standard blueprints, and metaphor. In reverse engineering, we start with

TABLE 5.3 | Principal value-related internal functions of a system, listing the operand and process parts of the function, and the system form

Principal Internal Operands	Internal Processes	System Form
Gain	Setting	Operational amplifier
Voltage	Increasing	
Requirements	Finalizing	Team X
Concepts	Developing	
Design	Approving	
Oxygen	Absorbing	Circulatory system
Blood	Pumping	
Oxygen	Delivering	
Flow	Accelerating	Centrifugal pump
Flow	Diffusing	
I index	Looping	Bubblesort code
J index	Looping	
Array entries	Testing	
Array entries	Exchanging	
Dough	Mixing	Kitchen
Bread	Baking	
Slices	Cutting	

the elements of form and reverse-engineer their functions. In the op-amp circuit there are only three elements. We reason, "What do these elements do?" A modest understanding of active analog circuits would reveal that the voltage divider in a feedback loop sets the effective gain for the op-amp, and the op-amp amplifies a signal.

Table 5.3 lists the internal functions of Team X. How did we identify these internal functions? This is a case in which we can apply our understanding of a *standard blueprint of process*. There are some combinations of processes that often go together to achieve an emergent function. For example, Box 5.6 describes a standard blueprint for making a decision.

Box 5.6 Insight: Standard Blueprints of Internal Processes

Sometimes functions just naturally unfold into a set of internal functions that are stable over many years. We call these standard blueprints. For example, when you compare the internal functions associated with riding a horse from your house to town 150 years ago (get the horse out of the barn, mount the horse, ride to town while navigating, arrive, and stable the horse) with the internal functions associated with driving a car from your house to the office today get in the car, (get the car out of the garage, drive while navigating, arrive, and park the car), you discover remarkable stability in these blueprints for internal process.

This may be because this is inherently a stable blueprint or because we use the blueprints of prior internal functions when we develop new systems. In either case, the blueprint is useful in understanding new systems.

Some examples of standard blueprints are:

- Transporting something with mass: overcome gravity, overcome drag, guide the mass
- Transferring information: encode the information, transmit while directing, decode the information
- Engaging an employee: recruit a candidate, conclude an agreement, train them, assign tasks, and periodically evaluate
- Making a decision: gather evidence, develop options, develop decision criteria, evaluate options, and decide
- Assembling parts: bring parts together, inspect them, assemble them, test the assembled system

---

Because developing a design is basically making a set of decisions, we would not be surprised if the internal processes of Team X therefore followed something of the blueprint of a decision process: Finalizing the requirements (Table 5.3) is gathering evidence and developing decision criteria (Box 5.6); developing design concepts is developing options; and approving the design is evaluating options and making a decision.

For the circulatory system, Table 5.3 indicates the internal functions. How did we identify these? In this case, you might consider a home heating system as a metaphor for the human circulatory system. In each there is a working fluid (water/blood) that absorbs "something" (heat/oxygen), and is pumped to a delivery point.

Other approaches to analyzing internal functions include the three techniques we developed in Chapter 2 for predicting emergence: precedent, analysis, and experiments. Often, just observing operations is valuable too. In applying these approaches, we must first *focus* on the internal processes that lead to the creation of value. We must also apply domain knowledge and experience.

We can now apply this procedural guidance to analyze making sliced bread. We could just observe someone making bread, or we could infer the internal processes from a standard blueprint for cooking: chop/mix food, heat, and break into chunks. It is not much of a jump from this blueprint to bread: mix to produce dough, heat to produce bread, and cut to produce slices. Figure 5.10 shows the OPM diagram for this set of internal processes.

Turning our attention to the pump, we focus on the value-related external function of pressurizing the water. It requires significant domain knowledge to identify the internal function of the system. The pump first increments kinetic energy and then trades it for the potential energy associated with pressure. Thus, in addition to the inflow and outflow functions, the two principal internal functions are *accelerating* to produce internal high-velocity flow and *diffusing* to produce internal high-pressure flow.

The matrix approach to representing internal function is to create an array that has the process on one axis and the operands on the second. We will call this the PO array (for process-operand). Looking only at the "create" steps of Table 5.4 shows the internal processes for the pump. Adding the "destroy" steps indicates the functional architecture, as discussed below.

For the bubblesort code, focus on the external value-related operand, the sorted array. Imagine a metaphor: You are at a gambling table sorting a hand of cards from smallest to largest. You might exchange them pairwise until they are sorted. We can also reverse-engineer the pseudocode, identifying what the lines of code do. We can consider the algorithm as a blueprint.

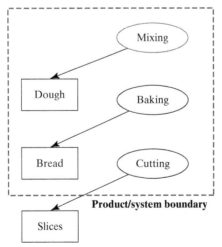

FIGURE 5.10 Internal functions of bread slice making.

TABLE 5.4 | The process-operand (PO) array of the functional architecture of the pump is made up of internal processes and operands, as well as interface operands (c' means create, d means destroy).

Process List	Internal low pressure flow	Internal high velocity flow	Internal high pressure flow	External low pressure flow	External high pressure flow
Inflowing	c'			d	
Accelerating	d	c'			
Diffusing		d	c'		
Outflowing			d		c'

Applying these techniques, we identify four internal functions associated with looping + I index, looping + J index, testing + array entries, and conditionally exchanging + array entries.

The OPM diagram of the internal functions of bubblesort (Figure 5.11) shows the internal functions. Looping affects the index value, in that it increments the value (changing its state), but it does not create or destroy the index. Likewise, exchanging affects the location or value of an array entry but does not create or destroy the entry. The array entries are actually an instrument of testing, and testing does not change anything about the array element; it just uses the information to make a decision. This is an example of an operand object being used as an instrument. Objects that are operands appear dynamically during operations and usually are transformed, but they can also act as instruments of subsequent processes.

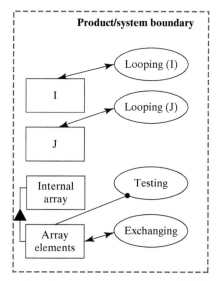

FIGURE 5.11 Internal functions of bubblesort.

98 PART 2 • ANALYSIS OF SYSTEM ARCHITECTURE

In summary:

- Within a system, there are internal functions, which include internal processes and internal operands.
- The procedure for identifying the internal functions includes these steps:
  - Focus on the primary externally delivered function, the value-related operand that leads to the creation or destruction of, or has an effect on, the value-related operand.
  - Involve domain experts; there is no substitute for experience.
  - Use the reverse engineering of key elements of form, standard blueprints in your field, and metaphors to understand internal functions.

5.5 Analysis of Functional Interactions and Functional Architecture

Functional Interactions and Functional Architecture

Having identified the entities of internal function, we can move to Question 5c of Table 5.1: "What is the functional architecture?" Answering this question requires understanding the relationships among the entities of function, which in Chapter 2 we called the interactions. This is the key step in analyzing architecture, because it is through the interaction of the internal functions that the externally delivered function emerges. The key idea is that the *exchanged or shared operands are the functional interactions.* The functions plus the functional interactions are the functional architecture.

A graphical OPM representation of the functional architecture of bread slice making is shown in Figure 5.12. Comparing Figure 5.12 to Figure 5.10 reveals that additional input

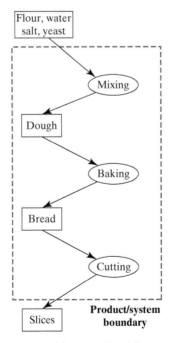

FIGURE 5.12 Internal functions and functional architecture of bread slice making.

operands have been identified: flour, water, salt, and yeast. The inputs are consumed by the mixing process; after mixing they no longer exist as entities. Mixing produces dough that is consumed by baking, after which the dough no longer exists. Likewise, the bread is consumed by cutting, and slices are produced. The operands of flour, water, yeast, salt, dough, bread, and slices *are the interactions between the processes*, being destroyed or created by mixing, baking, and cutting. The processes now connect to more than one operand, and the operands connect the processes. The external value-related operand (slices) cross the system boundary. A matrix-like representation called the process-operand array can represent the same information (Box 5.7).

> **Box 5.7 Methods: Creating the PO Array from OPM Diagrams**
>
> OPM diagrams of internal process and process-operand (PO) arrays enable us to visualize and analyze the internal functional architecture. Most people find it easier to sketch the OPM diagram and then convert it into a PO array, rather than trying to create the PO array from scratch. Both are useful representations, and each has a role to play in understanding systems. The procedure for translating an OPM diagram of internal function into a PO array is as follows:
>
> - Create an array in which the rows are processes and the columns are operands.
> - Isolate each process on the OPM diagram, and encode the connection between the process and connected operands in the row for the process, using the notation of Figure 5.2 and Figure 5.3.
> - Operands that are *created* by the process go into the PO array with a prime ('), and a notation (such as c') is placed in the column for the operand. This prime becomes important later when we are tracking causality in the system.
> - Operands that are *affected* by the process (a), are *destroyed* by the process (d), or are *instruments* of the process (I) exist before the process and go into the PO array with no prime and with the appropriate notation placed in the column for the operand.

Such a *simple flow-through* representation of interactions between processes is pleasing in its simplicity, and this kind of representation appears in many different types in engineering: pipe and flow software, control system block diagrams, SysML activity diagrams, and so on. In such simple flow-through systems, there is a unique operand that is created by the upstream process and destroyed by the downstream process, which constitutes the interaction.

Identifying Functional Interactions

Having worked through the sliced bread case, we can now generalize the procedure for identifying the functional interactions.

- Start with the diagram of the internal functions, which must include the value-related operand.
- Identify processes that are obviously missing from the diagram, such as inputting and outputting processes.
- Identify any operands that are obviously missing from the diagram of internal processes, such as inputs.

- For each process, ask whether there any other operands that are needed for the function to be completely represented.
- For each operand, ask what other processes it interacts with.
- Trace the path from the value-related output, considering whether the desired emergence is likely to occur based on the processes and operands represented.

Figure 5.13 shows an OPM diagram of the internal functions of the pump, fit into a simple flow-through pattern, in which each process destroys an operand and creates another. We can apply the procedure to derive this diagram. We start with only the "create" information in Table 5.4, which identifies two internal functions (accelerating and diffusing), and of course

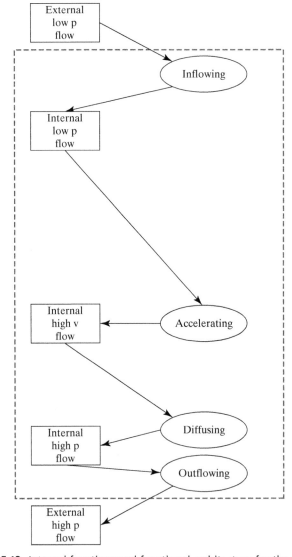

FIGURE 5.13 Internal functions and functional architecture for the pump.

we have the value-related operand (external high-pressure flow). There must also be an inflowing process and an outflowing process, as well as an external low-pressure flow. Between the processes of inflowing and accelerating, there must be an internal operand labeled internal low-pressure flow. With all the operands present, we connect the operands to processes as shown. We can now identify the *interactions* between the processes. For example, the interaction between the inflowing and accelerating processes is the exchange of the internal low-pressure flow operand.

The interpretation of the abstractions for the internal functions of a system is not unique, unlike decomposition of form, which is more tangible and therefore definite. There are often several equally valid ways to define the internal processes and operands. The real question centers on whether emergence is easily understood and predicted.

For example, Figure 5.13 gives the impression of the pump as a simple flow-through system. But we recognize that the water is not really destroyed by a process, and new water is not really created. What really happens is that the state of the water changes. Figure 5.14 captures this

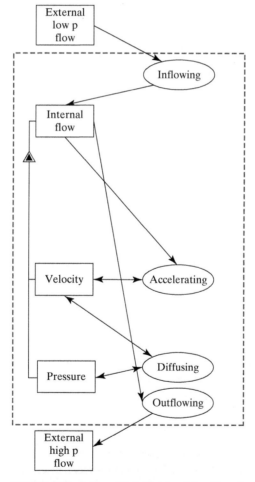

FIGURE 5.14 Internal functions and functional architecture: Alternative representation.

TABLE 5.5 | The process-operand (PO) array of the pump, made up of internal processes and operands, as well as interface operands, for the alternative representation with the state of the internal operands explicitly shown (c' means create, d means destroy, a means affect)

Process List	Internal flow — Internal flow existence	Internal flow — Velocity state	Internal flow — Pressure state	External low pressure flow	External high pressure flow
Inflowing A	c'			d	
Accelerating B		a			
Diffusing C		a	a		
Outflowing D	d				c'

alternative representation, showing that the internal flow has two important attributes: pressure and velocity. The accelerating process changes the velocity (from low to high), and the diffusing process changes both the velocity (from high to low) and the pressure (from low to high). Such a representation of state change loses the sense of simple flow-through but more accurately captures the physics.

The PO array that is the equivalent of Figure 5.14 is shown in Table 5.5. Now the operands at the top of the array explicitly show the states. The "a" under "velocity state" implies that the accelerating process affects (hence "a") the velocity state.

The Value Pathway

The exchanges of operands are the interactions among the processes, and the combination of the operands and internal processes is the *functional architecture*. One of the main features of the functional architecture is the *value pathway* along which value develops (Question 5c). It is along this pathway that benefit and ultimately value develop. It often starts with some input (low-pressure water, perhaps, or an unsorted array) and progressively moves through internal operands and processes until it reaches the value-related output. Figure 5.12, Figure 5.13, and Figure 5.14 show such internal value pathways. Note that in flow-though functional architectures, the value pathway is often readily evident, whereas in other types of architecture, more careful reasoning may be necessary to spot it.

One might ask, "What is not on the value pathway?" There are several classes of operands, processes, and objects of form that are not on the value pathway. In Figure 5.15, an elaboration of the bread slicing system, the value pathway starts with dough and ends with slices, as shown by the lines.

Things not on the value pathway include

- Entities of form, because form is not part of functional architecture and only provides an instrumental role to the processes (the oven and slicer), discussed in Chapter 4.
- Processes and operands that support secondary externally delivered functions (frying and fried dough), discussed in Section 5.6.

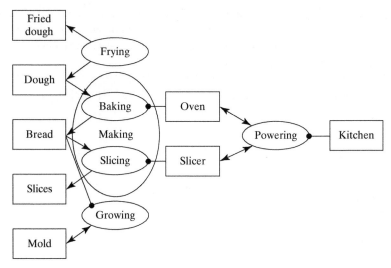

FIGURE 5.15 Entities not on a value pathway for bread slicing making.

- Processes and operands that don't contribute to the emergence of any desired externally delivered function. These are sometimes outcomes of unwanted side-effects, of poor design (growing mold), or of legacy, and they contribute to gratuitous complexity, as will be discussed in Chapter 13.
- Supporting processes and form. These are even further from the value pathway (the powering process enabled by the kitchen), discussed in Chapter 6.

Emergence and Zooming

Knowing the internal value pathway is a key step in effectively predicting emergence using any of the approaches discussed in Section 2.6. Figure 5.16 suggests the emergence of function of the pump. The pre-process operand is the low-pressure water, and the post-process value-related operand is the high-pressure water. In the view on the left, all of the internal operands are shown connecting to their associated internal processes. The right side of Figure 5.16 shows the pressurizing process emerging, with the internal processes suppressed and the internal and external operands simply attached to the emergent process.

Just like entities of form, process also has a whole–part relationship. The emergence of a system is the "smaller to larger" whole–part relationship. Zooming is the "larger to smaller" relationship. If we read Figure 5.16 from right to left, we see how the pressurizing process is zoomed. In zooming, we must identify not only the internal processes themselves but also how the external and internal operands interact with those processes.

A final approach to analyzing the functional architecture and predicting emergence is to project the functional architecture onto only the processes. This makes the way we view the operands as interactions more implicit. For simple flow-through systems, this can sometimes be done by inspection, but for more general systems it is best to apply the methodology of Box 5.8.

104 PART 2 • ANALYSIS OF SYSTEM ARCHITECTURE

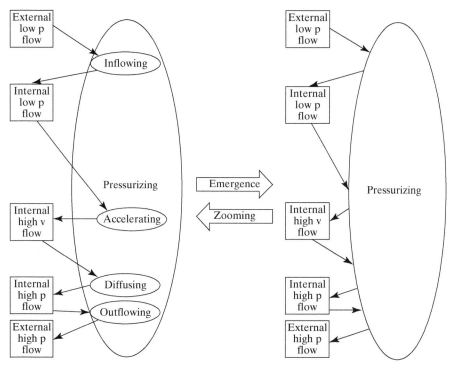

FIGURE 5.16 Emergence and zooming of process.

Figure 5.17 is an OPM showing the projection of the pump onto the processes. The method of Box 5.8 has been used, and the outcome has been converted back into an OPM diagram. Now the arrows between processes are open-headed (as they were in Chapter 4) and are labeled with the operand that links the processes. But these links now represent interactions among functions, not structure of form.

Functional Architecture in Software Systems

In principle, the functional architecture of software systems can be represented in the same way as any general system. Software engineers are comfortable with the functional domain, and they implement code (the form) only because it meets the demands of function. In practice, however, information systems are dynamic in structure and behavior and are therefore challenging to conceive as static architectures.

In software, the functions (as we define them) are of two types. There are computational statements (A = B + C), but there are also functions (as we define them) that dynamically allocate control (if..., then...). If there are functions, there must be operands: explicit variables used for computation, and more implicit control tokens used for control functions. The interactions produce the data interaction and control interactions among processes, respectively.

We can reverse-engineer bubblesort to show the functional architecture (Figure 5.19). As suggested by the procedure, we start with the internal processes and operands identified in Figure 5.11. This includes the value-related operand, which is the array in the final sorted state.

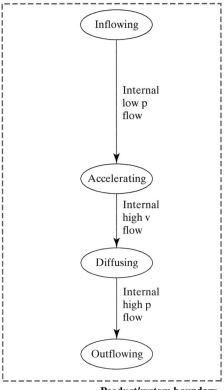

FIGURE 5.17 Projection of operands onto processes for the pump.

Box 5.8 Methods: **Projecting Operands onto Processes**

One of the advantages of matrix-like representations is that it is possible to compute on them. For example, the DSM community has created algorithms to sort and band the DSM. [3] Here we will introduce the ideas that the DSM can be broken into arrays and that projection operators can be developed.

The first step in this projection is to create the DSM:

$$\begin{vmatrix} PP & PO \\ OP & OO \end{vmatrix}$$

That is, the array is broken into four blocks, with the process-process (PP) block in the upper left and the operand-operand (OO) block in the lower right. The PP array is diagonal and simply contains an identifier for each process on the diagonal. Likewise, the OO array is diagonal with identifiers for each operand.

(continued)

The process-operand (PO) array has been developed (Box 5.7). The operand-process (OP) array is created by the following rules, as shown in Figure 5.18:

- Form the transpose of the PO array.
- Remove the primes from the create operator (c).
- Add primes to the destroy operator (d').
- Leave the affect (a) and instrument (I) operators the same.

Then form the array

$$PP_{operand} = PP \times OO \times OP$$

by symbolic matrix multiplication. This will create a projection of the operands onto the processes.

This will produce a nearly symmetric $PP_{operand}$ matrix, which indicates no causality. If you are interested in causality (for example, in what precedes and causes what), simply set to zero any term with a prime (') in it. This will leave a causal matrix that is not symmetric and can be interpreted by the normal DSM convention that an entry suggests flow from the column heading to the row heading.

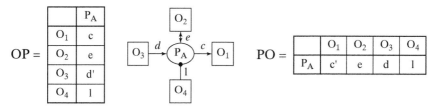

FIGURE 5.18 DSM notation for operand objects interacting with a single process.

We then identify missing processes: only importing and exporting, as well as operands that are missing, which include the internal variable called "array length," and the external operands: the external unsorted array, the external length of the array, and the external sorted array, which is delivered across the system boundary to create benefit. We can then trace the pathway from the external operands, through importing, looping, testing, and exchanging to exporting. This defines the functional architecture.

Some new features appear in the functional architecture. The operand for internal array length is not destroyed by the looping (I) or looping (J) processes, but rather becomes an instrument of looping. In general, operands that are created during operation can later become instruments. We also see the first appearance of a decomposition of an operand into its parts. In any given exchange, only two elements of the array are affected, not the entire array.

This analysis also reveals another new issue: There must be an operand that originates in the testing process, which otherwise has only instruments attached to it. The missing operand is part of the command flow of the routine. Figure 5.19 suggests that a control token created by the testing process is used by the exchanging process to decide whether to exchange the two array elements. Now that we have introduced the idea of a control token, we realize that the two looping processes must also create such tokens, in addition to their more explicit roll of incrementing

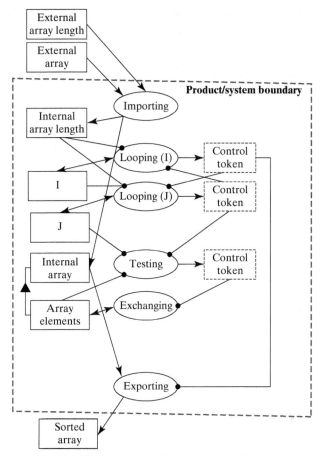

FIGURE 5.19 Functional architecture of bubblesort with command operands.

the I or J variables. Figure 5.19 shows the J looping process creating a token that passes control to either testing or I looping. I looping passes control to either J looping or exporting.

This example reveals the nature of the command interactions that exist in software, in parallel with the data interactions. In general, information systems of all kinds have some equivalent parallelism between control and data interactions.

In summary:

- The entities of internal function, the internal processes, and the interactions of operands among processes establish the functional architecture. Representation of internal function is not unique and can depend both on interpretation and on the style of representation.
- In general, the functional architecture will contain a value pathway of internal operands and processes along which the value-related external function emerges. We zoom the external function to identify the internal processes and operands.
- Diagrams and tables showing functional interactions can be simplified using projections, in which interactions between processes are indicated by shared operands.

5.6 Secondary Value-Related External and Internal Functions

Once a system delivers its primary value-related function, there is no reason why it should not also be designed to deliver other secondary value-related functions. Team X could produce the design and secondarily create reports to document the decision. The circulatory system could return carbon dioxide to the lungs, in addition to delivering oxygen. The operational amplifier could filter high-frequency noise, in addition to amplifying.

It is important to identify these secondary functions, because they are expected by customers and are a source of competitive advantage. Question 5d of Table 5.1 is "What are the other important secondary value-related external functions?" These other value-related external functions must also emerge from the internal processes. The procedure to address this question is just to repeatedly apply the reasoning for the primary value-related function to these secondary functions.

The centrifugal pump could have secondary value-related functions. Many water pumps have a "slinger" to sling water that leaks through the rotating seal away from entering the electric motor. The pump could also include a pressure sensor that indicates the outlet pressure of the water, a measure of the property associated with delivery of the primary value. Figure 5.20 shows

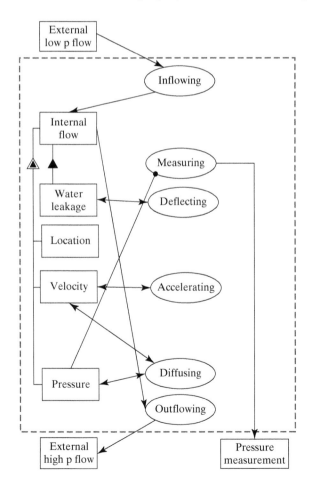

FIGURE 5.20 Internal functions associated with secondary value-related functions: Alternative representation.

TABLE 5.6 | Summary of the features of form and function

Form	Function
What a system is (a noun)	What a system does (a verb)
Objects + formal structure	Operands + processes
Aggregates (and decomposes)	Emerges (and zooms)
Enables function	Requires instrument of form
Specified at an interface	Specified at an interface
Source of cost	Source of external benefit
When transaction is a good	When transaction is a service

the two secondary value-related functions. One appears along the principal value pathway, the other on its own pathway.

In summary:

- Systems deliver secondary external value-related functions, in addition to the primary value-related function for which the system exists.

5.7 Summary

Like form, function is a system attribute conceived by the architect. It is the actions, activities, operations, and transformations that cause or contribute to performance. Function is made up of a process, which is pure activity, acting on an operand, which is an object changed by the process.

Built systems have a value-related operand, whose change is associated with the delivery of benefit and eventually value. The value of a system occurs when the externally delivered function acts on an external operand across the system boundary.

Externally delivered value-related function and other, secondary value-related functions emerge from the internal functional architecture, which usually features a value-related pathway. A complete description of functional architecture includes internal processes and internal operands.

A summary of features that contrast form and function appears in Table 5.6. In the next chapter, we will present the core idea of system architecture: the allocation of physical/informational function to elements of form.

References

[1] Tom Allen et al., *Lean Enterprise Value: Insights from MIT's Lean Aerospace Initiative* (New York: Palgrave, 2002).
[2] Gerhard Pahl and Wolfgang Beitz, *Engineering Design: A Systematic Approach* (New York: Springer, 1995).
[3] T.R. Browning, "Applying the Design Structure Matrix to System Decomposition and Integration Problems: A Review and New Directions," *IEEE Transactions on Engineering Management* 48, no. 3 (2001): 292–306.

Chapter 6
System Architecture

6.1 Introduction

We are now poised to understand the synthesis of form and function known as architecture. Form, a system attribute consisting of elements and structure, was developed in Chapter 4. Function, a system attribute consisting of entities of function and interactions through operands, collectively forming the functional architecture, was discussed in Chapter 5. Now we will explore system architecture, the combination of form and function.

Up to this point, we have agreed that system architecture is an abstract description of the entities of a system and the relationships between those entities. We will now adopt the more descriptive definition given in Box 6.1. The definition includes five key terms. *Function, form, relationship,* and *context* were discussed in Chapters 2 through 5. *Concept* is a mental image, notion, or system vision that maps form and function; it will be discussed in Chapter 7.

Box 6.1   Definition: *System Architecture*

System architecture is the embodiment of *concept,* the allocation of physical/informational *function* to the elements of *form,* and the definition of *relationships* among the elements and with the surrounding *context.*

As Figure 6.1 suggests, architecture is not an independent attribute, but the mapping between form and function.

The importance of architecture is expressed in the Principle of Value and Architecture (Box 6.2). In essence, if we are to deliver value with the system we build, it must have good architecture. Section 6.2 takes on the central question of how form and function map to each other. Section 6.3 discusses three additional ideas: what form appears in architecture to deal with non-idealities along the value pathway; what functions and forms must be present to support the value pathway; and the form and function of interfaces. Section 6.4 discusses the implications of operational behavior on system architecture. Finally, in Section 6.5, we survey the ways in which architecture can be represented. In this chapter, we will be guided by the questions in Table 6.1.

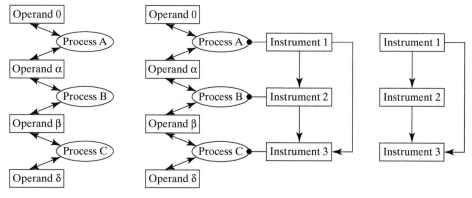

Functional architecture *System architecture* *Formal structure*

FIGURE 6.1 System architecture as the combination of functional architecture and elements and structure of form.

> **Box 6.2 Principle of Value and Architecture**
>
> *"Design is not just what it looks like and feels like. Design is how it works."*
>
> Steve Jobs
>
> *"If it is asserted that civilization is a real advance in the condition of man—and I think that it is, though only the wise improve their advantages—it must be shown that it has produced better dwellings without making them more costly; and the cost of a thing is the amount of what I will call life which is required to be exchanged for it, immediately or in the long run."*
>
> Henry David Thoreau, *Walden*
>
> *"Form and function should be one, joined in a spiritual union."*
>
> Frank Lloyd Wright
>
> Value is benefit at cost. Architecture is function enabled by form. There is a very close relationship between these two statements, because benefit is delivered by function, and form is associated with cost. Therefore, developing good architectures (desired function for minimal form) will be nearly synonymous with the delivery of value (benefit at minimal cost).
>
> - Benefit derives from the emergence of primary and secondary value-related function.
> - An axiom of lean manufacturing is that parts (form) attract cost.

6.2 System Architecture: Form and Function

Mapping of Form and Function

How would we describe the architectural differences between the two bridges shown in Figure 6.2? Both have the same external function (carrying vehicles), and they look surprisingly similar in form (both have two central towers, a roadbed, and cables).

TABLE 6.1 | Questions for defining architecture

Question	Produces
6a. How are the instrument objects mapped to the internal processes? How does the formal structure support functional interaction? How does it influence emergence?	The formal object-process-operand relationships that define the idealized architecture of the system
6b. What non-idealities require additional operands, processes, and instrument objects of form along the realistic internal value creation path?	The formal object-process-operand relationships that define the realistic architecture of the system
6c. What supporting functions and their instruments support the instrument objects on the value creation path?	The architecture one or two layers away from the value delivery pathway that is necessary to support the value delivery functions and their instruments
6d. What are the interfaces at the system boundaries? What operands are passed or shared? What are the processes at the interface? What are the instrument objects of the interface, and how are they related (identical, compatible)?	A formal and functional definition of the interfaces
6e. What is the sequence of execution of processes involved in delivering the primary and secondary functions?	An ordered set of actions the system undergoes in delivering function
6f. Are there parallel threads or strings of functions that execute as well?	A sequence diagram of the system
6g. Is actual clock time important to understand operations? What timing considerations or constraints are active?	

Yet the system architecture of these two bridges is different because their form–function mappings are different. The roadbed (form) of the cable-stay bridge carries load in compression (function), whereas the suspension bridge does not. The anchors (form) for the suspension bridge main cables are reacting to loads from the main cables (function). Since the cable-stay bridge carries tension in the roadbed, it cannot span as long a distance as a suspension bridge.

Question 6a of Table 6.1 is the crux of Part 2 and of this entire text. "How are the instrument objects mapped to the internal process? How does the formal structure support functional

FIGURE 6.2 A suspension bridge (left) and a cable-stay bridge (right). (Source: (Left) JTB MEDIA CREATION, Inc./Alamy (Right) CBCK/Shutterstock.)

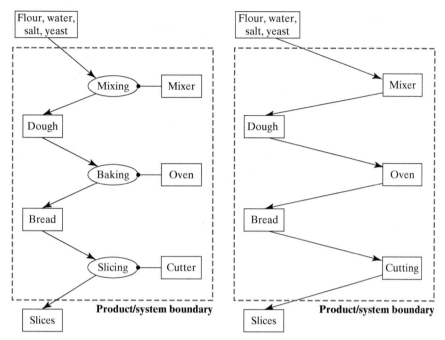

FIGURE 6.3 Simple system architecture of sliced bread making.

interaction? How does it influence emergence?" These questions define the important features of the architecture of a system—the relationship between form and function.

The center diagram of Figure 6.1 represents the simplest, and most commonly assumed architecture of a system: a simple flow-through functional architecture with distinct operands created at each stage. It further assumes that there is only one instrument linked to each internal process, and only one internal process linked to each instrument. This is a system that obeys the *independence* axiom. [1] If only life were so simple!

Figure 6.3 illustrates this simple architecture in a system for the production of sliced bread. As shown on the left, there is an instrument that does the mixing (a mixer), another that does the baking (an oven), and a third that does the slicing (a cutter). Each object of form is instrumental in one and only one process. Each process produces a distinct operand, so the function can be easily stated as something like *produce + outgoing operand*.

In such systems, the mental leap to the image on the right in Figure 6.3 is quite easy. The form is the proxy for the process it executes. [2] Dough is thought of as going into an oven, and bread emerges. One can temporarily suppress the processes altogether and imagine the operands passing through the instrument objects. When it works, it is quite useful, but is this paradigm generalizable?

One need only look at a slightly more complex representation of the architecture of producing slices of bread (Figure 6.4) to see the problem. The abstraction of a "mixer" has been replaced by a stirrer (such as a spoon), a bowl, and a stir agent (the cook). Likewise, the oven and the cutter have been shown in more detail. We no longer have the one-to-one mapping of elements of form and processes. And there is one instrument mapped to two processes; this becomes clear when we realize that the "stir agent" and the "slice agent" are the same person—the cook—as indicated

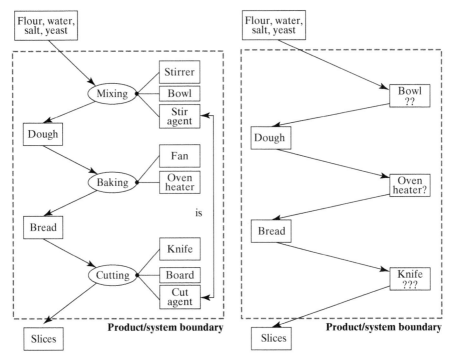

FIGURE 6.4 A not-so-simple system architecture of sliced bread making.

by the structural link. Now the diagram on the right does not make much sense. What does it mean to go into and out of a bowl or a knife? Using the instrument as the proxy for the process no longer makes sense.

Clearly, the simple model of one instrument for each process and one process for each instrument is not general. In reality, operands are created and destroyed not by elements of form but by processes. The mapping of form to function cannot be expected to be one-to-one. To gain perspective on various patterns in system architecture, we will examine fragments of architecture for some simple systems.

NO INSTRUMENT, OR COMBINED OPERAND AND INSTRUMENT OBJECT Figure 6.5a is a view of the architecture of ice melting, a case where no instrument object is shown. Is it really possible to have a process with no instrument? What does the melting? We would argue that there must be an instrument of form for each process. When you encounter such a process, think about what really is the enabler of the process.

Figure 6.5b shows a case where the operand, a person, is *also* the instrument of the process, walking. People "walk themselves." When it appears that there is no instrument for a process, consider that an operand might also be the instrument of the process.

To understand this apparent paradox, we have to look carefully at our definitions. We defined form (Box 4.1) as "the...embodiment of a system that exists [an object]...and is instrumental in the execution of function....Form exists prior to the execution of function." The person in Figure 6.5b satisfies this definition; she or he is the instrument of walking and exists prior to walking. We defined an operand (Box 5.2) as "an object...that *need not exist* prior to the

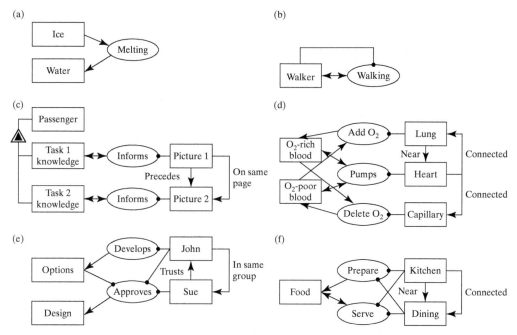

FIGURE 6.5 Fragments of system architecture of simple systems: (a) no instrument, (b) operand instrument, (c) one-to-one mapping affecting the same operand, (d) one-to-one mapping with multiple operands, (e) one-to-many mapping of form to processes, (f) many-to-many mapping of form to process.

execution of function and is in some way acted upon by the function." The person in Figure 6.5b also satisfies this definition; he or she is an object that happens to exist prior to the execution of function and is acted upon by the function. Thus, the definitions of form objects and operand objects are not mutually exclusive. An object can be both an element of form and an operand.

ONE-TO-ONE MAPPING WITH INCREASING OPERAND COMPLEXITY Figure 6.3 is an example of a simple flow-through architecture with one-to-one mapping of form and process. A slightly more complex architecture, the fragment from the emergency instructions supplied to passengers on airplanes, is shown in Figure 6.5c. Each picture informs the passenger and affects the passenger's state of task knowledge. The form-to-process mapping is the simple one-to-one case. The operands are not independent but, rather, are attributes of the same operand, a source of extra coupling.

In Figure 6.5d the heart, lung, and capillaries that make up the circulatory system all map one-to-one to a process. But on the operand side, the pumping process affects both oxygen-rich and oxygen-poor blood, while the adding and deleting processes create and destroy abstractions of oxygen-rich and oxygen-poor blood. From these three examples, we see that even in systems with one-to-one mapping of instrument to process, complexity can grow on the process-operand side.

ONE-TO-MANY AND MANY-TO-MANY MAPPING OF PROCESS AND OPERAND The bread slicing system of Figure 6.4, if we assume that the stir agent and the cut agent are different people, is the simplest case where one-to-one mapping breaks down. Here there are many instruments for each process, but only one process for each instrument. This is a very common occurrence. Very often, even simple processes need several instruments.

Moving up in complexity, Figure 6.5e depicts a fragment of Team X from Chapter 2. Team member John develops the design options, but he might help choose the design as well. Sue works only on the design decision and approval. John is now an instrument of two processes. This is also an example where an operand (the option) is subsequently used as an instrument. We also encountered this in the discussion of functional architecture in Section 5.5.

Finally, a fragment of the house example of Chapter 4 is shown in Figure 6.5f. Food is prepared primarily in the kitchen of a house, but some "preparation" can occur in the dining room (tossing salad, carving meat). Likewise, serving occurs primarily in the dining room, but some serving can occur in the kitchen (laying food out buffet-style, for instance). For these many-to-many mappings, the architect has to really focus on understanding all of the roles of each of the instruments.

Box 6.3 illustrates how the architecture for sliced bread making is represented in SysML.

> ### Box 6.3 Methods: **Architecture in SysML**
>
> To represent the system architecture of sliced bread making in SysML, we might use a combination of diagrams. Figure 6.6 shows the activity diagram in SysML, centered on the actions (processes) of mixing, baking, and cutting, with parameters (instrument objects and operands) shown as activity boundary nodes on the perimeter. To represent the baker, we could illustrate the baker (stir agent and cut agent) in a use case diagram (not shown).
>
>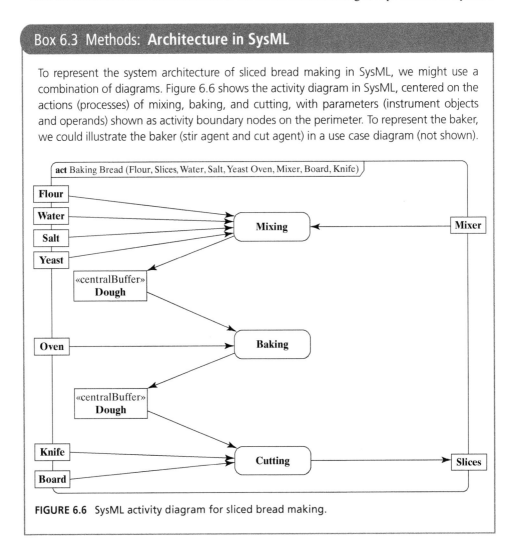
>
> **FIGURE 6.6** SysML activity diagram for sliced bread making.

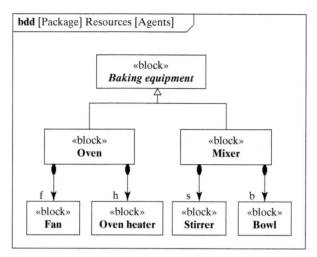

FIGURE 6.7 SysML Block definition diagram for the oven and mixer in sliced bread making.

To show decomposition of the mixer or of the oven, we could use a SysML structural diagram such as the block definition diagram (BDD) shown in Figure 6.7.

Identifying Form-to-Process Mapping

We are now ready to outline a procedure to address part of Question 6a in Table 6.1: "How are the instrument objects mapped to the internal processes?"

- Identify all of the important elements of form (Chapter 4).
- Identify all of the value-related internal processes and operands (Chapter 5).
- Ask what element of form is needed to execute each process (it may be among the identified elements of form, or it may lead to a new one).
- Map the elements of form to the internal processes.
- Identify the remaining unassigned elements of form, and try to reason about what process they might map to and whether this process is important to represent at this time.

After this procedure, reflect on the complexity of the form-to-process mapping and emergence.

Now we can apply this procedure to the analysis of the system architecture of the pump (Figure 6.8), which was developed from information in Figures 4.6 and 5.20. We will go process by process and identify the link between the internal processes and instrument objects. On the primary value pathway, the impeller is the instrument of accelerating the flow, and the housing is an instrument of diffusing as well as outflowing. Among the secondary value delivery functions, the water slinger is the instrument of deflecting the flow. There is no instrument of form shown in Figure 6.8 for the measuring function, because none was on the parts list of the motor in Chapter 4. Clearly we missed an instrument, such as a sensor. The mapping of form to process is nearly one-to-one, but the housing carries two processes. To obtain the desired emergence, the

118 PART 2 • ANALYSIS OF SYSTEM ARCHITECTURE

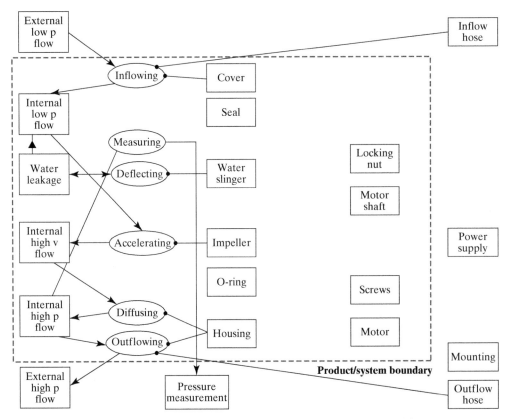

FIGURE 6.8 Pump system architecture with primary and secondary value-related functions.

design of this part must skillfully blend these two functions. We still have many elements of form unassigned to processes, which we will return to below.

The system architecture of the bubblesort routine is shown in Figure 6.9, which was developed from Figures 4.20 and 5.19. The importing, looping, testing, and exporting processes are associated with a single element of form (a line of code), assuming that we count the *end*...statement as part of the *for* or *if* statements. The exchanging process is spread over several lines of code.

The emergence of function form is clear in bubblesort. At its core are the three lines of code that do the exchanging. The "if" statement creates conditional exchanging. Adding the two "for" statements gives sorting, the desired emergent function.

Structure of Form Enables and Informs Functional Interactions

The externally delivered function emerges as form is assembled, and this emergence is enabled and informed by the structure of the form. Recall that structure—the formal relationships among the elements of form—can be of three types (Section 4.4). *Connections* indicate how the elements of form are linked or interconnected. Location and placement, including *spatial/topological relations* as well as address and sequence, indicate where elements are located. Intangible *relationships that simply exist* include membership, ownership, and human relationships.

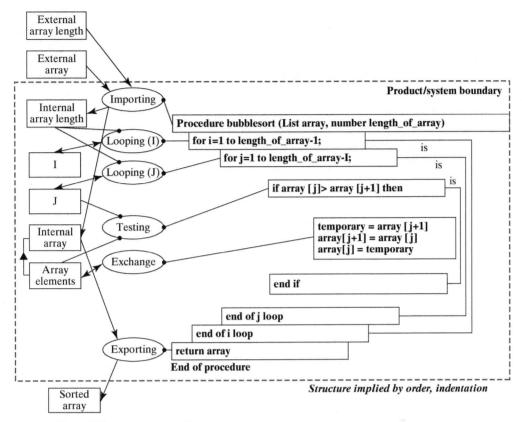

FIGURE 6.9 Bubblesort system architecture.

In general, *functional interaction takes place because there is a connection of form.* But be forewarned: Some functional interactions can take place in the absence of any connection whatsoever; examples include gravitational and electromagnetic interactions and some types of "ballistic" interactions among particles and fields.

In general, location and intangible relationships do not directly enable interactions and emergence of function but, rather, inform and influence the nature of the interaction or the degree of performance. In software, however, location, sequence, and address can be enabling of function.

We can illustrate this important role of structure by reexamining some of the very simple fragments in Figure 6.5.

In the circulatory system of Figure 6.5d, the connections enable the functional interaction: the movement of blood. The fact that the lung and heart are nearby informs performance but is not essential to function. It is easier to pump large quantities of blood because the heart and lungs are close together.

The passenger safety information card of Figure 6.5c is an information system, where the important feature of structure is sequence. The sequence of pictures is implicitly the sequence of instructions. If the pictures on the instruction card were reversed in order, the instructions would be invalid or at least confusing, so sequence enables emergence. The other spatial/topological relationship—that the individual pictures are on the same page (or, for software, in the same routine)—is a convenience. It may improve the reliability of the system (an "-ility") or how fast a passenger can execute the instructions (performance), but it is not essential to emergence of function. In Team X (Figure 6.5e) the fact that Sue trusts John (an intangible human relationship) probably improves performance.

Both the OPM and matrix representations have a way of showing which element of form is an instrument of which process, but neither has such an explicit way of showing which formal structural relationship enables which functional interaction. This is left to the architect to reason about and remember.

Identifying How Formal Structure Enables Function and Performance

Question 6a of Table 6.1 includes the questions "How does the formal structure support functional interaction?" and "How does it influence emergence?" We propose the following procedure to answer these questions.

- Identify which among the elements of form are likely to be important structural relationships (Chapter 4).
- Identify the functional interactions among the related internal processes (Chapter 5).
- Ask what formal structure is necessary to enable each of the functional interactions.
- Ask what formal structures would inform the other emergent properties of performance and the "ilities."
- Add, remove, or modify the structural relationships as necessary to support important functional interactions.

The spatial/topological structure of form among the lines of code of bubblesort is explicit, represented by sequence and the relationship of what is "within" other lines (informally represented by indentation). In Chapter 4, we summarized which lines share access to variables (this is only static information; the lines share the variables, but we are not sure yet whether they actually pass the variables—the interaction). In Chapter 5, we summarized the functional interactions. How well do the structure and the functional interactions match up?

Clearly, the sharing of variables allows the exchange of variables, so the data flow is relatively simple. The control flow is enabled in part by the static sequence. In this case, sequence is essential to emergence; if any two of the lines of code in the procedure were reversed, sorting would fail to emerge. The dynamic aspects of control flow, the movement of the control token in conditional statements, is a bit unresolved. Our model of system architecture would suggest that there must be some enabling structure that allows the flow of these control tokens. but none is evident in the code. There must therefore be, buried somewhere in the compiled version of the code, a sharing of some "control space" that enables the flow of the control tokens.

The important functional interactions for the pump revealed by Figure 6.8 are that the internal low-pressure flow is passed from the inflowing process, whose instrument is the cover, to the accelerating process, whose instrument is the impeller. A second important interaction is that the internal high-velocity flow is passed from the accelerating process, whose instrument is the impeller, to the diffusing process, whose instrument is the housing.

This suggests that there should be some sort of connectivity formal structure between the cover and the impeller, and again between the impeller and the housing, that enables these important functional interactions. Examining our connectivity structure diagram for the pump (Figure 4.16), we find no such connections. The connections in the pump support the interactions among elements of the mechanical parts, not the fluid flow. If we look at the summary of the structural relationships in Table 4.7, we observe that the only formal structure that exists between the impeller and the cover is a spatial relationship, which is listed as "close to" in Figure 4.14.

Can a spatial relationship of "close to" support interaction in a fluid flow? In fact it can, as was discussed in Section 4.4. Imagine the water stream from a garden hose emerging from the nozzle. If the nozzle were close to and pointed at a bucket, the water would be transferred from the hose to the bucket. In certain cases of such "ballistic" flows, proximity and alignment, which are spatial structural relationships, allow interaction as well. The formal structure of "close to" also informs performance. A smaller gap will allow a higher net rise in pressure.

In summary:
- The system architecture consists of instrument objects mapped onto internal processes. The processes interact through the exchange of operands, or through action on common operands. The relationship between the objects of form is the structure of the system (Figure 6.1).
- The mapping between the formal instruments and processes can be one-to-one (the simple case), many instruments to one process, or one instrument to many processes. In some cases, an operand that exists prior to the process can act as the instrument.
- The structure of the form supports emergence of function. Some aspects of structure actually are instruments of the interactions between the associated processes.
- The structure also informs the performance of the system—how well the system executes its functions.

6.3 Non-idealities, Supporting Layers, and Interfaces in System Architecture

Non-idealities in System Architecture

Looking at Figure 6.8, we observe that the seal, O-ring, motor, motor shaft, screws, locking nuts, power supply, and mounting are not yet associated with an internal process. What do these other instrument objects do?

The answer is that up to this point, we have analyzed only the idealized value delivery aspects of the pump, and there is a lot more to making a successful system. Getting the value delivery pathway right is necessary, but not sufficient. Other aspects of system function are necessary as well. These include non-idealities in the value pathway, supporting functions, and interface functions.

Non-idealities are the subject of Question 6b of Table 6.1. There is a whole class of these non-idealities that are associated with managing the operands: moving them, containing them, or storing them while the value processes are taking place. In the circulatory system, there are check valves to ensure that there is no reverse flow of blood. In a team, documents must be stored and controlled by a document system. At the physical level of computation, there are lots of bits stored and moved.

Another, related class of internal functions provides extra performance or robustness. For example, we might improve the performance of John in Team X (Figure 6.5e) by adding electronic design tools and rendering. We might improve the reliability of bread baking by monitoring the moisture content of the dough (Figure 6.5). And sometimes we need to offset biases, such as for the operational amplifier of Chapter 2.

If we are reverse-engineering, we can examine the instruments that are not associated with the principal and secondary value flows, and reason how these either help manage the operands or improve performance or robustness.

Applying this procedure to the pump of Figure 6.8, we identify the seal and O-ring as unassigned instruments of form that are "close" to the value pathway; they touch the water. The O-ring helps to contain the water, while the seal reduces leakage of the water by the shaft. The cover and housing already support ideal internal functions, but they have an additional non-ideal function: They guide and contain the water as it moves through the impeller. These non-idealities are shown in Figure 6.8.

It is interesting to note that bubblesort in Figure 6.9 does not appear to have any non-idealities. Digital information systems are supposed to be deterministic and not to have to deal with uncertainties. However, the run time implementation of this code would have additional instructions for moving and storing data that can be considered management of the operands. In communications systems, non-idealities include checking on correct transmission with processes such as error detection and correction codes.

Supporting Functions and Layers in Architecture

The objects of form on the value delivery pathway (such as the impeller and cover) themselves have to be *supported*. In general, physical objects have to be supported against gravity or other applied loads, powered and controlled. In organizations such as Team X, support comes from human resources, information technology, and management. Question 6c asks us to examine these supporting functions and their instruments.

A classic case of supporting layers is found in networked information systems. Bubblesort may run in an application layer, but if data are sent from bubblesort out on a network, layer upon layer will be supporting the application layer. In both the OSI seven-layer model and the Internet four-layer model, all of the value is created in the application layers. The remaining layers are all supporting. In Chapter 7, we will explore the architecture of networked information systems to illustrate this point.

Figure 6.10 shows the full architecture of the centrifugal pump. In general, the architecture of a system can be modeled in these *layers*: value operands, value processes, value instruments, and then several alternating layers of supporting processes and instruments. When identifying processes and instruments in the supporting layers, reflect on how the value instruments are supported mechanically, energetically, biochemically, informatically, and so on.

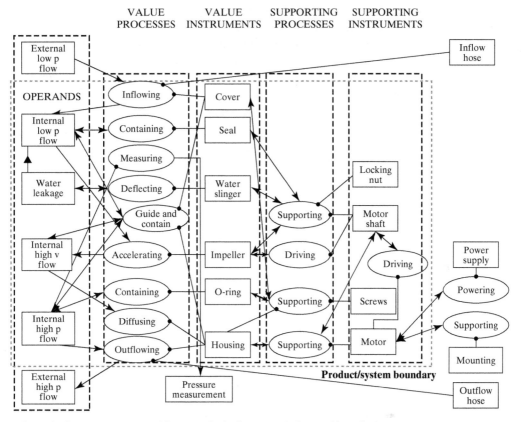

FIGURE 6.10 Pump system architecture, including supporting and interfacing processes.

The details of Figure 6.10 show some interesting features of supporting processes and form. Note that the motor drives the shaft but also supports the shaft. The shaft in turn drives and supports the impeller. The housing, which is in the value instrument column, also supports the cover and O-ring, which are also in the value instrument column. Thus the assignment of an instrument to one column or another is not unique, but in general, instruments should be identified as close to the value processes as possible.

System Interfaces in Form and Function

Boundaries are central to the definition of a system (Section 2.4). The importance of clearly communicating where the system boundary is located cannot be overstated. The system boundary divides the entities in the product/system from the accompanying system and defines the entities under the control of the architect. It identifies the interfaces that must be controlled. Question 6d of Table 6.1 focuses our attention on boundaries interfaces.

When something is passed into or out of a system, there is a system interface at the boundary crossing. For the pump, Figure 6.10 shows five potential interfaces: inflowing fluid, outflowing fluid, external powering, external supporting, and one to represent the pressure measurement leaving the product/system.

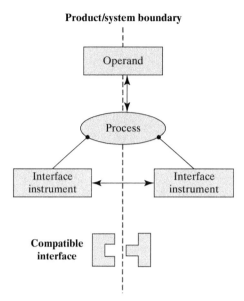

FIGURE 6.11 Model of an interface as a system boundary.

It is the obligation of the architect to *specify* the interfaces. Whether this is done by citing a standard or writing an interface control document, there is certain information that must be specified, as shown symbolically in Figure 6.11. Like all entities in a system, the interface has form and function (operand and process). These include:

- The operand that passes through the boundary, which will be the same on both sides.
- The process of passing the operand, which is usually shared or common at the boundary.
- The two interface instruments, which have some formal relationship. They can be androgynous, meaning that the interface form is identical on both sides, or they can be compatible, meaning that they are different but they somehow fit together.

Four representative interfaces from the pump of Figure 6.8 are shown in the interface diagrams of Figure 6.12. For the inflowing process the operand is the water, which passes through the boundary unchanged. The instruments are a hose and the cover, which contains some fitting for the hose (outflowing is similar). The powering interface process passes current and involves the power cable of the motor connecting to a socket. The measurement is transmitted from an internal wire to an external one. The mechanical supporting process passes load from the motor legs to a mounting plate.

A similar analysis of interfaces for bubblesort can be done by examining Figure 6.9. At the start of the procedure, two operands are passed by the importing function: the external array and the length. The instruments of the passing are a call statement in the main routine and the procedure statement. The exporting function passes the shared value-related operand back, with the return statement serving as the instrument, although the sorted array may just be left at the global address of the initial array.

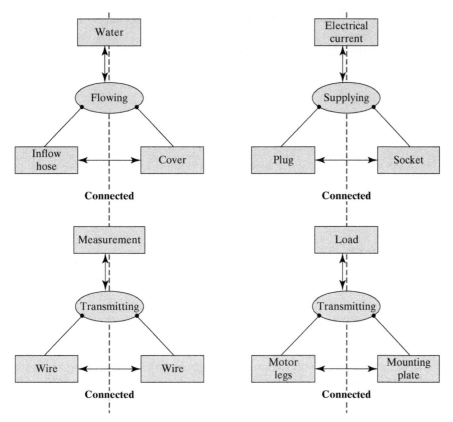

FIGURE 6.12 Models of an interface as a system boundary.

In summary:

- Even though there is nominally an idealized value delivery pathway along which primary and secondary value are delivered, in reality there are non-idealities that must be accommodated in the architecture of a system. These are often associated with the management of the operands—containing them, moving them, or storing them.
- Architecture can be represented in layers of operands, processes, and instruments of form. The supporting processes and the additional instruments often appear in one or two additional layers.
- Interfaces have form and function. At a minimum, the definition of an interface should include the common operand, the process, and the compatible instruments associated with the product/system and the accompanying system.

6.4 Operational Behavior

Our analysis of architecture has until now focused on form and function. But function is a somewhat quasi-static view of a system. Function is more about what the system *can do* than about what actually happens when the system runs. The operational or run time environment is more dynamic; things in the system happen in certain sequences, and the system interacts dynamically with surrounding systems and people. It is the difference between saying that an ATM can dispense money (its function) and listing, in order, all the steps necessary to actually get money out of an ATM (its operational behavior). In this section, we will examine three aspects of operational behavior: the operator, behavior, and operational cost.

Operator

The operator is the person who will use the system. In some cases (such as a riding a bicycle or playing a video game), the operator is so important that the system simply will not operate without his or her active involvement. For other products, the operator exercises supervisory control over operations, such as changing the channel on a television. All systems must accommodate humans any time they touch it, which is normally informed by human factors and industrial design. The human operator is so important that we will consider the human operator a product/system attribute.

In our simple systems, the pump and the bubblesort code have an operator at some level of supervision. Someone turns the pump on and off, or at least supervises the control software that does so. Code execution does not directly involve the operator, but someone ran the routine that called bubblesort.

Because these systems are not rich examples of operator interaction, we will introduce one more simple system: the two-levered corkscrew, commonly called a butterfly corkscrew (Figure 6.13), that is used to remove the cork from a wine bottle. The operator is essential to the operation of this device. The operator places the device on top of the bottle, twists the screw handle, pushes down the levers, and then removes the cork from the screw.

FIGURE 6.13 Butterfly corkscrew. (Source: Ekostsov/Fotolia)

Behavior

Behavior is the sequence of functions (and associated changes in state) that form executes in order to deliver value, and it is a product attribute. It is important to represent the sequence of events that contribute to external function. We will distinguish between the operations sequence and the dynamic behavior, or timing, of the system. Timing includes specific reference to clock time, whereas sequence is based on relationships among actions.

The *operations sequence* is the total progression of actions or processes that the system undergoes. Sequence will include actions associated with the primary and secondary externally delivered function, as well as with supporting and interfacing functions. From another perspective, sequence is the array of changes in state of the operands of the system. Sequence includes not only what step *follows* what, but also whether there are *overlaps*. Sometimes steps must follow in sequence (you must start the car before driving the car). Sometimes sequence is uncertain or optional (you can adjust the mirrors before or after starting the car).

There are various ways to represent the operating sequence of the corkscrew as it engages the cork, but the simplest is the sequence line diagram at the right in Figure 6.14, which indicates the starting and ending states for the processes. This is a variant of the SysML sequence diagram.

Even in this simple case, we immediately realize that the cork is not the only important object involved; the bottle is being restrained, and the corkscrew is being used. A more complete representation of the corkscrew sequence is shown by the other two sequences of Figure 6.14. Processes are represented on the right side of the line, and states are shown to the left of the line. In particular, the corkscrew undergoes several processes and state changes to achieve the desired value function. As indicated for the corkscrew (far right), these start from storage and end up in maintenance (cleaning) as will be discussed in Chapter 8.

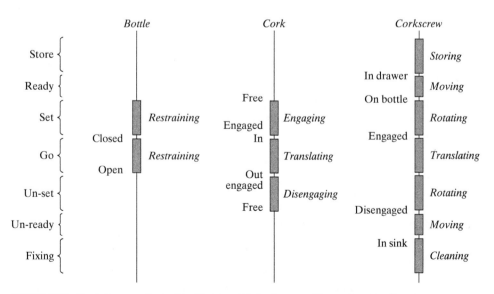

FIGURE 6.14 Corkscrew, cork, and bottle operations shown with a sequence diagram.

There are a variety of other diagrams that can be used to represent the states and processes of the system, and their conditional interaction. These include state control diagrams (used in control theory) and control flow diagrams (such as functional flow block diagrams). In SysML's, activity diagrams and state machine diagrams are used. OPM uses these diagrams or an animated simulation of tokens passing through the system to represent sequence. [3]

In contrast to sequence, *dynamic behavior* is the detailed timing of steps, start time, duration, overlap, and so on. Such behavior is important in real-time systems. For example, in applying the brakes of a car there would be a sequence, but there is also a timing constraint: The brakes must deploy within some number of milliseconds of pedal movement. Timing is also important for systems closely linked to clock and calendar, such as passenger train operations.

The issues that typically arise in dynamic behavior include start-up transients and latency of information or elements in a system. One of the most vexing aspects of real-time systems is timing constraints associated with multiple parallel sequences or threads. Multi-thread issues are not unique to software; a car can skid while turning, for example. In real-time software, multi-thread issues are particularly complex because of indeterminacies in timing in interactions with other applications or the operating system, including interrupts. Dynamic behavior (including multi-threads, non-synchronous events, and non-deterministic events) is, in our experience, best represented in the specific sub-domain of the architecture and is left to the architect to explore. The architect is advised to learn about these issues if they are relevant to the architecture of his or her product/system.

Operations Cost

Forecasting operations cost is frequently a consideration when architecting systems. The operational behavior, the operational concept (Chapter 7), and the details of system operations (Chapter 8) all influence the operational costs. Operational cost is typically expressed per event, per day, or in terms of usage (per mile in a car). Like operational behavior and operator, operational cost is a *product/system attribute.*

Operations cost is built up of a number of components. Chief among them is often the cost of the operator and other personnel needed to run the system. Consumables are another major category of operational cost. Indirect costs of operations include maintenance, nominal upgrades, and insurance.

The architect must carefully consider architectural decisions that affect the operational costs, because these costs are an important factor in the long-term competitiveness of the product/system.

In summary:
- Every system has an operator, whether that operator is closely involved in the operation of the system or plays a more supervisory role. The interface to the operator is a key part of the architecture.
- Behavior is the sequence of functions necessary for the value to emerge from the system. Behavior includes sequence, the execution of series of processes, and associated changes in states. Dynamic behavior or timing specifically references relative or absolute time in the operations.
- Operating cost is one of the key drivers in eventual system competitiveness. It is built up of direct costs (operational personnel, consumables) and indirect costs (maintenance, upgrades, and the like).

6.5 Reasoning about Architecture Using Representations

Simplified System Representation

The view of the system architecture that we have developed may be too complete for many needs. Simplified or condensed versions of the representation may be adequate or even preferable, providing the right amount of information or providing it in a way more broadly understandable. Starting with the full representation of the system architecture (Figure 6.10), there are several approaches to producing a simpler rendition: Hide detail in abstractions, leave out detail, or project the system. The first approach is to create larger-scale abstractions to conceal details. Examining Figure 6.10, for example, we could create an abstraction for the motor assembly that includes the motor and shaft.

Another option is to simplify the representation of the system by leaving out some details. The procedural guidance here is to reflect on the important details in the hierarchy, applying the Principle of Focus (Box 2.6), and preserving insight into the value pathways (operands and processes) while simplifying elsewhere.

A simplified pump architecture created in light of these guidelines is shown in Figure 6.15. Comparing Figure 6.15 with Figure 6.10 suggests that it may be easier to understand the pump architecture with the simplified drawing. The value pathway is preserved, but there is still a suggestion that the cover and housing manage the non-ideal movement of the operand. The role of

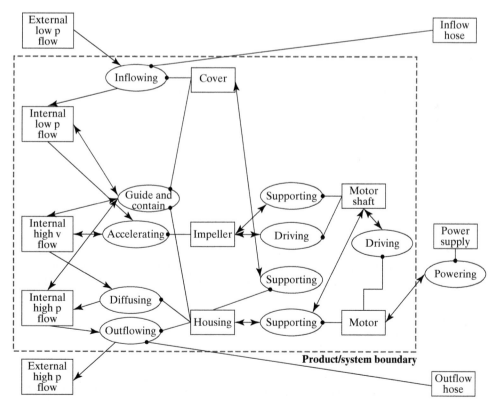

FIGURE 6.15 Simplified pump system architecture.

TABLE 6.2 | System architecture of the simplified pump. Internal and external processes are shown on the left, internal and external operands and internal and external instruments on the top.

		O1 internal low pressure flow	O2 internal high velocity flow	O3 internal high pressure flow	O4 external low pressure flow	O5 external high pressure flow	cover	impeller	housing	motor	motor shaft	pipe (in)	pipe (out)	power supply
P1	inflowing	c'			d		I					I		
P2	accelerating	d	c'					I						
P3	diffusing		d	c'				I						
P4	outflowing			d		c'		I					I	
P5	guiding/containing	a	a	a			I	I						
P6	supporting (rotating)								a		I			
P7	driving (shaft)								a		I			
P8	supporting (housing)							a	I					
P9	supporting (motor)								a	I	a			
P10	driving (motor)									I	a			
P11	powering										a			I

the motor shaft in supporting and driving the impeller stands out more. The key operand interfaces have been retained, as well as one of the supporting interfaces. We will use this simplified pump example for the remainder of the chapter.

The simplified architecture of the pump can also be represented by an array. In Box 5.10, the process-operand array (PO) was introduced. By extension, we can define a process-form array (PF), which represents the relationships between the processes and their instruments. Combining these two ideas, we develop the array |PO PF| of Table 6.2, which contains exactly the same information as Figure 6.15.

In Table 6.2, the shaded section represents the value-related processes. The graphical representation (Figure 6.10) has the advantage of showing the partitioning of the system into layers a bit more visually, whereas the matrix representation (Table 6.2) is more compact and allows for computation, as we will see shortly.

Projected System Representations

If there is still too much information in the simplified representation, another option is to condense the system by projection, as discussed in Section 3.6. Box 5.11 presented a projection of operands onto processes. Here we explore two additional types of projections: onto objects and onto form.

CHAPTER 6 • SYSTEM ARCHITECTURE   131

TABLE 6.3 | The full system DSM of a system architecture, an N-Squared matrix with the processes, operands, and objects of form shown on both axes

	Process	Operand	Form
Process	PP	PO	PF
Operand	OP	OO	OF
Form	FP	FO	FF

The starting point of this procedure is to construct a DSM as shown in Table 6.3. Box 5.11 defined the arrays PP, PO, OP, and OO. The PF array is defined above, and the FP array is derived from the PF array by analogy, using the procedure in Box 5.11. The FF array is generally a diagonal with an identifier for each element of form. This array *does not* contain the structural connections discussed in Section 4.4. The OF and FO arrays are generally zero, because the operands do not usually have a connection directly to objects of form; there is always an intervening process.

Projection onto Objects

In the projection of processes onto objects, the operands and instruments are shown explicitly, and the processes become links. The advantage of this projection is that to many viewers, the objects are more concrete. However, those less familiar with system architecture may require some explanation of the idea of *operand objects.* Such a projection onto objects for bread slicing is shown in Figure 6.16.

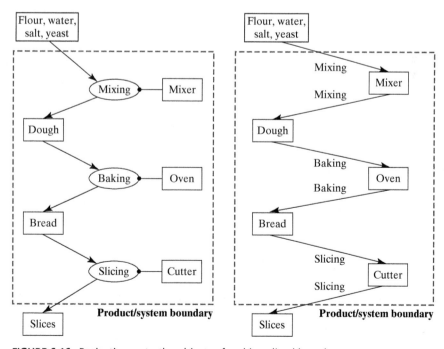

FIGURE 6.16 Projection onto the objects of making sliced bread.

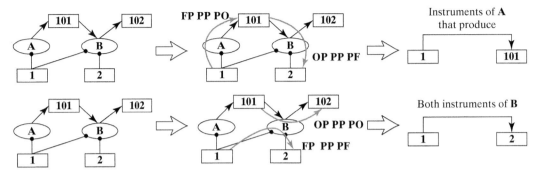

FIGURE 6.17 Schematic projection of the system architecture onto the objects.

Figure 6.17 shows the schematic for "hand drawing" the projection onto objects from an OPM diagram of the full architecture:

1. Isolate each object.
2. Follow links from the object (operands or instruments) to the connecting process and then to the next connecting object.
3. Represent this object-process-object path as a structural relationship between the two objects, and label it.

The formalism of matrix projection for this case is to compute the relationships between the operands (OP × PP × PO), the elements of form (FP × PP × PF), and the cross terms (OP × PP × PF) and (FP × PP × PO). The way in which they would be constructed if we were "hand drawing" the projection is shown in Figure 6.21. The final array representing the projection would read

$$\begin{vmatrix} OP \times PP \times PO & OP \times PP \times PF \\ FP \times PP \times PO & FP \times PP \times PF \end{vmatrix}$$

The terms with primes could be retained, or the primed terms could be dropped to more explicitly show causality.

The OPM diagram projected onto objects is shown for the value stream objects of the simplified pump in Figure 6.18. The five operands are connected by the processes. These same processes connect the operands to the elements of form.

All of the architectural information is captured in Figure 6.18 by the objects and the linking processes. You can reason from operands through the process links to instruments to understand function. You can also reason from form object to form object to understand the amount of *functional coupling* between the elements of form.

Projection onto Form

The entire system architecture can also be projected onto the most tangible elements, the objects of form. This projection makes the entities appear very simple but stores a lot of information on

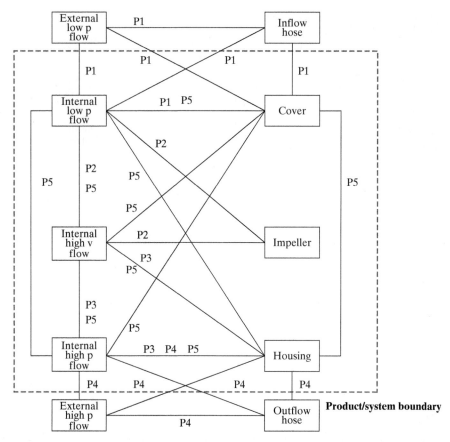

FIGURE 6.18 Projection of the value stream of the simplified pump system architecture onto the objects. (See Table 6.2 for legend of the processes.)

the links. Figure 6.19 shows the projection onto form of the sliced bread architecture. Now the links between objects of form encode the previous process, the connecting operand, and the following process.

The schematic for developing this information is shown in Figure 6.20. Now there are two parts of the projection. The links between elements for form through a single process are simple. However, in order to reach the operands, you have to go from an object of form to a process, then to an operand, then back to a process, and then back to an element of form (FP × PP × PO × OO × OP × PP × PF). The sum of these two resulting arrays will form an N-Squared matrix that corresponds to the most common DSM used for products, with elements of form on two sides.

The projected arrays for the simplified pump can be calculated from the information in Table 6.2 and are shown in Figure 6.21. The links between objects of form through processes

134 PART 2 • ANALYSIS OF SYSTEM ARCHITECTURE

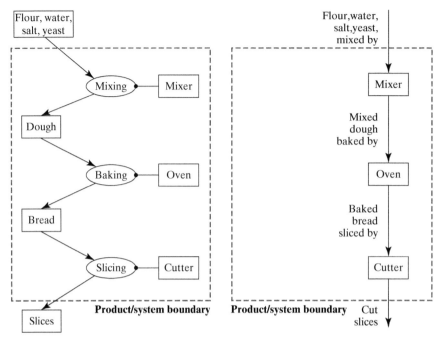

FIGURE 6.19 Projection onto the form of making sliced bread. (Compare with Figure 6.16.)

are quite simple to understand. The operand links are more complex and are marked with the operand that forms the link, but not the two processes.

In some ways, Figure 6.21 is the most compact representation of a system. The instrument objects are very concrete, and the information is sufficiently abstracted that most non-experts can understand it. Something similar to Figure 6.21 can be constructed with a SysML internal

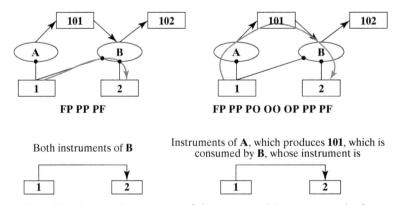

FIGURE 6.20 Schematic projection of the system architecture onto the form.

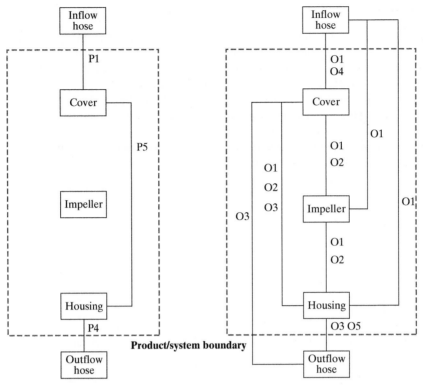

FIGURE 6.21 Projection of the value stream of the simplified pump system architecture onto the form. FP x PP x PF is shown on the left, and FP x PP x PO x OO x OP x PP x PF on the right. (See Table 6.2 for more information.)

block diagram if item flow links are used. Sometimes in making models, the operand interactions between the elements of form are represented by classes of interactions such as those shown in Table 6.4.

TABLE 6.4 | Classes of interactions for operand interactions

Matter	Mechanical	Mass exchange	Passes flow to
		Force/momentum	Pushes on
	Biochemical	Chemical	Reacts with
		Biological	Replicates
Energy		Work	Carries electricity
		Thermal energy	Heats
Information	Signal	Data	Transfers file
		Commands	Triggers
	Thought	Cognitive thought	Exchanges ideas
		Affective thought	Imparts beliefs

In summary:

- The representation of an architecture that explicitly contains operands, processes, and objects of form has the advantage of capturing and explicitly showing all of the information about the system, but it produces an image or matrix that is quite complex. Sometimes the representation can be simplified or abstracted.
- The architecture can be projected onto the operand and form objects. This produces a diagram that is simpler and is very useful to the architect, but it requires an understanding of the role of operands.
- The architecture can be projected onto just the objects of form, producing the most compact diagram, understandable by all, but with a lot of information stored on the links.

6.6 Summary

In Chapters 4, 5, and 6 we discussed the core ideas of system architecture: form, function, how function is allocated to form, and how function emerges as form is assembled. Although form and function are system attributes, the allocation we call architecture is not a system attribute, but a mapping between the two.

The structure of the form is quite important to consider when trying to understand emergence. There are aspects of structure that enable emergence, and aspects of form that inform the performance associated with the emergent function.

Primary and secondary value emerge along the value pathway, but non-idealities often appear as a consequence of the need to manage the operands and deal with uncertainties. The value pathway instruments must be supported by supporting processes and instruments, tracing through to interfaces with the context.

The final representation of architecture is quite complex (a lot of information about entities and relationships) and therefore may appear complicated (hard for a human to understand). Various abstractions, simplifications, and projections can be used to allow the architect to reason about the architecture, and to facilitate communication.

Measuring our progress by returning to the definition of system architecture in Box 6.1, we have presented the essence of system architecture in this chapter. Earlier we had discussed form and its relationships (structure) and with function and its relationships (interaction). What remains to discuss from our definition of architecture, is *concept*.

References

[1] Suh, Nam P. "Axiomatic Design Theory for Systems." *Research in Engineering Design* 10.4 (1998): 189–209.

[2] Ulrich, Karl T., and Steven D. Eppinger. *Product Design and Development.* Vol. 384. New York: McGraw-Hill, 1995.

[3] A software application has been developed to generate six of nine SysML views from an OPD, but an independent attempt at conducting the mapping is educational (Grobhstein and Dori, 2010).

Chapter 7
Solution-Neutral Function and Concepts

The authors would like to thank Peter Davison of MIT for his substantial contribution to this chapter.

7.1 Introduction

Forward Engineering and More Complex Systems

Up to this point, we have pursued an approach of reverse engineering; we have assumed that we have the system in front of us and we are trying to understand what the system is (form) and how it works (function). In Chapters 4 through 6 we did this with simple systems (pump, bubblesort, bread slicing, and so on) to work through the approaches of system architecture before complicating the picture with systems of realistic complexity. But frankly, using these techniques of system architecture on systems as simple as bubblesort and a pump is like taking a sledge hammer to a thumb tack.

It is not possible to understand all of system architecture from such simple systems examined from a reverse engineering perspective. Therefore, we will now shift to an approach of "forward engineering," in order to introduce two important ideas: solution-neutral function (Section 7.1) and concept (Section 7.3). In addition, we will introduce two systems of higher complexity: an air transportation service and a home data network. The air transportation service was chosen specifically to demonstrate how system architecting applies to the design of services. The home data system is a microcosm of global information networking.

In this chapter, we will introduce one new guiding question and begin to collate the questions from Chapters 4 through 6 into a method. Note that we modify some previous questions with new information, as shown by the brackets in Table 7.1.

An Introduction to Solution-Neutral Function and to Concept

We begin by introducing *concept* as a notion or shorthand that explains the system in brief. Let's use the simple system of a corkscrew from Chapter 6. A corkscrew opens a wine bottle, but can we name other concepts for opening a wine bottle?

How would we describe a corkscrew in a general fashion? A corkscrew uses a screw to pull the cork. Surely, this is not the only way to open a bottle. If we get creative, we might think of other methods to "pull" the cork. We might glue a tab to the top of the cork and pull on the tab. We could potentially open the bottle without even touching the cork, by creating suction on the cork: surrounding it with a lower-pressure environment. While we're thinking about using a pressure differential to remove the cork, we might consider pressurizing the air *inside* the bottle.

138 PART 2 • ANALYSIS OF SYSTEM ARCHITECTURE

TABLE 7.1 | Questions for defining concept. This table begins to build the "forward engineering" view. We have amended questions from Chapter 5 (5a and 5b) with square brackets [ ] to describe new ideas presented in Chapter 7.

Questions	Produces
7a. Who are the beneficiaries? What are their needs? What is the solution-neutral operand whose change of state will meet these needs? What are the value-related attribute and the solution-neutral process of changing the states? What are the other attributes of the operand and process?	A solution-neutral framing of the desired function of the system
5a. What is the primary externally delivered value-related function? The [specialized] value-related operand, its value-related states, and the [specialized] process of changing the states? What is the abstraction of the instrumental form? [What is the concept? What are several other concepts that satisfy the solution-neutral function?]	An operand–process–form construct that defines the abstraction of the system
5b. What are the principal internal functions? The internal operands and processes? [What are the specializations of those processes? What are the concept fragments? What is the integrated concept? What is the concept of operations?]	A set of processes and operands that represent the first-level and potentially second-level downward abstractions of the decomposed system.

This solution is actually sold as a household product, complete with a thin, hollow needle to pierce the cork, and a hand air pump. Alternatively, the wine steward in fine restaurants will sometimes use a fork-like device to pull on the sides of (shear) the cork.

Possible concepts for removing a wine bottle cork are shown in Figure 7.1, where we describe the concepts in a structured way as an operand-process-instrument set— for example, cork (operand) pushing (process) injected gas (instrument object).

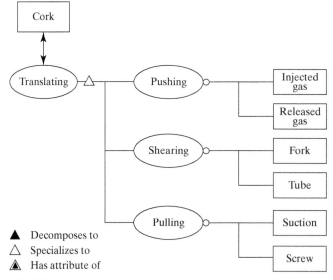

FIGURE 7.1 Concepts for removing a wine bottle cork.

This simple example illustrates a profound idea that we call solution-neutral function. [1] *Solution-neutral function* is the function of a system stated without reference to how the function is achieved. Box 7.1 explains this idea and its benefits.

> **Box 7.1 Principle of Solution-Neutral Function**
>
> *"We cannot solve our problems with the same thinking we used when we created them."*
>
> *Albert Einstein*
>
> Poor system specifications frequently contain clues about an intended solution, function, or form, and these clues may lead the architect to a narrower set of potential options. Use solution-neutral functions where possible, and use the hierarchy of solution-neutral statements to scope how broad an exploration of the problem is to be undertaken.

In Figure 7.1, we described the solution-neutral function of our system as a "cork translating" and then *specialized* the cork translation function into cork pushing, shearing, and pulling. Note that there is more than one possible instrument object per specialized function; "pulling" alone is insufficient to describe the concept. A concept that specializes cork translating is "cork pulling with a screw"—a corkscrew.

There are a number of advantages associated with this structured representation of concepts, and chief among them is that it stimulates us to think about alternative specialized functions. For example, pulling led us to think about pushing the cork. Drawing the diagram of Figure 7.1 is one approach to structured creativity, which we will discuss in Chapter 12.

Solution-neutral function exists in a hierarchy, as shown in Figure 7.2. Translating the cork generalizes to removing the cork, but we could also destroy the cork by burning or melting it. Removing the cork generalizes to opening the bottle, which also specializes to breaching the cork (drilling a hole in it) or breaking the bottle to get at the wine. The highest level shown is accessing the wine. In Figure 7.2, we have omitted the associated instrument object in order to fit all the solution-neutral functions on one diagram: the corkscrew for pulling, the blowtorch for melting the cork, the table edge for breaking the bottle, and so on.

The breadth of concepts we generate is heavily dependent on the functional intent we pose. In this case, *cork removing* leads us to a narrower set of solutions than *wine accessing*. All else being equal, the more solution-neutral our expression of the functional intent of the system, the broader the set of concepts we will develop.

We have to choose where in the hierarchy of solution-neutral function to be. We could have continued upward to *drink providing* in the diagram, but there is the matter of practicality to consider. This is a question of functional intent. If we have a wine bottle on the table, it does not make sense to consider concepts that sit above *wine accessing*.

Having introduced the three key ideas of concept, solution-neutral function, and functional intent, let's proceed to define each idea more rigorously in the context of higher-complexity systems.

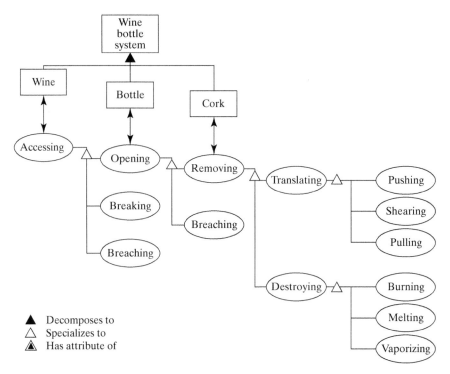

FIGURE 7.2 A hierarchy of broader concepts for accessing wine in a bottle.

7.2 Identifying the Solution-Neutral Function

In order to move to more realistic systems, we will now introduce two higher-complexity examples: an air transportation service and a home information network. We'll begin by identifying the functional intent expressed as solution-neutral function. We use *intent* to designate goals for a system. In order to reason about the architecture of a system, we need to understand something about the goals.

We will focus here on the functional intent derived from the primary need of the primary beneficiary. A more detailed and comprehensive approach to prioritizing stakeholders and goals derived from their needs will be presented in Chapter 11.

For the air transportation service, we'll focus on the traveler as the primary beneficiary. The traveler's primary need might be to develop business in another city by traveling to see a client. For our home data network, the primary beneficiary is the person using the network, whom we will call the *surfer*. His or her primary need might be to buy an interesting book online.

With the ingredients discussed so far, we are building a procedure for deriving a solution-neutral statement of a functional intent, which is summarized in Question 7a of Table 7.1. The procedure is as follows:

- Consider the beneficiary.
- Identify the need of the beneficiary you are trying to fill.

- Identify the *solution-neutral operand* that, if acted upon, will yield the desired benefit.
- Identify the *attribute* of the solution-neutral operand that, if changed, will yield the desired benefit.
- Perhaps identify *other relevant attributes* of the solution-neutral operand that are important to the statement and fulfillment of the goal.
- Define the *solution-neutral process* that changes the benefit-related attribute.
- Perhaps identify *relevant attributes* of the solution-neutral process.

The results of following these steps are shown in Table 7.2 for the air transportation service and home data network. These examples illustrate how carefully you have to think about the beneficiary's needs in order to write a solution-neutral statement of goals.

In the first case, we have to understand that the traveler needs to visit another city. Visiting is a conceptually rich idea, which involves a traveler traveling to someplace, staying for a while, and returning. So the operand in this endeavor is the *traveler*. The value-related state of the traveler is *location*, and the value-related process is *changing*, which give us *changing location of the traveler* as the value function. Changing location can also be called traveling or transporting. It is also useful to know other attributes of the traveler (that the traveler is alone, with light luggage) and attributes of traveling (it needs to be done on demand and safely).

Here is a summary version of the *solution-neutral function*: "safely and on demand transporting to a new location a traveler with light luggage." Note that we have not said anything about how the transporting happens. We are in the solution-neutral domain.

Table 7.2 also lists the results for the home data network case, where the surfer wants to buy a book. *Book* is the operand, and *ownership* is the value-related attribute. The other attribute has something to do with a book you want to read. *Buying* is the process of changing ownership by exchange of a good for money, and online is an attribute of how the buying will take place. The solution-neutral function is "buying online a book that the buyer likes."

These solution-neutral functions become the functional intent of the systems. Meeting these value-focused goals will probably meet the need of the beneficiary and deliver value. There will certainly be other intents for the system based on secondary value delivery: the needs of enterprise stakeholders and so on. Examining all the influences that impact goals is discussed in Chapter 10, and the process of deriving a complete set of goals is presented in Chapter 11.

TABLE 7.2 | Formulation of the solution-neutral functional intent for the transportation service and the home network

Question	Transportation Service	Home Network
Beneficiary?	Traveler	Surfer
Need?	"Visit a client in another city"	"Buy a cool book"
Solution-neutral operand?	Traveler	Book
Benefit-related attribute?	Location	Ownership
Other operand attributes?	Alone with light luggage	Consistent with tastes
Solution-neutral process?	Changing (transporting)	Buying
Attributes of process?	Safely and on demand	Online

When the system operates, an interesting thing happens. The intent vanishes! When people fly in an airplane, it hard to tell why they are traveling. When we see a room of servers, it is hard to tell for what purpose these servers are being used. In designing a system, it is critical that the architect state the solution-neutral function. However, once a system is built, it can be difficult to determine what the intent is.

This may seem elementary, but this formalism captures a key idea: Intent is often discussed but rarely written down. Intent may be discussed casually in the hallway ("I thought we were building a retro-inspired minivan.") or it may be the centerpiece of product concept renderings ("Here's the artist's sketch of our retro-inspired minivan."). The architect must identify and record the main functional intent of the system and use it to guide design.

In summary:

- The functional intent for a system should be stated as a solution-neutral function.
- The solution-neutral function contains an operand and a process (and their attributes) that are linked to the delivery of value; It contains no reference to solution.

7.3 Concept

The Notion of Concept

The intellectual distance from the solution-neutral function to the architecture of the system is a large gap to jump. In order to help, we create an intellectual construct known as concept. This intellectual construct is not strictly necessary, but it is very useful in reasoning about complex systems.

Concept is the high-level mapping of function to form. It is defined in Box 7.2. Concept is a way to conveniently and concisely convey the vision of the system. To put it another way, the concept

- Is the transition point from the solution-neutral to the solution-specific.
- Must allow for value-related functions to be executed, enabled by form.
- Establishes the vocabulary for the solution and is the beginning of the development of architecture.
- Implicitly sets the design parameters of the system.
- Implicitly sets the level of technology.

Box 7.2 Definition of Concept

Concept is a product or system vision, idea, notion, or mental image that maps function to form. It is a scheme for the system and how it works. It embodies a sense of how the system will function and an abstraction of the system form. It is a simplification of the system architecture that allows for high-level reasoning.

Concept is *not* a product/system attribute but a notional mapping between two attributes: form and function.

Both concept and architecture contain mappings of function to form. However, concept explains how function is mapped to form in a general way. Figure 7.3 suggests this notional

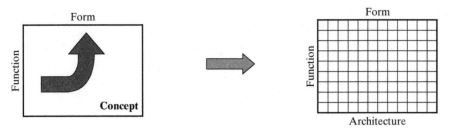

FIGURE 7.3 Relationship between concept and architecture.

mapping of function and form by concept, and by the much more elaborate information that is contained in the architecture. If you have a concept in mind, it is a guide to building the architecture. If you have an architecture in front of you, *concept rationalizes the architecture.*

We can use simple systems to illustrate these main points. The solution-neutral function of the pump is "moving fluid." The concept is water (specific operand) pressurizing (specific process) using a centrifugal pump (specific instrument of form). Here *specific* means "in the solution-specific domain," as opposed to the solution-neutral domain. So now we have transitions from the solution-neutral to the solution-specific by choosing the concept. Referring to a centrifugal pump immediately establishes a vocabulary of motor, housing, and impeller; sets design parameters such as impeller speed and pressure rise; and implicitly sets the level of technology (higher than moving water with buckets, but lower than high-temperature turbo machinery). In contrast to one simple phrase describing concept (a centrifugal pump pressurizing water), the definition of the architecture would require a detailed description of the parts, their structure, and their mapping to internal functions, as was developed in Chapter 6.

Likewise, bubblesort is actually the name of a concept. The solution-neutral function is "sorting the array." The concept is array (specific operand) exchanging sequentially (specific process) using bubblesort (specific instrument of form). Bubblesort allows for the sorting but is not as fast as other algorithms such as quicksort. Mentioning bubblesort to anyone who has studied algorithms immediately conjures up a simple algorithm, a solution vocabulary, and a set of instrument objects that must be specified to define the architecture (the code). From these two examples, it is clear how much more is defined about a system when the concept is chosen than is defined before the concept is chosen—and how much more must be specified to define the architecture.

Framework for Developing Concept

Figure 7.4 shows the elements of a rigorous definition of the concept for a system. Outside the concept box in Figure 7.4, we show the solution-neutral function as a precursor in the column labeled "Intent." The column labeled "Function" shows that we should identify the solution-specific operand and the specific operating function of the system, along with their attributes. This is a *specialization* of the solution-neutral function. The column labeled "Form" shows that we identify the specific form and attributes of the form.

There are five key ideas in Figure 7.4, implied by the five shapes with thicker borders. Table 7.3 shows the analysis of the systems we have been discussing. In the case of the pump,

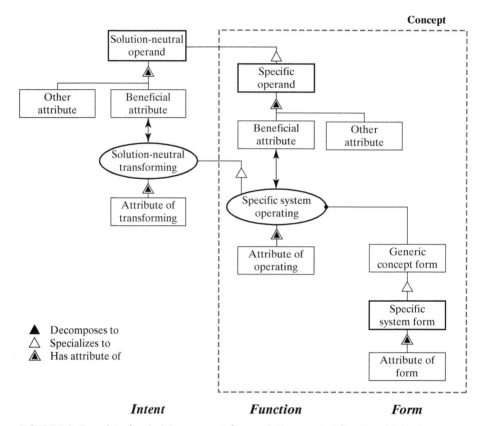

FIGURE 7.4 Template for deriving concept from solution-neutral functional intent.

moving fluid is the solution-neutral function. Water, a type of fluid, is the specific operand, and pressurizing, which is one concept of operations of a pump, is done with a centrifugal pump. For bubblesort, the array is the solution-neutral operand, entries of the array are the specific operand, and sequentially sorting (in contrast with inserting) using bubblesort (rather than quicksort) is the concept.

TABLE 7.3 | Solution-neutral function and solution-specific concept for five systems

Solution-Neutral Operand	Solution-Neutral Process	Specific Operand	Specific Process	Specific Instrument
Fluid	Moving	Water	Pressurizing	Centrifugal pump
Array	Sorting	Array entries	Sequentially exchanging	Bubblesort
Cork	Translating	Cork	Pulling	Screw
Traveler	Transporting	Traveler	Flying	Airplane
Book	Buying	Internet	Accessing	Home DSL connection

CHAPTER 7 • SOLUTION-NEUTRAL FUNCTION AND CONCEPTS **145**

Looking at all of the examples in Table 7.3 makes it clear that there is no single relationship between the specific operand and the solution-neutral operand, or between the specific process and the solution-neutral process, as discussed in Box 7.3.

Box 7.3 Methods: Specializing the Solution-Neutral Function to Concept

The step from solution-neutral function to concept is not in any way automatic: rather, it requires careful consideration and creativity. This is especially true in specializing the operands. Any of the following can occur in the specialization of the operand (shown in *italics*). This listing is not comprehensive but indicates some of the changes that can occur.

- The solution-neutral operand and the specific operand are completely different, and the solution-neutral process and the specific process are different as well:
  - Entertaining a *person* by watching a *DVD*
  - Preserving a *memory* by capturing an *image*
  - Serving a *customer* by dispensing *cash*
- The solution-neutral operand and the specific operand are completely different, but the process remains the same:
  - Choosing a *leader* by choosing the *president*
- The specific operand is a part of the solution-neutral operand:
  - Open the *bottle* by removing the *cork* (the cork comes with the bottle)
  - Sorting an *array* by exchanging *array entries*
  - Making a *sandwich* by cutting bread *slices* (slices are part of the sandwich)
- The specific operand is a type of the solution-neutral operand:
  - Moving *fluid* by pressurizing *water*
  - Fixing *cars* by fixing *sports cars*
- The specific operand is an attribute of the solution-neutral operand, or an attribute is added:
  - Amplifying a *signal* by increasing *signal voltage*
  - Controlling a *pump* by regulating *pump speed*
- The specific operand is an information object that describes an attribute of the solution-neutral operand:
  - Evacuating *people* by informing the *task knowledge of the people*
  - Checking *equipment status* by communicating *status signal*

Similar patterns occur in specializing the solution-neutral process, even for the same operand (processes in *italics*).

- The solution-neutral process is specialized to a different process:
  - *Sheltering* people by *housing* people
- The solution-neutral process is specialized to a type of process:
  - *Transporting* a traveler by *flying* a traveler
  - *Preserving* wood by *painting* wood
  - *Cooking* potatoes by *boiling* potatoes
- The solution-specific process is formed by adding or changing an attribute to the solution-neutral process:
  - *Powering* the tool by *powering (electrically)* the tool

Steps for Developing Concepts

The steps for developing the complete description of the concept are implicit in Figure 7.4: Identify the specific operand, whose change will satisfy the functional intent; identify the beneficial attribute of the specific operand whose change is associated with value; and so on. These points are summarized in Question 5a of Table 7.1.

The analysis of the concept for the two higher-complexity systems is shown in Table 7.4, using the approach of Figure 7.4. Transporting has now been specialized to the specific process of flying. "Flyer" is the dummy form, which is specialized to an airplane that is commercial. This is a case where the solution-neutral and specific operands are the same, but the process specializes. In simple language, the concept is as follows: a single traveler with light luggage flying in less than 2 hours on a commercial airplane.

The home network case proceeds somewhat differently. The solution-neutral operand is book, but the specific operand jumps to network. Gaining access, or simply accessing, is the process, and reliable is the attribute of accessing. This is done by a dummy "accesser," which specializes to an inexpensive home DSL network. In simple language, the concept is as follows: reliably accessing a high-speed Internet connection using an inexpensive home DSL network.

As we noted in Chapter 2, the operand-process-instrument construction we used to describe a system can be likened to the noun-verb-noun structure of human language. Notice that when the concept is described using the template of Figure 7.4, the concept is described by a full sentence. The instrument is the noun subject, and the attributes are adjectives. The process is the verb, and attributes of the process are adverbs. The operand is the object (grammatically speaking), and the attributes of the operand are adjectives in the predicate. Thus, a concept boils down to a sentence.

The idea of concept is very close to the idea of *patterns* in software. A software pattern is a "problem–solution pair in a given context." [2] Looking at Figure 7.4, the "problem" is what here is called the solution-neutral functional intent. The "solution" is the concept.

TABLE 7.4 | Formulation of the specific function and form defining the concept of the transportation service and the home network. See Table 7.2 for the solution-neutral function.

Question	Transportation Service	Home Network
Specific operand?	Traveler	Internet
Benefit-related attribute?	Location	Access
Other operand attributes?	Alone with light luggage	High-speed connection
Specific process?	Flying	Gaining (Accessing)
Attributes of process?	In less than 2 hours	Reliably
Generic concept form?	"Flyer"	"Accesser"
Specific form?	Airplane	Home DSL connection
Attributes of form?	Commercial	Inexpensive

Naming Concepts

There is no simple convention for naming concepts. Rationally, concepts should be named operand + process + instrument, and a few are, such as light + emitting + diode in Table 7.5. Examples of other conventions are shown in Figure 7.5 as well. Concepts are not normally named for just the operand, because it doesn't contain enough information to describe a concept. In English, the ending "er" shows up to make a process sound like an instrument, but we don't really express much more by calling something a mower than by calling it a mow.

Sometimes a concept is named for an attribute of the process or instrument, as "wireless" was used for two different concepts at the beginning of the twentieth century and at the beginning of the twenty-first.

Sorting Alternative Concepts

Developing concepts is an open-ended creative process, and it is in the interest of the architect to have a rich set of ideas to select from.

Once the structured creativity or brainstorming phase of developing concepts is concluded, then the sorting and down-selecting begin. The process of deriving selection criteria based on stakeholder needs and product/system goals will be deferred to Chapter 11. In this section, we will discuss the process of sorting and organizing the possible concepts.

Figure 7.5 shows an OPM graphic that suggests an organization scheme for alternative concepts. First, all of the concepts that operate on the same specific operand would be organized on

TABLE 7.5 | Naming conventions for concepts

Concept-Naming Convention	Operand	Process	Instrument
Operand-Process-Instrument	light	emitting	diode
	data	storage	warehouse
Operand-Process	lawn	mow	(er)
	hair	dry	(er)
Operand-Instrument	cork	(removing)	screw
	fire	(burning)	place
	hat	(storing)	rack
	suit	(carrying)	case
Process-Instrument	(food)	dining	room
	(stuff)	carrying	case
Process	(TV)	control	(er)
	(image)	project	(er)
Instrument	(head)	(covering)	hat
	(food)	(serving)	table
	(person)	(carrying)	bicycle

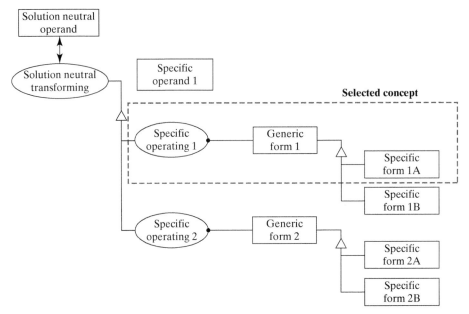

FIGURE 7.5 Tree showing the options among concepts, with different specific operating and specific instruments of form, but all with the same specific operand 1.

one page. Different concepts may in fact operate on different specific operands, while producing the same solution-neutral result. For example, if the solution-neutral statement for the pump is to move fluid, one option is *pump water*, but it could also be *circulate air* (increasing the rate of evaporation). The solution-neutral operand fluid could specialize to the specific operand water or air, as discussed in Box 7.3.

The graph in Figure 7.5 organizes the concept options into an operand layer, a process layer, and an instrumental form layer. In general, there are fewer operand options than process options, and fewer process options than form options, so this decision tree representation makes sense.

Figure 7.6 shows an example of a tabular version of the same decision tree as in Figure 7.5, specialized to the "centrifugal pump," which is now just one of seven concept options in this non-exhaustive list. For water as an operand, there is a principle of operation of "pressurizing" that could be accomplished by a centrifugal pump or an axial flow pump. Alternatively, there is a different operating principle called accelerating, which can be done by a different kind of centrifugal pump or a jet ejector. Displacement pumps work on yet another principle. The same reasoning, applied to sorting in Figure 7.7, reveals three common principles of operation in sorting, each with several specific algorithms, which are one of the abstractions of form for computation.

Transporting the traveler is developed in Figure 7.8, where traveler is the only likely operand. There are a wide variety of specific processes for transporting, and it is a long-standing human endeavor. We limited the specific process list to flying, rolling, and floating, and then linked each to common instruments.

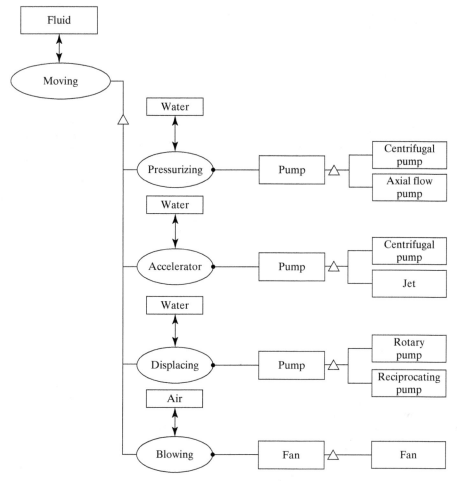

FIGURE 7.6 Solution-neutral function and solution-specific concept options for "pump" system.

Figure 7.9 shows the more limited option space for the "home network" example. In buying a book online, it is hard to avoid accessing the Internet, so here the options are only how you access: through various types of home connections, in public spaces, or at work.

It is important for the architect to develop a comprehensive view of the concept space prior to launching into the details of architecture, and thus answering the last item in Question 5a: "What are several other concepts that satisfy the solution-neutral function?"

Broader Concepts and Hierarchy

The analysis of concept alternatives focused on looking "downward" to increasing detail from the solution-neutral statement of the intent. However, we can also look "upward" to increasingly more general intent, just as we did with wine accessing. Why do we move fluid? Sort arrays? Travel? Buy books? This is a question of functional intent, and we have learned that solution-neutral intent is a hierarchy.

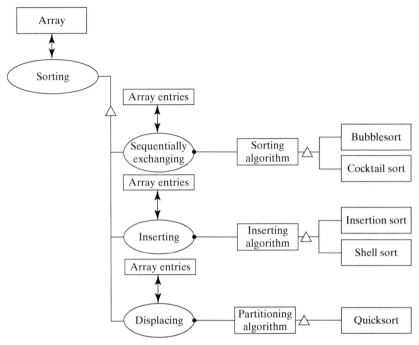

FIGURE 7.7 Solution-neutral function and solution-specific concept options for "bubblesort" system.

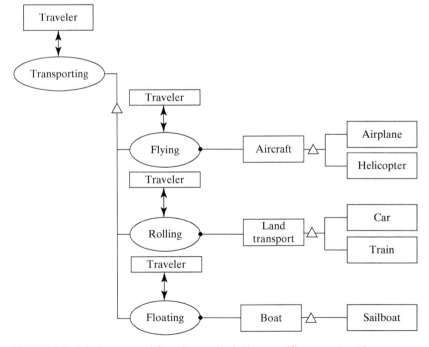

FIGURE 7.8 Solution-neutral function and solution-specific concept options for "transportation service" system.

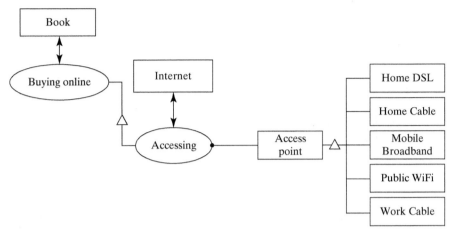

FIGURE 7.9 Solution-neutral function and solution-specific concept options for "home network" system.

Maybe we move fluid to dry a basement. Why dry a basement? Maybe to improve the living conditions in a house. We should not be surprised that there is hierarchy in functional intent, just as there is hierarchy in built systems.

Figure 7.10 shows a hierarchic concept tree for the traveler. We have focused (for example, in Tables 7.2 and 7.4) on the level of a traveler traveling, specifically flying. But looking at the need statement in Table 7.2, we see a hint of why the traveler travels: something about visiting a client. In Figure 7.10 we notionally explore what led to the need to travel: to close a deal, by learning more about client preferences, by meeting the client (at her or his office), by traveling, by flying. At all the levels there are alternative concepts that could be pursued. For example, one might ignore the needs of the client and try to change the client's mind. One might have a teleconference or study the client's materials. One might meet the client at a conference or trade show.

This hierarchy shows that the *specific function at one level becomes the solution-neutral functional intent at the next level* down the hierarchy. For example, learning client preferences is

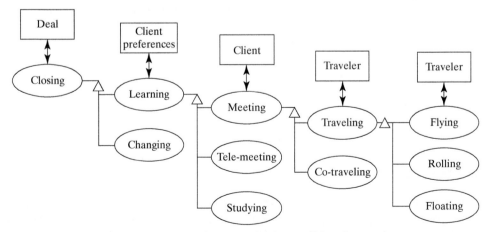

FIGURE 7.10 A hierarchy of intents and concepts leading to flying the traveler.

152 PART 2 • ANALYSIS OF SYSTEM ARCHITECTURE

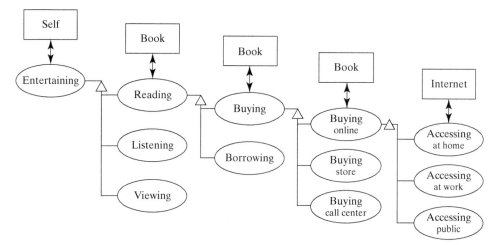

FIGURE 7.11 A hierarchy of intents and concepts leading to accessing the Internet at home.

a specific function compared to the solution-neutral intent "deal closing," but it then becomes the solution-neutral intent compared to the more specific "client meeting."

Which of these is the right level at which to start examining the goals of the system? There is no right answer, and in fact it is very useful for the architect to understand the hierarchy of intent for the system "up" through one or two levels.

Figure 7.11 shows the similar hierarchy of intents and concepts for the home network. It begins with entertainment, specializes to reading books, which requires borrowing or buying books, which can be bought online, which leads to accessing the network. Understanding this hierarchy might identify new options, such as selling electronic books.

In summary:
- Concept is a system vision that maps function to form. Concept establishes the solution vocabulary and the set of design parameters. Concept rationalizes the details of the architecture and guides its development.
- Concept is derived by specializing the solution-neutral operand and process to the specific operand and process, and to an abstraction of form (the specific instrument), as suggested in Figure 7.4.
- There is no common convention for naming concepts, and the operand, process, and/or instrument may appear in the name.
- Concept options should be developed and sorted to form a set that spans the possible functional concepts and instrument concepts.
- Concepts sit in a hierarchy. The specific function at one level becomes the solution-neutral functional intent at the next level down the hierarchy. It is important for the architect to understand several levels of this hierarchy.

7.4 Integrated Concepts

It is often the case that the process part of the concept is rich in meaning and can almost immediately be expanded or "unpacked" into several more detailed internal functions. We encountered one such example when we learned that the traveler wants to "visit" the client. Visiting implies going to the place of the client, spending some time there, and returning. If any of these three internal processes were missing, it would fail to meet the definition of "visiting."

An *integrated concept* is made up of these smaller *concept fragments,* each of which identifies how one of the internal processes is specialized. When we encounter a rich process, we expand it into internal processes and then identify concept fragments for each. It is really just the concept development process used recursively. Question 5b of Table 7.1 focuses on concept fragments and integrated concepts.

Returning to the transportation service case, we can identify at least three important internal processes in any transportation process: lifting, or overcoming gravity; propelling, or overcoming drag; and guiding. Without these three internal processes, effective transporting is not possible on or near Earth. Transporting is a rich concept.

In the case of a car, for example, lifting is associated with wheels, which is why in Figure 7.8 we called the process rolling. In a car, wheels also provide the propulsion, through torque on the road surface. Wheels also provide the guiding function for cars.

Using this idea, we can enumerate a surprising number of concepts by combining one instrument for each of the three important internal functions. A representation of integrated options is shown in Table 7.6. In such a *morphological matrix,* each column selects one of the choices of a concept fragment for each of the internal processes. We see that a car is wheels-wheels-wheels, whereas a train is wheels-wheels-ground. The only conceptual difference between cars and trains is how they are guided. Morphological matrices are covered in detail in Chapter 14.

TABLE 7.6 | Expanded morphological matrix of instrument alternatives for the three internal processes associated with transporting. The integrated concepts are named at the top.

Internal Process	Instrument	Car	Train	Jet Aircraft	Propeller Aircraft	Helicopter	Airship	Glider	Boat	Submarine	Jet Boat	Hydrofoil
Lifting	Wheels	X	X									
	Propeller					X						
	Wings			X	X			X				X
	Closed hull						X			X		
	Open hull								X		X	
Propelling	Wheels	X	X									
	Propeller				X	X	X		X	X		X
	Jet			X							X	
	Gravity							X				
Guiding	Wheels	X										
	Propeller					X						
	Rudder			X	X		X	X	X	X	X	X
	"Ground"		X									

We can observe some patterns from Table 7.6. Aircraft and gliders are held aloft by wings guided by rudders and differ only in propulsion. However, helicopters are all about propellers. Airships and submarines are conceptually the same and differ only in the medium in which they operate. Such an expanded morphological matrix allows the architect to view how the concept fragments for the internal processes come together to define the integrated concepts.

The morphological matrix for the home data network system is shown in Table 7.7. We identify five key internal processes: connecting local network and ISP; modulating ISP carrier signal; managing data on local network; connecting user devices and local network; and interacting with user. For each internal process, there are several possible instruments of form, which, together with the process, define the concept fragment. Here we have used the physical device (say, the gateway) as a placeholder for the electronics, software, and protocols for each device.

If we select one specific instrument of form for each function, we can obtain an integrated concept for the home data network. Figure 7.12 shows an integrated concept that corresponds to Integrated Concept 1 of Table 7.8. This integrated concept combines a separate DSL modem and a box containing a gateway and switch with both WiFi and Ethernet connections. Common variants would have the DSL modem in the box, or other user devices.

Note that this integrated concept is not, strictly speaking, made up of exactly one assignment for each function; some flexibility is employed in combining concept fragments. For example, managing the data network is done by both a residential gateway and a switch, plus a separate wireless access point (WAP). Likewise, the morphological matrix lists the various user devices separately, while this expanded morphological matrix combines four different user devices (laptop, smartphone, desktop, and printer). When various combinations of such choices for concept fragments are possible, the architect can use some creativity in expressing the integrated options. When computational searches are performed, the representation must be more rigorous, as discussed in Chapter 14.

The second integrated option shown in Table 7.8 is a common offering for homes with cable, wherein VOIP (Voice over Internet Protocol) phone, cable, and Internet are bundled. The system includes a combined device that comprises the cable modem, gateway, and Ethernet switch. A hard-wired network of local devices (desktop and printer) is connected via Ethernet. The TV connects directly to the cable, and the VOIP phone connects via WiFi.

Although it is more likely to be used outside of the home while on the road, the option shown as Integrated Concept 3 in Table 7.8 is quite different. It uses mobile broadband to connect a smartphone to a laptop via a local WiFi network.

Integrated concepts are developed by identifying internal function, specializing, and then combining. Some integrated concepts have common names (car and train), but many are easier to describe by enumerating the component instruments.

In summary:

- Often, concepts unfold to reveal a set of internal processes that must be executed. Choosing a specific operand, process, and instrument for each of these unfolded processes defines the concept fragments.
- The architect should explore and combine concept fragments to build desirable integrated concepts.

CHAPTER 7 • SOLUTION-NEUTRAL FUNCTION AND CONCEPTS   155

TABLE 7.7 | Morphological matrix of instrument alternatives for the five internal processes associated with a home network

Function	General Form	Specific Form
Connecting local network and ISP	Physical connection	Fiber
		Coaxial cable
		Coaxial cable
		Phone (dial-up)
	Radio connection	Mobile broadband
		Satellite
Modulating ISP carrier signal	Embedded modem	Embedded mobile broadband modem
		Embedded dial-up modem
	External modem	DSL modem
		Cable modem
		Fiber modem
		External mobile broadband modem
		Satellite modem and antenna
Managing data on local network	Single-function hardware	None
		WAP
		Residential Gateway
		Switch
		Tethering Device (hotspot)
	Multi-function hardware	Residential Gateway + Switch
		Switch + WAP
		Modem + Residential Gateway
		Modem + Res Gateway + Switch
		Modem + Res Gateway + Switch + WAP
Connecting user devices and local network	Homogeneous	None
		Wi-Fi
		USB
		Bluetooth
		Ethernet
	Hybrid	Wi-Fi + Ethernet
		Wi-Fi + USB
Interacting with user		Smartphone
		Television
		Laptop
		Home server
		VOIP phone
		Desktop
		Printer

156 PART 2 • ANALYSIS OF SYSTEM ARCHITECTURE

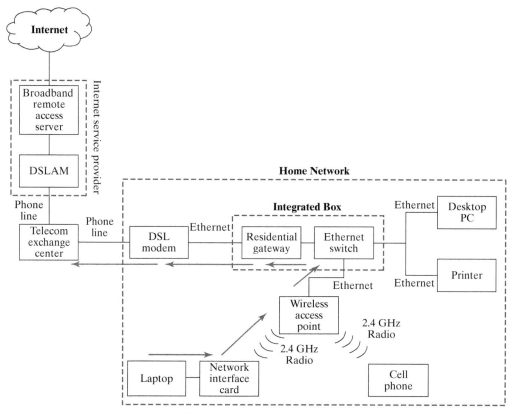

FIGURE 7.12 An integrated concept for the home data network.

TABLE 7.8 | An informal expanded morphological matrix for the home data network

Function	Integrated Concept 1	Integrated Concept 2	Integrated Concept 3
Connecting local network and ISP	DSL	Coaxial cable	Mobile broadband
Modulating ISP carrier signal	Dedicated DSL modem	Cable modem in integrated box	Embedded mobile broadband modem
Managing data on local network	Integrated gateway and switch + WAP connected to switch by Ethernet	Integrated cable modem, gateway, Ethernet switch	Integrated modem + cell phone as tether
Connecting user devices and local network	WiFi (to laptop and cell phone) + Ethernet (to desktop and printer)	Local cable to TV + Ethernet (to desktop, printer, and VOIP phone)	WiFi
Interacting with the user	Laptop + cell phone + desktop + printer	VOIP phone + TV + desktop + printer	Laptop

7.5 Concepts of Operations and Services

We have defined the concept of the system, but one aspect of conceptual development remains in Question 5b of Table 7.1: *concept of operations,* or "conops." Operational behavior of a system was introduced in Chapter 6. There we learned that operation is broader than function. Function is a somewhat quasi-static view of what a system *can do.* Operation is the sequence of things leading to the delivery of the primary function, and it indicates what the system actually *does.*

The relationship between concept of operations and a detailed sequence of operations is the same as the relationship between system concept and system architecture. The concept of operations sketches out how the system will operate: who will operate it, when, and coordinated with what else.

As an example, Figure 7.12 represents Integrated Concept 1 from Table 7.8. Both the table entry and the figure make the system look rather static. How does it actually operate? The concept of operations would identify three main operations executed by this network: data are moved from the user device (laptop) to the ISP; data are moved from the ISP to the user device; and the local network is managed, to prevent congestion. A complexity of data networks is that all of these can be going on simultaneously.

The concept of operations for the outgoing data is suggested by the arrows in Figure 7.9. The sequence of operations is that the data are moved from the laptop through the WAP to the switch, then the gateway, the modem, and on the telephone line toward the ISP. Incoming data from the ISP to the laptop follow the same path in reverse. The traffic management performed by the switch and gateway will be discussed in Chapter 8. With this concept of operations in hand, it is much easier to understand how the home data network actually operates and delivers value.

As a second example, the left side of Figure 7.13 sketches out the concept of operations for an air transport aircraft. Weeks ahead of a flight, there is some scheduling of aircraft and crews, followed by flight planning on the day of the flight. Just before the flight, the aircraft is loaded, and then actual flight operations begin. The plane arrives and is unloaded. At some designated interval, the aircraft is maintained. This concept of operations contains much more information than the statement of the function of the aircraft—to transport passengers and baggage. The concept of operations is very focused on the aircraft, and the traveler appears only as "baggage."

This is an operational view of the aircraft. What does it look like if we view this as a transportation service? If the enterprise transfers the instrument, it is called a *good.* If it transfers the function, it is a *service.* In this case, an aircraft manufacturer sells airplanes (a good), and an airline sells transportation services.

The right side of Figure 7.13 illustrates the consumer-centric view of the *concept of service.* The traveler plans the trip and buys the ticket. After arriving at the airport, the traveler checks in and checks baggage. By different processes, each is loaded on the aircraft. This is a point where the concept of operations of the aircraft and that of the transportation service are identical. The details of taxi, takeoff, and the like are relatively unimportant to the traveler. After unloading, the traveler and baggage are unloaded, and the traveler "checks out" of the airport.

During operations, there is a very important relationship concerning the aircraft. From the perspective of the concept of operations of the aircraft, the aircraft is the operand: It is loaded, it is flown, and so on. From the perspective of the concept of the service, the aircraft is the instrument: It transports the traveler.

158 PART 2 • ANALYSIS OF SYSTEM ARCHITECTURE

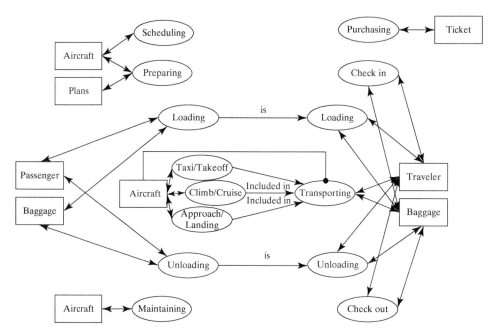

FIGURE 7.13 Concept of operations of an aircraft (left) and the concept of the service of air transportation (right).

In retrospect, we see that a service can be architected the same way as a system; it is more process-focused than a product. A service is a system! In addition, we see that the concept of operations contains information vital to understanding the system architecture.

In summary:

- The concept of operations defines at a conceptual level how the system actually operates when it delivers value.
- Concepts of operations can be defined for the system, and for the service built upon the system.

7.6 Summary

In this chapter, we transitioned from simple to more complex systems, and from a purely analytical approach to the beginning of a synthetic approach to architecture. Table 7.1 introduced a new guiding question (Question 7a) and refined some of the questions developed in Chapter 5 to distinguish the ideas of solution-neutral function and specific concept.

This synthetic process begins by examining the needs of the stakeholder and identifying the solution-neutral function. Actual design starts by specializing the operand and process, and adding an instrument—the steps to define a concept. For almost any solution-neutral function (we considered the example of transporting a person), there are many possible concepts (such as driving in a car and flying in an airplane). The architect can creatively develop these options, sort them, and then choose one to develop into an architecture, the subject of Chapter 8. It might be desirable to expand the selected concept into concept fragments, which are then recombined to form an integrated concept. Before leaving the conceptual development task, it is useful also to develop a concept of operations for the system.

In the next chapter, we will build upon the selected system concept and concept of operations to develop the architecture of the system.

References

[1] Nam P. Suh, "Axiomatic Design: Advances and Applications," *The Oxford Series on Advanced Manufacturing* (2001).

[2] Erich Gamma, Richard Helm, Ralph Johnson, and John Vlissides, *Design Patterns: Elements of Reusable Object-Oriented Software* (Pearson Education, 1994).

Chapter 8
From Concept to Architecture

The authors would like to thank Peter Davison of MIT for his substantial contribution to this chapter.

8.1 Introduction

We are almost ready to begin examining the synthetic process of creating system architecture, which starts in earnest in Chapter 9. The remaining steps we need to understand to complete Part 2 is how concept expands to architecture. Concept is a notional mapping between function and instruments of form, whereas architecture is a fairly comprehensive description of the relationships between internal functions and instruments of form.

As discussed in Chapter 7, the amount of information required to describe a system is defined by the solution-neutral functional intent and by the concept—the amount of information needed to fill in the template of Figure 7.4. The amount of information needed to describe the architecture is orders of magnitude larger. For example, the concept of a pump or that of bubblesort can be described by one line, but to describe the architecture, we need to present all of the information from Chapter 6.

Table 8.1 lists all the key questions from Chapters 4 through 7. These questions approximately summarize the information needed to document an architecture. The questions are listed in the order in which they are addressed during the synthesis of an architecture. Table 8.1 includes questions about form (Chapter 4), function (Chapter 5), form-to-function mapping (Chapter 6), and solution-neutral function and concept (Chapter 7). Two new questions, 8a and 8b, are added in this chapter. These questions relate to extending an architecture from Level 1 to Level 2, and possible modularization of Level 2 objects.

The main tasks in developing the Level 1 architecture are summarized in Questions 5c through 6g and discussed in Section 8.2. Then we need to recursively apply these steps to develop the Level 2 architecture and a scheme for modularization. This is the topic of Questions 8a and 8b and is discussed in Sections 8.3 through 8.5.

In this chapter, we will use the air transportation system as the running example to illustrate the tasks. In Section 8.4, we will summarize the process by developing an architecture for the home data network.

TABLE 8.1 | Questions for defining a system in synthesis. Items in square brackets [ ] have been added to clarify the task's relationship to *concept*, as introduced in Chapter 7.

7a.	Who are the beneficiaries? What are their needs? What is the solution-neutral operand whose change of state will meet these needs? What are the value-related attribute and the solution-neutral process of changing the states? What are the other attributes of the operand and process?
5a.	What is the primary externally delivered value-related function? The [specialized] value-related operand, its value-related states, and the [specialized] process of changing the states? What is the abstraction of the instrumental form? [What is the concept? What are several other concepts that satisfy the solution-neutral function?]
5b.	What are the principal internal functions? The internal operands and processes? [What are the specializations of those processes? What are the concept fragments? What is the integrated concept? What is the concept of operation?]
5c.	What is the functional architecture? How do these internal functions connect to form the value pathway? How does the principal external function emerge?
5d.	What are the other important secondary value-related external functions? How do they emerge from internal functions and pathways?
4a.	What is the system?
4b.	What are the principal elements of form?
4c.	What is the formal structure?
4d.	What are the accompanying systems? What is the whole product system?
4e.	What are the system boundaries? What are the interfaces?
4f.	What is the use context?
6a.	How are the instrument objects mapped to the internal processes? How does the formal structure support functional interaction? How does it influence emergence?
6b.	What non-idealities require additional operands, processes, and instrument objects of form along the realistic internal value creation path?
6c.	What supporting functions and their instruments support the instrument objects on the value creation path?
6d.	What are the interfaces at the system boundaries? What operands are passed or shared? What are the processes at the interface? What are the instrument objects of the interface, and how are they related (identical, compatible)?
6e.	What is the sequence of execution of process involved in delivering the primary and secondary functions?
6f.	Are there parallel threads or strings of functions that execute as well?
6g.	Is actual clock time important to understand operations? What timing considerations or constraints are active?
8a.	How does the architecture of Level 1 extend to Level 2?
8b.	What is a possible modularization of the Level 2 objects?

162 PART 2 • ANALYSIS OF SYSTEM ARCHITECTURE

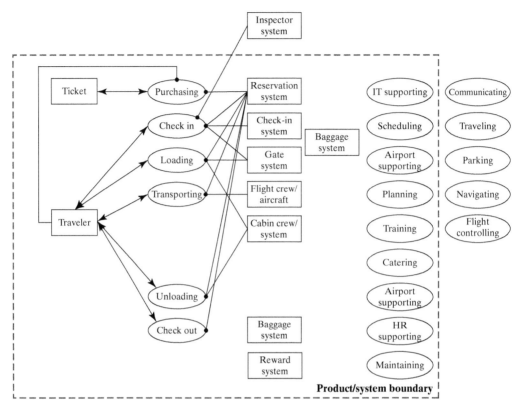

FIGURE 8.1 Partial architecture of the primary value delivery of the air transportation service.

8.2 Developing the Level 1 Architecture

Expanding Concept to Functional Architecture

Our first task in expanding from concept to architecture is to identify the value pathway that delivers the primary value. This is done by starting with the key internal functions identified in the integrated concepts (Question 5b) and linking them together from inputs or start points to outputs or end points.

The solution-neutral function of the air transportation service (transporting a traveler) and concept (flying a traveler with an airplane) have been identified. The left two columns of Figure 8.1 show the primary value delivery pathway, created by linking the key internal functions from Chapter 7. This figure identifies the primary operand as the traveler and leaves the checked baggage as a secondary value operand to be discussed below. Interestingly, all of the processes act on the traveler except the purchasing of the ticket. The state of the traveler progresses from checked-in in the departure city to checked-out in the arrival city. The only other operand is the ticket, which is essentially the information object that encodes itinerary, traveler status, reservation, and fees. This functional architecture represents a simple "no frills" airline that just gets the traveler from A to B. This figure responds to Question 5c.

In contrast to this fairly simple functional architecture, Figure 8.2 adds secondary value delivery: transportation of checked baggage, the provision of entertainment and food on board,

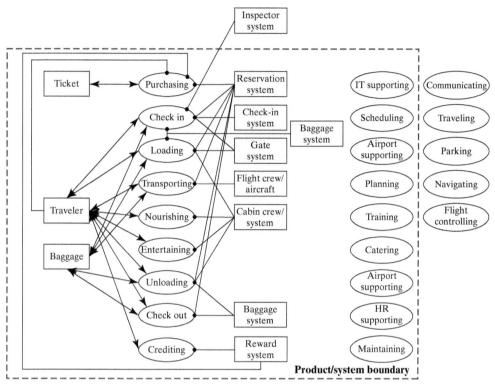

FIGURE 8.2 Partial architecture of the primary and secondary value delivery of the service of air transportation.

and the involvement of a frequent-traveler incentive program. This answers Question 5d and makes the architecture noticeably more complex. For example, many of the operations for the traveler must now be replicated for the baggage.

Defining the Form

As introduced in Chapter 3, *zigzagging* is the idea that we reason in one domain as long as is practical and then switch to the other. For example, we can only decompose the function *transporting* so far before it becomes necessary to specify something about the form part of the concept, in order to understand what further functions are required.

We have gone about as far as we can go in the functional domain, so it is time to "zig" into the form domain and begin to address Questions 4a through 4f.

QUESTION 4A asks us to define the abstraction of the form that constitutes the system. The answers within the dashed line in Figure 8.2 are the employees and equipment of the airline, as well as the leased entities of the airport that are directly associated with processing passengers and baggage (such as baggage conveyers and ticket desks). Excluded from the form of the system are the non-airline services of an airport (such as arriving by car and parking) and the air traffic control, navigation, and security services provided by the government.

The form at Level 1 decomposition is depicted in Figure 8.2 (which shows only the instruments of the value delivery processes). The structure of form (where it is located and how the form is connected) is not shown. We have answered part of Question 4b and deferred answering Question 4c.

QUESTION 4D A complete analysis of the form of the accompanying systems is not shown in Figure 8.2, but the entities of form that are instruments of the accompanying processes can be inferred from it. For example, navigating requires ground-based radio transmitters and/or space-based navigation satellites. Flight controlling would require controllers, towers, and so on.

QUESTION 4E Likewise, a complete analysis of the form of the interfaces has not been shown in Figure 8.2, but one explicit interface appears there: the federally supplied security inspection personnel and equipment required for the check-in process. Other interfaces are implicit in the supporting functions and the provision of accompanying functions shown outside of the product/system boundary, such as air traffic control and navigation services that interact with the aircraft and flight crew.

QUESTION 4F The use context is that of a traveler traveling on a scheduled commercial flight from an airport near one urban area to another. It is a single leg of a domestic flight, or a flight within a zone of common passport control. Significant infrastructure exists around the airport that is not shown in Figure 8.2, including hotels, roads, and rental cars.

Mapping the Function to Form

Next we can "zag" back to the functional domain and examine the system architecture that results from mapping the function to form. At Level 1, this is shown by the links between form and process in Figure 8.2. The reservation system, which contains the itinerary, seat assignment, and travel program status, is an instrument of virtually all of the processes. The check-in system and gate system get the traveler checked in and on board. The flight crew are only responsible for transporting. These observations answer Question 6a.

If this architecture begins to look complicated, consider for a moment the non-idealities of a more realistic transportation service (Question 6b). The traveler transiting the airport and waiting at gates has not been included in the model. We have also not considered making a connection in a hub airport on the way from the departure area to the arrival area, or traveling internationally, with passport control.

The supporting functions (Question 6c) are enumerated in Figure 8.2 but are not shown as connected to the value-related instruments. This is because a deeper understanding of the mapping needs to be made at Level 2, and because the mapping would be very dense. For example, nearly all supporting instruments involve personnel who need to be trained and supported by human resource (HR) processes.

The sequence of operations (Question 6e) can be inferred from the way Figure 8.2 is drawn: Processes are vertically listed in the order in which they occur. The passenger purchases a ticket, checks in, is loaded onto the aircraft, and so on. There is a strong suggestion that there are parallel threads or strings of operations (Question 6f). For example, along with the operations on the passenger, there is almost an entire parallel path for processing the

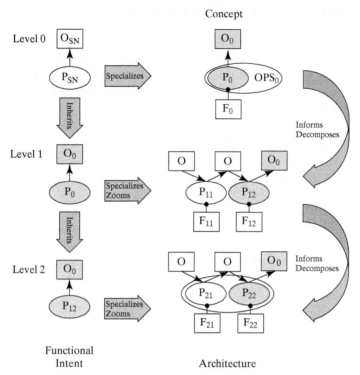

FIGURE 8.3 Reasoning about architecture down through the levels.

checked baggage, which separates from the passenger at check-in and is (we hope) reunited with the passenger at check-out. Also, the passenger is fed and entertained while being transported. We know from experience that clock time is important (Question 6g)—there are actual schedules that aircraft keep—but this information is not shown in Figure 8.2 and would be captured in additional diagrams.

In summary, the tasks of developing the Level 1 architecture are presented in Table 8.1. The starting point is the Level 0 information: the solution-neutral function (O_{sn} and P_{sn}), the solution-specific concept (O_0, P_0, and F_0), and the concept of operations (OPS_0), as shown in Figure 8.3.

In order to develop the Level 1 architecture, the specific function at Level 0 (the functional part of the concept) becomes the functional intent at Level 1. This is a key point: *The solution at one level becomes the problem statement at the next level.* We call this *function–goal reasoning*. The intent at Level 1 also inherits other aspects of the Level 0 intent.

This Level 1 functional intent is next specialized and zoomed to form the functional architecture at Level 1: the operands (including the external operand O_0) and the internal processes at Level 1 (P_{11} and P_{12}). The Level 0 form is next decomposed to define the Level 1 entities of form and mapped to the Level 1 internal function. This is also informed by the Ops Concept (OPS_0) at Level 0. Finally, the remaining detail of the architecture at Level 1 is filled in: non-idealities in function, supporting processes and form, interfaces, and so on.

8.3 Developing the Level 2 Architecture

Intent and Recursion at Level 2

The complete Level 1 architecture that would be developed from Figure 8.2 would have at least 18 internal processes (counting the 9 value-related processes and at least 9 more supporting processes), plus the associated form, plus interfaces with some accompanying systems. This is already a great deal of complexity.

Why develop more detail at Level 2 in this OPM representation? We want to make sure that Level 1 is reasonably accurate, and we want to make sure the boundaries of internal modules (and internal interfaces) are reasonably placed. These internal boundaries drive the next step: actually distributing the responsibility for design to other groups (internal and suppliers). If the architecture is not well decomposed at Level 1, this distributed effort and subsequent integration will suffer.

The principal question is therefore "Is *this* the right decomposition at Level 1, or at least a good one?" In Chapter 13, we will discuss the choice of decomposition in detail. The fact is that the real information about how the entities at Level 1 should be clustered or modularized is hidden at Level 2. We need to dive down one more level to see how the details really work, and then make decisions on how Level 1 is best structured.

We therefore employ the hierarchic decomposition (Chapter 3) to develop the Level 2 architecture, which is summarized in Question 8a of Table 8.1: "How does the architecture of Level 1 extend to Level 2?"

We will *recursively* apply the procedure from Table 8.1 to the task of expanding from Level 1 to Level 2, as suggested by Figure 8.3. The central procedure we employ is that *each function at Level 1 becomes a statement of functional intent at Level 2,* allowing us to recursively employ the analysis structure already developed. We are recursively applying what we called function–goal reasoning above.

For example, if "purchasing tickets" is the process we focus on in the Level 1 view of the air transport service (Figure 8.2), "purchasing tickets" becomes the functional intent at Level 2, and we have to make a choice at Level 2 among many specialized processes to accomplish the intent of purchasing (such as purchasing online) and then to zoom the purchasing process, as will be discussed below.

The intent at Level 2 retains by inheritance the overall intent for the system at Levels 0 and 1. Additionally, we often want to capture intent from operations. For example, how is the ticket purchasing sequenced in the air transport service? Did our operational concept create any intent as to how late the ticket can be purchased before the scheduled departure?

Developing the Level 2 Architecture

Given an understanding of the intent at Level 2, we can now derive the architecture at Level 2. We will recursively follow the questions of Table 8.1. Starting with Question 5b we identify the expanded operands, attributes, and processes.

We illustrate the development of Level 2 architecture by focusing on the ticket purchase process. The Level 1 function ("purchasing tickets") has now become the Level 2 intent. As shown in Figure 8.4, this has been specialized to an online system, primarily supported by the

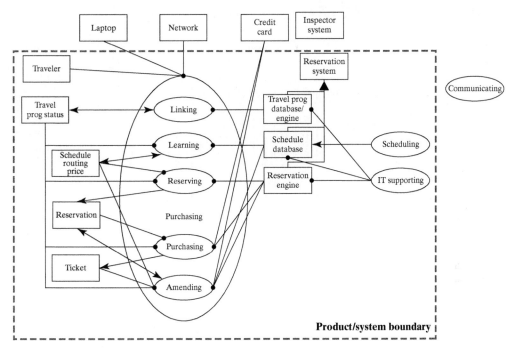

FIGURE 8.4 Zooming to Level 2 for the ticket-purchasing process.

reservation system. Five internal processes have been identified, as well as three new internal operands (ticket exists as a Level 1 operand). The reservation system has decomposed to three instruments within the system, and three more accompanying instruments are identified outside of the product/system boundary.

If this zooming is applied to each of the 9 value delivery pathway processes of Figure 8.2, then we can produce the 28 internal processes at Level 2 as shown in Table 8.2. Note that 3 of the processes of Figure 8.2 (entertaining, nourishing, and crediting) have been "demoted" to Level 2 process. Such judgments are often made in architecting. Table 8.2 also indicates the 16 states of the 4 major operands. All operands are somehow affected by the internal processes, and some also act as instruments.

Table 8.3 maps the 28 internal processes onto the 22 instruments of form. These relationships appear to be approaching more of a one-to-one correspondence. The notable exceptions are the agents and flight attendants, who, as humans, are adaptable to many roles.

Together, Table 8.2 and Table 8.3 express the architecture at Level 2. The tables do not yet include the supporting systems, interfaces, or external accompanying systems present in the whole product/system.

The procedure to examine architecture at the next level could be *recursively* applied to derive Level 3. But do we need to do this? Consider the desired outcome. Examining Level 2 has two objectives: to confirm that the higher-level abstractions at Level 1 are reasonable, and to investigate the appropriate ways to cluster or modularize Level 1, as we discuss in Section 8.6.

TABLE 8.2 | An array showing the Level 2 processes and relationships with operands for the air transportation services (I stands for instrument, a for affect, and c for create).

			OPERANDS													
			passenger							carry-on		checked bags		itinerary		
			passenger	location	inspection	alert status	information	entertainment	nourishment	travel prog status	location	examination	location	checked status	schedule/price	reservation	ticket status	boarding pass
PROCESSES	ticketing	linking	I							a								
		learning	I							I					a			
		reserving	I							I					I	c		
		purchasing	I							I						I	c	
		amending	I							I					I	a	I	
	checking in	arriving at airport	I	a							a		a					
		issuing								I					I	I		c
		checking	I							I			a		I			
		inspecting	I		a					I					I			
		examining	I							I		a			I			
		alerting	I			a												I
		changing	I							I					a	I	a	
	loading	loading											a		I			
		embarking	I	a							a							I
		storing	I								a							I
	transporting	informing	I				a											
		entertaining	I					a										I
		nourishing	I						a									I
		transporting		a														
		conveying									a							
		shipping											a					
		evacuating	I	a														I
	unloading	collecting	I								a							
		disembarking	I	a							a							I
		unloading											a	I				
	checking out	collecting	I										a	a	I			
		departing airport	I	a							a		a					
		crediting								a						I		

TABLE 8.3 | Mapping between the Level 2 processes and form for the air transportation services (I stands for instrument, a stands for affect).

			INSTRUMENTS OF FORM																				
			traveler program database/engine	schedule database	reservation engine	computer/network	credit card	car	agent	baggage conveyor	inspectors	metal detector	x ray	posting boards	baggage handler	mobile conveyor	boarding pass checker	passenger database	flight attendants	video	food	flight crew	aircraft	carousel
PROCESSES	ticketing	linking	I			I																		
		learning		I		I																		
		reserving		I	I	I												a						
		purchasing				I	I																	
		amending		I	I	I	I																		
	checking in	arriving at airport						I																
		issuing							I															
		checking							I	I														
		inspecting									I	I												
		examining									I		I											
		alerting							I					I										
		changing							I															
	loading	loading													I	I								
		embarking							I								I	I	I					
		storing																	I					
	transporting	informing																	I	I				
		entertaining																		I				
		nourishing																	I		I			
		transporting																				I	I	
		conveying																				I	I	
		shipping																				I	I	
		evacuating																	I					
	unloading	collecting																						
		disembarking																	I					
		unloading														I	I							
	checking out	collecting													I									I
		departing airport						I																
		crediting	I																					

Consider the complexity. A typical Level 1 model will have about 7 +/− 2 primary value processes, as well as non-idealities and supporting processes, and about as many entities of form, for about 20–30 entities in all. A full model at Level 2 might have 50–100 entities, and at Level 3 there would be hundreds. Three levels of hierarchy are hard to develop and too much to comprehend easily. We don't need to go to Level 3.

In summary, the goal is to produce a model at Level 1 that is robust and well modularized, so that the architecture can be broken up and distributed to others to develop further. As suggested by Figure 8.3, the Level 2 model can be created by recursively applying the procedures to develop the architecture (Table 8.1), focusing successively on all of the internal processes in the initial Level 1 model. The key step is applying function–goal reasoning: We start the analysis of Level 2 by treating each internal function at Level 1 as a goal at Level 2.

8.4 Home Data Network Architecture at Level 2

We now present the home data network architecture at Level 2, as an example of what this method of architectural analysis can produce. Recall that the solution-neutral function of this system is to buy a book, and the concept is to do so by accessing the Internet using a DSL modem. The integrated concept involves a dedicated DSL modem, an integrated gateway and switch, a WAP connected to the switch by Ethernet, and a laptop connected to the WAP by WiFi (Table 7.8). A Level 1 architecture is shown in Figure 7.12.

Figure 8.5 shows the architecture at Level 2, following the route of an IP packet sent from a laptop on the home network until it leaves the local network bound for some external server or computer. We break the functional architecture into three layers: the Internet layer, the link layer, and the physical layer. The idea of layers in an architecture was introduced in Section 6.3.

As shown in the form representation of the home network, data flow from the user's laptop through its network interface card and then are broadcast via WiFi (2.4-GHz signal) and picked up by a wireless access point (WAP). From there, data travel through the Ethernet switch and are directed to the residential gateway/router. The router sends them to the modem, which then modulates the data onto a signal traveling out of the house via a phone line.

The functional path that the data follow (a description of what happens to the data at each step along their path out of the network) is much less straightforward. The source of this difficulty is that the network protocols are divided into layers. The purpose of a layered approach is to allow a network designer or IT manager to view the network through the lens of only a single layer, abstracting the functions of the other layers. This layered approach is common to network communication models, with the OSI seven-layer model and the five-layer Internet protocol suite being the primary examples. The three layers in our home network model are identical to the bottom three layers of the Internet protocol suite, with the application and transport layers excluded (the home data network does not play a major role in creating or managing data).

This layered approach gives modern networks many of their desirable emergent properties, such as scalability, robustness, and flexibility; unfortunately for us, it also makes it difficult to trace the functional pathway of data across multiple layers. However, we believe this "linearized" view of the network functional pathway is valuable, so we include the following description.

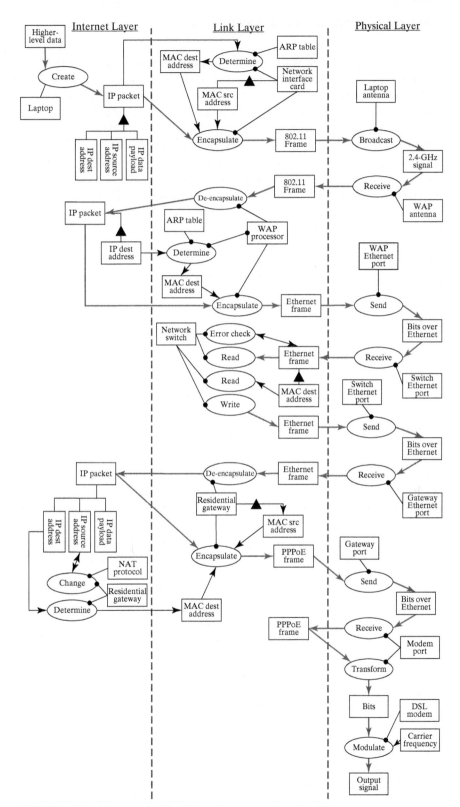

FIGURE 8.5 Architecture for the home data network at Level 2.

The first step along this path is the creation of an IP packet by the user laptop from some source of higher-layer data. The packet consists of a number of distinct elements, but we will focus on three: a source IP address, a destination IP address, and a data payload (error checking is another key element). The source address points to the laptop, and the destination address points to the server or computer where the IP packet is being sent (presumably to request data from a server or to upload data to be stored). The data payload contains the higher-layer data.

This packet (IP Layer) gets transformed into one or more frames (Link Layer), in a process known as encapsulation, to be transferred across the local WiFi link. Encapsulation into an 802.11 frame (WiFi protocol) adds media access control (MAC) source and destination addresses. MAC addresses encode the frame's location and destination on the local network. MAC addresses are similar to IP addresses, except that they encode location within the "neighborhood" of the local area network and do not enable the routing functions that are the core of IP. The source MAC address is a static property of the laptop's network interface card, whereas the destination address is retrieved from an address resolution protocol (ARP) table that maps destination IP addresses to destination MAC addresses.

The 802.11 frame(s) is (are) then sent wirelessly to the wireless access point (WAP) by modulating the bits of the frame onto a 2.4G-Hz radio signal and broadcasting this signal from the laptop antenna. This transfer occurs at the Physical Layer. Once received, the 802.11 frame is unpackaged by the WAP to reconstruct the IP packet. Because the WAP needs to know where to send the data next, it reads the IP destination address and sees that this destination is located through the residential gateway. It then encapsulates the IP packet into an Ethernet frame, setting the MAC destination address as that of the residential gateway.

The WAP then sends the frame toward the gateway by sending it via the Ethernet cable to the network switch. Again, this transfer of bits through a physical medium (in this case, copper wire) occurs at the Physical Layer. Once the frame reaches the switch, it is stored in a buffer, and the destination MAC address is read. Using this address and knowledge of the location of the residential gateway, the switch then writes the buffered frame to the link on which the gateway is located. In addition, the switch performs an error-checking function to ensure that the frame was read and transferred correctly.

Once the Ethernet frame reaches the residential gateway (again by way of a Physical Layer connection), it is again unpackaged to form the original IP packet. Here the gateway performs a function known as network address translation (NAT), wherein the source IP address is changed to the "public facing" IP address of the residential gateway from the "local" IP address of the laptop. This address change accomplishes two functions: It preserves IP addresses (with NAT, not every Internet-enabled device needs a unique IP address), and it adds an additional layer of security because the user device is not broadcasting its address across the wider network.

Once the IP address is set, the packet is encapsulated into a PPPoE (Peer-to-Peer Protocol over Ethernet) frame that allows the gateway to communicate with the Internet service provider on the Link Layer. This frame is then modulated by the DSL modem onto a high-frequency carrier signal that travels through the home phone line to the telephone company's exchange center and then on to the ISP to be routed to its final destination.

In summary, the home data network can be expanded to Level 2 in a way similar to what occurs in the air transportation service. The home data network architecture with 24 internal processes and associated operands is represented by an OPM diagram (Figure 8.5). The operands and processes give a sense of the flow of data through the system. The instruments include both physical devices (laptop, switch, antenna) and informational objects (software and data tables). The Level 2 model enables us to judge the robustness of the Level 1 model and associated modularization.

8.5 Modularizing the System at Level 1

We have now decomposed both examples to Level 2, but we have not yet asked whether we chose the correct decompositions at Level 1. We expanded the Level 1 representation in order to examine the relationships among the entities at Level 2. These relationships should inform the modularization or clustering at Level 1. This process is captured by Question 8b "What is a possible modularization of the Level 2 objects?"

In the air transportation service example, we initially presented the Level 1 processes according to the major events that happen along the timeline of air travel: ticketing, checking in, loading, transporting, unloading, checking out. How do we know this is a good way to decompose an air transportation service?

Our objective is to cluster entities of the architecture that are tightly interconnected. [1] This enables us to minimize interactions across modules and maximize internal coherence within modules. In order to examine how the Level 2 entities cluster, we first need to decide *how* we will cluster (Box 8.1). In this example, we use a process-centric clustering, with relationships based on operand interactions. Our assumption is that if two processes share a large number

> ### Box 8.1 Methods: **Clustering by Interactions in System Matrices**
>
> When we examine clustering, there are three important tasks: choosing the basis of the clustering, representing that information, and computing the clusters.
>
> The common ways to cluster are based on processes or form. If we choose processes, we will most likely use links through the operands, focusing on interactions (Box 5.8). If we choose form, we can use links through processes and operands, which emphasize interactions among the elements of form (Section 6.5), or through structure of form (Section 4.4), which stresses static relationships.
>
> When we represented a system with matrix views (PO and PF), in Chapters 4 through 6, we chose to capture the type of interaction (create, destroy, affect) that is represented on the corresponding OPM diagram. When we do clustering, we suppress the *type* of interaction, representing only the *number* of connections. When we replace any interaction symbol in the PF and PO matrices with a 1 and set the PP, OO, and FF matrices to the identity matrix, the resulting matrix gives only the count of the interactions between entities indexed by the row and column.
>
> The final clustering is identified by a clustering algorithm. A clustering algorithm rearranges the DSM entities (rows and columns) to group together sets of entities that have a high degree of connectivity, resulting in blocks of entities that are more highly coupled and are more decoupled from other blocks. [2]

of operands, they are tightly coupled. The nature of the interaction is suppressed, and the only information left is the number of interactions. For example, it is easy to see in Table 8.4 that the "arriving at airport" process interacts with four different operands and that three of these operands are shared with "embarking."

We then cluster the processes using the Thebeau algorithm. [3] Table 8.4 shows the $PP_{operand}$ matrix for the air transportation service architecture, grouping Level 2 processes into clusters.

TABLE 8.4 | A DSM array showing the coupling through operands of the processes of the air transportation service with suggested Level 1 clustering. Matrix entries show the number of operands linking the row, column processes.

Cluster	Process	Reservation					Ticket			Passenger							Checked Bags					Carry Bags			Secondary			
		linking	learning	reserving	purchasing	amending	changing	issuing	crediting	arriving at airport	inspecting	embarking	transporting	disembarking	evacuating	departing airport	checking	loading	shipping	unloading	collecting	examining	storing	conveying	collecting	alerting	informing	entertaining	nourishing
Cluster 1	linking	2	2	2	2	2	2	1	1	1	1	1	0	1	1	1	2	0	0	0	1	1	1	0	1	1	1	1	1
	learning	2	3	3	2	3	2	1	1	1	1	1	0	1	1	1	2	0	0	0	1	1	1	0	1	1	1	1	1
	reserving	2	3	4	3	4	3	2	2	1	2	1	0	1	1	1	3	1	0	1	2	2	1	0	1	1	1	1	1
	purchasing	2	2	3	4	4	4	3	2	1	2	1	0	1	1	1	3	1	0	1	2	2	1	0	1	1	1	1	1
	amending	2	3	4	4	5	4	3	2	1	2	1	0	1	1	1	3	1	0	1	2	2	1	0	1	1	1	1	1
Cluster 2	changing	2	2	3	4	4	5	4	2	1	2	2	0	2	2	1	3	1	0	1	2	2	2	0	1	2	1	2	2
	issuing	1	1	2	3	3	4	4	2	0	1	1	0	1	1	0	2	1	0	1	1	1	1	0	0	1	0	1	1
	crediting	1	1	2	2	2	2	2	2	0	1	0	0	0	0	0	2	1	0	1	1	1	0	0	0	0	0	0	0
Cluster 3	arriving at airport	1	1	1	1	1	1	0	0	4	1	3	1	3	2	4	1	1	1	1	2	1	2	1	2	1	1	1	1
	inspecting	1	1	2	2	2	2	1	1	1	3	1	0	1	1	1	2	1	0	1	2	2	1	0	1	1	1	1	1
	embarking	1	1	1	1	1	2	1	0	3	1	4	1	4	3	3	1	0	0	0	1	1	3	1	2	2	1	2	2
	transporting	0	0	0	0	0	0	0	0	1	0	1	1	1	1	1	0	0	0	0	0	0	0	0	0	0	0	0	0
	disembarking	1	1	1	1	1	2	1	0	3	1	4	1	4	3	3	1	0	0	0	1	1	3	1	2	2	1	2	2
	evacuating	1	1	1	1	1	2	1	0	2	1	3	1	3	3	2	1	0	0	0	1	1	2	0	1	2	1	2	2
	departing airport	1	1	1	1	1	1	0	0	4	1	3	1	3	2	4	1	1	1	1	2	1	2	1	2	1	1	1	1
Cluster 4	checking	2	2	3	3	3	3	2	2	1	2	1	0	1	1	1	4	1	0	1	3	2	1	0	1	1	1	1	1
	loading	0	0	1	1	1	1	1	1	1	1	0	0	0	0	1	1	2	1	2	2	1	0	0	0	0	0	0	0
	shipping	0	0	0	0	0	0	0	0	1	0	0	0	0	0	1	0	1	1	1	1	0	0	0	0	0	0	0	0
	unloading	0	0	1	1	1	1	1	1	1	1	0	0	0	0	1	1	2	1	2	2	1	0	0	0	0	0	0	0
	collecting	1	1	2	2	2	2	1	1	2	2	1	0	1	1	2	3	2	1	2	4	2	1	0	1	1	1	1	1
Cluster 5	examining	1	1	2	2	2	2	1	1	1	2	1	0	1	1	1	2	1	0	1	2	3	1	0	1	1	1	1	1
	storing	1	1	1	1	1	2	1	0	2	1	3	0	3	2	2	1	0	0	0	1	1	3	1	2	2	1	2	2
	conveying	0	0	0	0	0	0	0	0	1	0	1	0	1	0	1	0	0	0	0	0	0	1	1	1	0	0	0	0
	collecting	1	1	1	1	1	1	0	0	2	1	2	0	2	1	2	1	0	0	0	1	1	2	1	2	1	1	1	1
Cluster 6	alerting	1	1	1	1	1	2	1	0	1	1	2	0	2	2	1	1	0	0	0	1	1	2	0	1	3	1	2	2
	informing	1	1	1	1	1	1	0	0	1	1	1	0	1	1	1	1	0	0	0	1	1	1	0	1	1	2	1	1
	entertaining	1	1	1	1	1	2	1	0	1	1	2	0	2	2	1	1	0	0	0	1	1	2	0	1	2	1	3	2
	nourishing	1	1	1	1	1	2	1	0	1	1	2	0	2	2	1	1	0	0	0	1	1	2	0	1	2	1	2	3

Now, rather than processes being organized according to the original "function timeline," the architecture is grouped more functionally. The first cluster, which might be called the Level 1 reservation process, deals with the creation and amendment of the reservation. The second cluster, closely related to the first, includes processes that affect the ticket.

The next three clusters contain processes that control the path of the passenger, of checked bags and of carry-on bags, respectively. All three of these groups follow a similar pattern of checking in, moving, and checking out, as they move from the departure location to the arrival location. Finally, the sixth cluster contains the secondary value processes that are aimed at the passenger's travel experience.

This clustering is not perfect—there are interactions among the blocks—but this will be the result for any clustering. By performing this clustering analysis, we are not saying that this Level 1 modularization (according to shared operands) is better than the first attempt in Figure 8.2, which was based on a progression of time. The two different decompositions simply represent two ways in which the airline service provider might be organized. If running time were important, you might focus on the time-based organization. If you were thinking of end-to-end system reliability, the second might be more appropriate.

In summary:

- In order to modularize the system well at Level 1, it is necessary to expand to Level 2, examine the relationships based on processes or form, and then cluster to create a new Level 1 decomposition.
- Clustering algorithms can arrange the entities into an order that maximizes the local interaction and minimizes the interactions between blocks. The resulting clustering might be better than the original one, and it better informs the architect about how to decompose the system, define internal interfaces, and then distribute the task of more detailed design.

8.6 Summary

In this chapter, we followed the development of an architecture from a simple concept through to Level 1 and then recursively to Level 2. The main point is that there is a process to go from Level N to Level N+1, which can be applied in expanding the concept to Level 1, and equally well in expanding Level 1 to Level 2. Because the really important information about the relationships among the Level 1 entities are hidden at Level 2, we must probe down two levels, identify the relationships of interest, and then cluster at Level 1.

We have come to the end of Part 2 of the text, in which we described the entire process of analysis of an architecture: defining the form and structure (Chapter 4); the functional architecture of processes and operands (Chapter 5); the mapping of form to function (Chapter 6); the identification of solution-neutral function and concept (Chapter 7); and the development of Level 1 and Level 2 architectures and their modularization at Level 1 (Chapter 8). This analysis was built upon the lessons from Part 1 on system thinking (Chapter 2) and thinking about complex systems (Chapter 3). We now have a strong foundation on which to discuss the creation of architecture. We will transition to a discussion focused on generating new architectures, in the context of real-world complexity.

References

[1] Steven D. Eppinger and Tyson R. Browning, *Design Structure Matrix Methods and Applications* (Cambridge, MA: MIT Press, 2012).

[2] See DSMweb.org for more information. <http://www.dsmweb.org/en/understand-dsm/technical-dsm-tutorial0/clustering.html>

[3] Ronnie E. Thebeau, "Knowledge Management of System Interfaces and Interactions for Product Development Process," 2001. MIT master's thesis. <http://www.dsmweb.org/?id=121>

Part 3
Creating System Architecture

In Part 2, we retrospectively analyzed the decisions of other architects. We worked backwards through existing architectures to identify the form, function, architecture, and concept. Now begins the harder work of real synthesis: defining architectures that do not yet exist, and doing so for complex systems.

Part 3: Creating System Architecture, takes a forward path, from early identification of system needs to concept selection and architecture. However, in complex systems, we can no longer represent examples in their entirety. Our emphasis switches from complete representations to choosing the right lens through which to view the architecture. It will become increasingly important that we choose lenses wisely, lest we become inundated with information.

In Chapter 9, we outline the role of the architect, in terms of tasks and deliverables, and then focus on the first task, reducing ambiguity. We then proceed to a tool to drive ambiguity from the system—a holistic product development process. Chapter 10 identifies the main organizational interfaces as potential opportunities for reducing ambiguity in architecture: corporate strategy, marketing, regulation, internal and external competence to manufacture, and operations.

In Chapter 11, we examine a systematic way to capture the needs of stakeholders, prioritize them, and convert them into goals for the system. Then, in Chapter 12, we propose approaches to focusing the creativity of the architect on inventing and selecting concepts. Finally, in Chapter 13, we review approaches to managing complexity during system development.

Chapter 9
The Role of the Architect

9.1 Introduction

Throughout this text, we have emphasized holism in the role of the architect, which places a great burden on the architect in terms of the number of considerations to be managed. This drive for holism can be misinterpreted as a failure to define a system boundary for the architect's role. The architect is not the design team, the financial controller, the marketing executive, or the plant manager. To paraphrase Eberhardt Rechtin, the architect is not a generalist, but rather a specialist in simplifying complexity, resolving ambiguity, and focusing creativity.

The architect crafts the vision of the system and communicates that vision among the stakeholders and the extended project team. We will specifically highlight three principal roles of the architect:

- **Reduce ambiguity:** Define the boundaries, goals, and functions of the system.
- **Employ creativity:** Create the concept.
- **Manage complexity:** Choose a decomposition of the system.

In this chapter, we introduce the role of the architect and define the deliverables of the role. We discuss different types of ambiguity and relate them to the role of the architect (Section 9.2). We discuss the product development process as a tool of the architect (Section 9.3). We conclude with a case study on civil architecture, which highlights some universal themes in architecture.

9.2 Ambiguity and the Role of the Architect

The Role of the Architect

Before the architect is engaged, there has been an "upstream process" that has defined issues, opportunities, and needs for the new system. This upstream is full of ambiguity. For example, what are the high-priority needs of the customer vs. the "nice to haves"? What are the regulations that must be considered and how will they be dealt with? And how is manufacturing competence aligned with the envisioned product?

The architect drives ambiguity from the upstream process. The architect is responsible for creating boundaries and concretizing goals. This includes:

- Interpreting corporate and functional strategies
- Interpreting competitive marketing analyses

- Listening to users, beneficiaries, customers, or their representatives
- Considering the competence of the enterprise and its extended supply chain
- Considering the operations and operational environment of the system
- Infusing technology where appropriate
- Interpreting regulatory and pre-regulatory influences
- Recommending standards, frameworks, and best practice
- Developing goals for the system based on the upstream influences

Once the goals for the system have been defined, there is the creative task of defining a concept for the system. A good concept does not ensure system success, but a bad choice of concept almost certainly dooms a system to failure.

The architect employs creativity to create the concept for the product, which captures the essential vision of the system, and differentiates the product relative to previous concepts. Several of the tasks in defining the concept are:

- Proposing and developing concept options
- Identifying key metrics and drivers
- Conducting highest-level trades and optimization
- Selecting a concept to carry forward, and perhaps a backup
- Thinking holistically about the entire product life cycle
- Anticipating failure modes and plans for mitigation and recovery

In the moment when the concept is chosen, there is a relatively small amount of information available to define the system. But very rapidly the information explodes, as team members start defining external interfaces, the first-level design and decomposition take place, and the considerations for downstream factors begin to be defined.

The architect manages investment in complexity and the evolution of complexity in the system, in order to ensure that the goals are delivered. A key component of the architect's role in managing complexity is ensuring that the system is comprehensible to all. This aim is achieved by:

- Decomposing form and function
- Clarifying the allocation of functionality to elements of form
- Defining interfaces between subsystems and the surrounding context
- Configuring the subsystems
- Managing flexibility vs. optimality
- Defining the degree of modularity
- Articulating vertical vs. horizontal strategies
- Balancing in-house vs. outsourcing design and manufacturing
- Controlling product evolution

These three principal roles of the architect all center on information: identifying the necessary, consistent, and important information by reducing ambiguity; adding new information through creativity; and managing the explosion of information in the final architecture.

Traditional systems engineering conceives of a project as a three-way tradeoff among performance, schedule, and cost. [1] The enduring wisdom is that the systems engineer must identify the tensions among these and, in the end, can set one of these three variables, can manage to a second variable, and must allow the third variable to float. The dynamics that couple performance, cost, and schedule have now been largely defined through project management models in System Dynamics, [2], [3], [4] but the endogenous estimation of the three variables is rare, given the difficulty of obtaining the data and the expertise required to calibrate these models.

Complexity, creativity, and ambiguity are not as directly coupled as performance, cost, and schedule, but they are far more amorphous and difficult to manage. The metaphor of tension in systems engineering is directly applicable to the role of the architect; one of the primary mechanisms by which the architect acts is recognizing, communicating, and resolving tensions in the system (Box 9.1 Principle of the Role of the Architect).

Box 9.1 Principle of the Role of the Architect

"Some single mind must master, else there will be no agreement in anything."

Abraham Lincoln

"Timing has a whole bunch to do with the outcome of a rain dance."

Cowboy saying

The role of the architect is to resolve ambiguity, focus creativity, and simplify complexity. The architect seeks to create elegant systems that create value and competitive advantage by defining goals, functions, and boundaries; creating the concept that incorporates the appropriate technology; allocating functionality; and defining interfaces, hierarchy, and abstractions to manage complexity.

- In view of the ambiguity and complexity present in most systems, and the need for balance, it is often desirable to have the architecture created by a single individual or a small group.
- The architect maintains a holistic view but always focuses on the small number of issues critical to the design.
- The architect doesn't follow a single method but adopts different frameworks, views, and paradigms as appropriate.

Reducing Ambiguity

Zhang and Doll [5] coined the term "the fuzzy front end" to express the idea that early development activities are challenging because the goals are unclear. These goals may be unclear because a defined customer base cannot articulate what it wants next, because the end customer is unclear, or because the underlying technology progress is unclear.

Leadership is not a management task unique to system architects. It is leadership that sets a vision for the firm and molds an amorphous context into an executable plan. However, we believe that the task of the architect is particularly challenging because the architect sits between firm strategy and product/system definition and therefore must competently bridge both worlds.

To these ends, we believe it is important for the architect to recognize and be able to deal with the various types of ambiguity. Strictly speaking, ambiguity is composed of two ideas: fuzziness and uncertainty. However, in common usage, ambiguity can also connote incorrect, missing, or conflicting information.

Fuzziness occurs when an event or state is subject to multiple interpretations. A color can be fuzzy; whether a color is blue or purple could be interpreted differently depending on the viewer. The color has a defined wavelength, but its adherence to the categories we call blue and purple is fuzzy. Fuzziness is rampant in the statements of needs of customers—for example, when they say they want a "smooth" finish or "good" gas mileage. Fuzziness is also influenced by context. What is good gas mileage will depend on the background of the customer. Customers in North America and Europe have different expectations. [6]

Uncertainty occurs when an event's outcome is unclear or the subject of doubt. The outcome of a coin flip is uncertain. We can clearly articulate the potential states that could happen, but we don't know which state will occur. For example, a new technology may or may not be ready at the date needed to support a new product.

Compounding the ambiguity that arises from fuzziness and uncertainty are the related phenomena of unknown information, conflicting information, and false information. In order to illustrate these variations, we use the shorthand (X, Y, Z), where X, Y, and Z are the true inputs to the system.

Unknown information (X, __, Z) occurs when information is under-determined or unavailable. This is a very common case early in the product development cycle; for example, you may know nothing about the product/market intentions of your competitors. You may know about the absence of information; an example of such a *known unknown* occurs when you don't know whether your existing competitor will launch a product. Even more dangerous, you may not know about the absence of information; this is often called an *unknown unknown*. For example, a *new* competitor could introduce a competing product.

Conflicting information (X, D, Z and X, B, Z) occurs when two or more pieces of information offer opposing indications. The problem is over-determined. For example, you may have information from one source that the government will impose a new regulation affecting your product, but information from another source may suggest that the government will not do so.

False information (F, Y, Z) occurs when you are presented with incorrect inputs. In this case you may believe you have all the information (your information is complete), but some of it is wrong. For example, you may have been told that a supplier will be ready on a certain date, when the supplier actually has no plan whatsoever to be ready on that date.

Consider the following statements. What types of ambiguity are present?

- Please make me a smooth cover for this phone. (Fuzzy)
- Will it be a boy or a girl? (Uncertain, Fuzzy)
- Make sure you meet your quarterly goals. (Unknown, Conflicting)
- Produce a low-cost, high-quality product. (Conflicting, Fuzzy, Unknown)
- Every fourth year is a leap year. (False)

Ambiguity is almost always present in the upstream influences, which we have roughly segmented below by corporate functions.

- **Strategy:** How much risk is corporate, the board, or the government willing to take?
- **Marketing:** Will it fit with marketing's plans for product line positioning and channels of distribution?
- **Customers:** What do the customers want/need? Will their needs change with time?

- **Manufacturing:** Will manufacturing be ready to produce it?
- **Operations:** What failures must the product endure and still operate?
- **R&D:** Is that technology infusible?
- **Regulations:** What are applicable regulations? Are they likely to change?
- **Standards:** What are the applicable standards, and will they change?

For each of these, the initial information that the architect receives could be fuzzy, uncertain, missing, conflicting, or false.

Given all these sources of ambiguity (Box 9.2 Principle of Ambiguity), the architect's responsibility is to provide more certain, less fuzzy inputs to the team. In some cases, the ambiguity can be pared down by analysis. In other cases, one may choose to impose constraints simply to make the problem more tractable. Sometimes it is necessary to make an assumption so that work can proceed, being careful to mark the assumption well for future verification.

> **Box 9.2 Principle of Ambiguity**
>
> *The best-laid schemes o' mice an' men*
> *Gang aft agley [often go awry]*
>
> Robert Burns
>
> The early phase of a system design is characterized by great ambiguity. The architect must resolve this ambiguity to produce (and continuously update) goals for the architect's team.
>
> - Development is possible only with the acceptance of uncertainty.
> - In general, no one designs or rigorously controls the process upstream of architecting, so there should be no expectation of unambiguity. There will be uncoordinated, incomplete, and conflicting inputs.
> - Uncertainty can create opportunities; it is not always bad.
> - Ambiguity contains known unknowns and unknown unknowns, as well as conflicting and false assumptions.
> - Ambiguity is especially apparent at the interface to the upstream influences, because no one designs the upstream processes.
> - Uncertainties should be identified and prioritized so that they can be managed.

The architect must realize that no one explicitly "designs" the upstream influences; rather, they just occur, often with incomplete, overlapping, or conflicting outcomes. It is easy to get trapped in the thinking that the upstream influences (corporate strategy, marketing, the board, the customer, the government, and so on) have knowledge that can reduce the ambiguity, or that they have a responsibility to inform the architecture by commenting on early drafts.

The architect must engage these upstream influences and drive the ambiguity out to create a vision and plan for a successful product. This requires knowledge of what the upstream influences are, who has "control" of them, and how they are best engaged. The architect must keep in mind the objective: a consistent, complete, attainable, clear, and concise set of goals for the product.

Removing, reducing, and resolving ambiguity is the main role of the architect at the interface to the upstream process. The tools that we present to help with this include a list of the deliverables of the architect and the use of a holistic product development process (Section 9.3), the examination of several of the principal upstream and downstream influences (Chapter 10), the ABCD Product Case (Chapter 10), and the stakeholder needs-to-goals process (Chapter 11).

Deliverables of the Architect

As a summary of the role of the architect, the list of deliverables of the architect is given below. Note that deliverables are very different from tasks in that they are the end result, not the procedure by which the state is achieved. The deliverables include:

- A clear, complete, consistent, and attainable (with 80–90% confidence) set of goals (with emphasis on functional goals).
- A description of the broader context in which the system will sit, and the whole product context (including legal and standards).
- A concept for the system.
- A concept of operations for the system, including contingency and emergency operations.
- A functional description of the system, with at least two layers of decomposition, including description of primary and secondary externally delivered function; process flow with internal operands and processes, including non-idealities, supporting processes, and interface processes; and a process to ensure that the functional decomposition is followed.
- The decomposition of form to two levels of detail, the allocation of function to form, and the structure of the form at this level.
- Details of all external interfaces and a process for interface control.
- A notion of the developmental cost, schedule, and risks, and the design and implementation plans.

Table 8.1 leads to the development of all but the last deliverable, which represents a summary of the project to be undertaken. However, one of the many challenges the architect faces is to represent these deliverables such that they can be easily absorbed and used. In Part 3, we focus on representing architecture in the context of complex development projects, where we recognize that diagrammatic languages such as OPM and SysML, while semantically exact, may not fit with existing process or culture. Therefore, part of the role of the architect is to capture the information from Part 2 however it can be most easily interpreted.

In summary:

- The role of the architect includes reducing ambiguity, employing creativity, and managing complexity.
- Architecture sits astride the upstream activities (things that are done before the architecture is created) and downstream activities (things that are done after the

architecture is created but which must be factored into the architecture). Ambiguity arises from both.
- In common practice, ambiguity is a combination of fuzzy information, uncertainty, missing information, conflicting information, and incorrect information. The principal job of the architect at the interface to the upstream (the fuzzy front end) is to drive as much of this ambiguity from the system as possible.
- A high-level way to do this is to start to prepare the deliverables of the architect.

9.3 The Product Development Process

Nearly every large firm today has defined an internal product development process (PDP). This captures the methodology of product development, including the terminology, phases, milestones, schedules, and lists of tasks and outputs. The intent of the PDP is to provide an enterprise framework that captures the wisdom of previous development efforts and provides standard processes and approaches (which upstream efforts reflect lessons learned from past developments, which test procedures are critical to regulatory success, and so on).

From our perspective, a principal advantage of the PDP is that it is a tool for reducing ambiguity by defining tasks and responsibilities. It is easy to become lulled into the notion that the PDP will work stepwise through the upstream ambiguities, leaving the architect with a clear understanding of strategy, market, customer needs, and technology. This is only partially true. The PDP can do a disservice to the architect if it presumes more certainty than is present. We will use the PDP as a starting point for examining upstream and downstream influences and as a method to drive early ambiguity from the system.

Similarities and Differences among Enterprise PDPs

A passing inspection of several enterprise-specific PDPs suggests that they appear surprisingly similar, even across very different industries and sectors. This suggests that there is an underlying effective practice in product development. The main purpose of this section is to develop a Generic PDP, using an intentionally different representation from a traditional stage-gate graphic. We will focus on the common activities of product development, rather than on their sequence, so that the model is applicable to both stage-gated, and non-stage-gated processes.

A second purpose is to develop the reader's skill in understanding and evaluating contemporary innovation in product development—methods such as Lean, Agile, Six Sigma, Scrum, and Voice of the Customer. Modern engineering culture is rife with methods and initiatives. Some of these methods advance propositions that can be tested for validity; others do not. Some of these methods attempt to capture intangible aspects of company culture, with varying degrees of success. Our goal is for the reader to become an educated consumer of these methods—to evaluate their utility, to understand their implications, and to apply them in the context in which they were intended, but not beyond.

From a scope perspective, many PDPs exclude the upstream process of conceiving the product (defining what is to be built). Additionally, many do not include actually operating the product but simply assume that the responsibility of the builder ends with shipment. This is true in some sectors, but often the need to consider operations and service is sufficiently great that operations through to retirement should be considered for inclusion in the PDP. In keeping with our

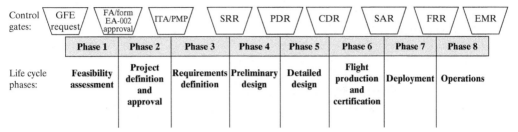

FIGURE 9.1 NASA JSC Flight Product Development Process [7], showing the expected linear progression and stage-gates. (Source: Project Management of GFE Flight Hardware Projects. EA-WI-023. Revision C, January 2002. Retrieved August 12, 2014)

holistic approach, we seek a PDP that is not only "birth to death" but also "lust to dust"; that is, it captures the activities from the first time a product is envisioned until the last instance is retired.

We begin with NASA (see Figure 9.1). NASA's PDP includes some aspects of the fuzzy front end (feasibility, approval, requirements), but much of the fuzziness is hidden in the initial government furnished equipment (GFE) request. The NASA process also includes operations, because NASA is the operator of most of its products.

The upstream influences on NASA's PDP are not captured in this PDP; where you might expect to find market analysis, NASA's PDP notes only a feasibility question. The reality is that NASA's market analysis combines budget decisions, public value, science goals, and industrial base questions. Notionally, NASA's PDP proceeds linearly from feasibility to operations, [8] without iteration and without explicit reference to the technology development process.

NASA's PDP reflects the realities of its stakeholders, and the products it builds: science satellites, the International Space Station, the Space Shuttle, and so on. These are predominantly low-volume, expensive space systems. They cannot be tested in situ before operations, nor can they easily be repaired. There is a high perceived cost for public failure, [9] especially for human spaceflight. Whether viewed through the lens of safety or of operations, NASA's PDP places significant emphasis on traceability of requirements and formal review, approval and change control processes, an attribute that has been both commended (the repair process for imperfections in the Hubble Space Telescope mirror) and challenged (the Space Shuttle *Columbia* accident investigation board).

Helicopter Inc.'s PDP is shown in Figure 9.2, where the name of the firm has been masked. Helicopter Inc. builds helicopters for military and civilian applications. It derives a majority of revenue from government contracts, and its products compete in a highly regulated market. By comparison with NASA's PDP, the iteration loops stand out in Figure 9.2. This is an acknowledgment that even in a NASA-like environment of aerospace development, iteration is inevitable and is not at odds with a stage-gated process. Helicopter Inc.'s process literally centers on a regulatory event, FAA certification. However, that certification is an instrument of the more important process, namely sales.

Interestingly, this diagram includes product feedback (implied from customers) but does not explicitly represent customers up front. Does Helicopter Inc. conceive products and then test them in the market, or does it gather customer needs and development grants, or a combination of both? At this level, all that can be ascertained is that input from customers and other stakeholders is implicit in the mission statement. In the Helicopter Inc. PDP, we can begin to see the four major

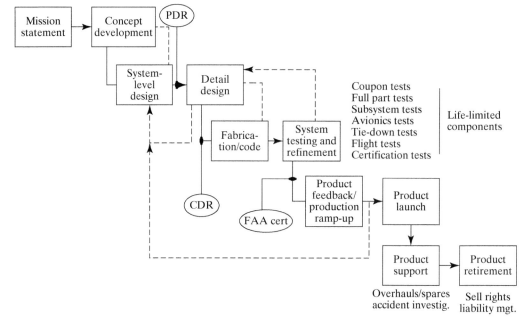

FIGURE 9.2 Helicopter Inc.'s product development process.

activities that are common to nearly all complete PDPs and will form the basis of our Generic PDP: conceiving (mission statement, concept development, system-level design); designing (detailed design); implementation (fabrication, testing, and ramp-up); and operations (implicit in support and retirement, although Helicopter Inc. may not explicitly operate its products).

A third PDP, for Camera Co., is shown in Figure 9.3, where the name of the firm has been masked. Camera Co. sells camera film and cameras for amateur and professional applications.

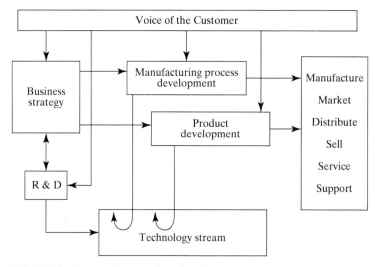

FIGURE 9.3 Camera Co's. product development process.

The level of competition in its markets is very different from that faced by NASA and Helicopter Inc., as is its distributed and diverse customer base. Camera Co.'s PDP explicitly shows upstream processes of strategy, voice of the customer, and R&D, all part of our generic phase of conceiving. The manufacturing development process actually precedes product development, given that Camera Co. is in a process-driven industry. This brings to the surface the idea that the implementation of products is an important driver of the overall PDP.

Although Camera Co. is often thought of as an R&D-driven company, it is interesting to note that the voice of the customer notionally precedes R&D and the technology stream. Is this an actual representation of the process, or is it an attempt to influence a company traditionally focused on technology? Manufacturing occurs in the PDP before product development. What does this suggest about the role of the architect in reducing ambiguity? Perhaps reducing ambiguity associated with manufacturing quality is a more important role than innovation in product offering, given Camera Co.'s stable, loyal customer base.

As a fourth example, the Agile PDP emphasizes iterative and incremental development. Using Agile, collaborative teams evolve requirements and solutions in a somewhat evolutionary approach. Its origins are in the rapid prototyping of software code for small and medium-sized applications. Agile emphasizes interactions among individuals who are building actual code over documentation, negotiations, and planning (Figure 9.4). Beyond software, Agile has been used as a by-word for shaking up development processes in more capital-intensive, longer-lifecycle industries. To be an educated consumer of PDPs, the reader should consider in what contexts Agile is appropriate and useful.

Based on an analysis of the four simplified product development processes, we identified many similarities and differences. Key to this discussion of PDPs is the question of whether differences are superficial or substantial: Did someone design in the differences to reflect the sectors in which the organizations operate? Similarly, we raised the question of whether the PDP represented the "actuals" of product development or a shifted image intended to motivate and transform the organization?

Here are some of the differences we observe among product development processes:

- Existence and number of phases and phase exit criteria (and degree of formality)
- Existence and number of design reviews

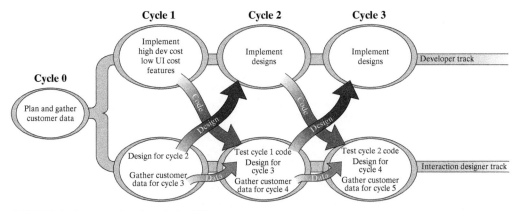

FIGURE 9.4 An example of an Agile PDP.

- Timing for committing capital
- Requirements "enforcement"
- Degree of customer input, feedback, sales, and aftermarket integration
- Degree of explicit/implicit iteration
- Amount of prototyping
- Internal testing and validation effort, importance of traceability
- Timing of supplier involvement

We would argue that several of these differences, such as the number of phases and the representation of customer input, are superficial. Which differences are important?

There are many factors that *could* drive differences between PDPs, such as hardware vs. software, existing product or new product, standalone vs. platform, production volume, capital intensity, and technology push vs. market pull vs. process-intensive. An informed user of PDPs judges which aspects of the process are well honed by experience, implementable, and necessary, and then applies them judiciously to the product under development.

Generic Product Development Process

Despite these differences among PDPs discussed above, there are enormous similarities. In fact, when engineers and architects from various backgrounds move past the superficial distinctions in their PDPs, they find a large body of common effective practice, which we codify in the *Generic PDP*.

Figure 9.5 provides a representation of the common activities of the Generic PDP. The emphasis is on activities of various types, and Figure 9.5 should *not* be read as a stage-gate

Conceive	Design		Implement		Operate	
Mission Conceptual design	Preliminary design	Detailed design	Element creation	Integration test positioning	Lifecycle support	Evolution
			Downstream influences			

• Business strategy • Goals
• Functional • Function
  strategies • Concepts
• Customer • Business/
  needs   product
• Segments   plan
• Competitors • Platform plan
• Technology • Supplier plan
• Regulation • Architecture
• Scope and   • Commitment
  plan
	• Requirements definition
	• Model development
	• Requirements flow down
	• Detail decomposition
	• Interface control
		• Design elaboration
		• Goal verification
		• Failure and contingency analysis
		• Validated design
			• Sourcing
			• Implementation ramp-up
			• Element implementation
			• Element testing
			• Element refinement
				• Product integration
				• Product, system testing
				• Refinement
				• Certification
				• Market positioning
				• Channels
				• Delivery
					• Sales distribution
					• Operations
					• Logistics
					• Customer support
					• Maintenance, repair, overhaul
					• Upgrades
						• Product improvement
						• Platform expansion
						• Retirement

Primary Domain of the Architect

FIGURE 9.5 Our Generic PDP. It represents activities of a complete PDP and is not intended to represent a stage-gate process.

approach. Studies have shown that the linear representations of the design process are deeply flawed. [10] They fail to represent iterations and feedback, presume a linear flow of time and effort across stages, and can mask a design's immaturity when gate criteria are compromised.

The Generic PDP of Figure 9.5 is intended to be a checklist of the activities that appear in complete PDPs, loosely bundled into four groups: conceiving (determining what will be built, based primarily on market needs and technology availability); designing (rendering the information object that defines what is to be implemented); implementing (converting the design into reality); and operating (the operation of the real system to deliver value). The word "implementing" is deliberately chosen to include the coding of software, the manufacturing of hard goods, and the integration into systems. "Operations" has a sense of both supporting existing instances of the product and releasing future products. Operations ends in retirement.

Figure 9.5 also represents the architect's system boundary. The architect often engages with a project after a small number of upstream activities, such as functional strategy decisions. If they are to be successful, the architect must be fully involved in the activities labeled "the primary domain of the architect."

One of the roles of the architect is *moving relevant downstream information into the architecting phase*. This is illustrated as a reverse flow in Figure 9.5. Experience would suggest that it is easy to move constraints upstream. For example, we could argue that repairability is an important attribute of the product, therefore all parts that wear should be colored differently, quickly accessible, and easily diagnosed. Information to move upstream is plentiful, but knowing *which information* will meaningfully differentiate between architectures is more difficult. This is a theme that we will revisit formally in Part 4.

In order to further support comparison among product development processes, we will move outside this more traditional representation, and construct a second PDP called the Global PDP. Our intent is to construct an extended baseline against which different PDPs can be analyzed, in the interest of identifying influences upstream and downstream of the architecture.

To begin our Global PDP, we develop three nested views of architecture influences. The first level, shown in Figure 9.6, centers on the architecture of the product, which we have represented by its seven product/system attributes. Function and form are discussed extensively in Part 2. The choice of the architecture of the system is a reflection of product goals, which are in turn set to respond to market and stakeholder needs. Beyond the architecture, we have captured downstream operations from the perspective of the product operator, including operations and cost, both of which are shaped by the architectural choices made and represent a source of downstream ambiguity for the architect.

Although it runs from left to right, this generic representation represents activities, not a stage-gated process. The double-headed arrows showing the principal dependencies imply the possibility of iteration.

At this level of magnification, the architect is trying to answer the canonical *W Questions* also shown in Figure 9.6: why, what, how, where, when, who, and how much. Note that in this representation, form answers where, since form exists and therefore has location. One of the authors' students once opined that he did not attend MIT to be brought back to the fourth grade. As with many fundamental truths, this list is sure to reappear in the search for holism in system development.

The W Questions share a common linguistic ancestor: *quo* from Indo-European. Many Indo-European languages use a common sound as the basis for these words. French uses "qu," as in

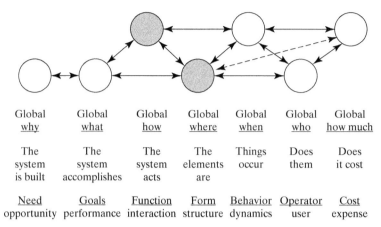

FIGURE 9.6 Holistic framework for the attributes of a product/system considered in product development. Note that interactions are two-directional and that flow from left to right is not implied. Architecture, consisting primarily of form and function, is highlighted at the center.

qui, quoi, quand, combien, ou, comment, and p*ourquoi.* Hindi uses "k," as in *kab, kya, kyon, kitne, kisne,* and *kahan*. The ancient Greek rhetorician Hermagoras defined seven circumstances at the heart of all issues: *quis, quid, quando, ubi, cur, quem ad modum,* and *quibus adminiculis*. The implication is that the architect who wants to have a holistic view of the product development processes must understand these attributes well.

Moving up a layer in our Global PDP (Figure 9.7), we include the activities of the design process and the implementation process, noting that each has its own instantiation of the canonical W Questions. The design process is likely to have its own set of form, notably design tools, which will shape and constrain the available system form (a draftsperson equipped with a straightedge will produce different renderings than one equipped with a French curve).

As an example of design costs influencing architecture, many automakers have a nonrecurring engineering (NRE) design cost policy for platform vehicles, whereby any vehicles occurring in the same year and produced at the same volume have to split NRE design costs evenly. This policy is not explicitly part of the technical evaluation of a concept, nor would it have been present in a stage-gate review, but it is clearly part of our Generic PDP because it will have a significant impact on the ability of a variant to deliver value.

Likewise, the implementation process will have capabilities and limitations. To the architect, this representation implies that the existing processes in implementation must be factored into the architecture of the product.

Any PDP or new design method can be critically evaluated by comparing it with the 21 questions implicit in Figure 9.7. Few methods will answer all 21 questions, so the purpose is to understand which areas a framework is strongest in. For example, the Lean movement began with studies of the Toyota Production System and evolved into principles, methods, and tools for improving manufacturing cost and quality. [11] Lean later moved upstream, asking how

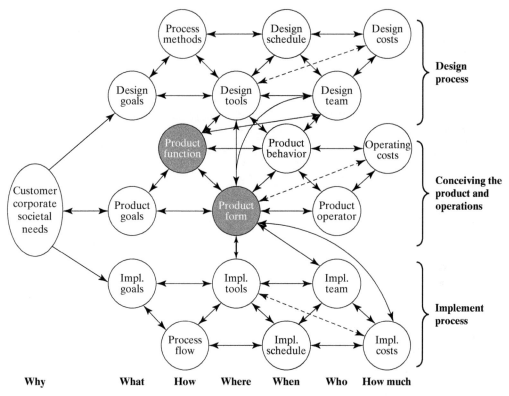

FIGURE 9.7 The second nested view of the Global PDP, our holistic framework for product/system development, explicitly showing the design and implementation efforts as parallel activities. Note that interactions are two-directional and that flow from left to right is not implied.

design decisions could enable leaner manufacturing. [12] However, Lean has also been applied directly to the design *process* [13], an application now far removed from the original intent.

The third nested view of our Global PDP (Figure 9.8) places the firm's PDP in the context of its overall activities. Several firm functions are shown inside the enterprise boundary. For example, R&D and corporate functions are primarily upstream. Public relations, sales, and distribution are downstream. Human skills, information systems, and engineering tools permeate the diagram. Some of these functions are well represented as upstream or as downstream, whereas others (such as corporate strategy) may play a role in several places.

External to the firm, the diagram illustrates an incomplete list of actors and attributes that affect the enterprise. These range from the availability of capital in the market to the competition and the social impact. The corporate functions near the external interactions are meant to facilitate the interactions.

The purpose of this broadest view of the PDP is to remind the architect that understanding the contents of the PDP is not necessarily sufficient. The architect must be aware of other issues that influence the firm and context, as shown in Box 9.3 Principle of the Stress of Modern Practice.

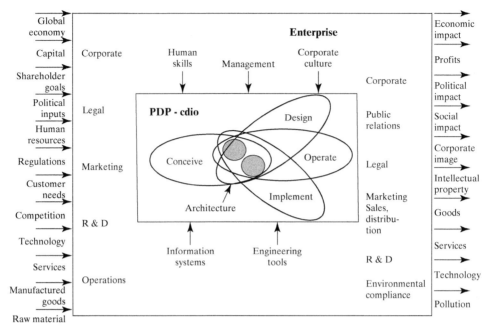

FIGURE 9.8 The Global PDP in the context of the producing enterprise, showing contributions from within the enterprise and interactions at the enterprise boundaries.

> **Box 9.3 Principle of the Stress of Modern Practice**
>
> *"Things which matter most must never be at the mercy of things which matter least."*
>
> Johann Wolfgang von Goethe
>
> Modern product development process, with concurrency, distributed teams, and supplier engagement, places even more emphasis on having a good architecture. Recognize this trend and the impact it has on the architecting process.
>
> - Accelerating the development process by concurrency increases the importance of initial conceptual decisions.
> - Driving decision making to the lowest appropriate level (empowerment) and the use of distributed or non-collocated teams increase the importance of well-coordinated and visible high-level design guidance.
> - Involving suppliers early in the process increases the importance of having an architectural concept and baseline; suppliers can define or defy decomposition.

9.4 Summary

This chapter focused on the role of the architect, particularly his/her role in reducing ambiguity. The main finding is that the architect is not a generalist but a specialist in resolving ambiguity, focusing creativity, and simplifying complexity. The architect must understand the types of ambiguity: fuzziness, uncertainty, and missing, conflicting, or incorrect information.

The architect is responsible for producing certain deliverables, which are central to the role of reducing ambiguity and serving as a bridge between corporate functions and the design environment.

To highlight this role, we examined where it fits within the context of a product development process. Although many PDPs appear different, they share underlying features. Architects must be able to discern which segments of a PDP are centrally linked to their role, actions, and outcomes, just as the architect must become skilled in interpreting new product development initiatives.

Box 9.4 *Case Study: Civil Architecture and System Architecture*

By Steve Imrich, Architect and Principal at Cambridge Seven Associates
With thanks to Allan Donnelly

The term "architect" arises from civil architecture, but what do the two fields share?

When one compares civil architecture with system architecture, one realizes that there is an interpretive-interactive-performance aspect of building design that makes civil architecture different than many other forms of industrial or product design. More than many other kinds of products, buildings inherit complex cultural and symbolic qualities that affect space planning, material quality, and user experience. Ideally, the "special problems" of a building inform the "design." Certain devices like helicopters, sailboats, and computers (among many others) are good examples of the success in resolving "special problems" and how the concept, design, and engineering may introduce an elegance, simplicity, and beauty that transcend a style for these products. Because buildings can only be seen and experienced as part of their site, the "special problems" of a building always relate to a wider context of time, culture, existing infrastructure, human behavior, local environment, and re-interpreted use. Because buildings are not normally mocked-up beforehand there is a large element of risk in determining the success of a design … something that is generally associated with the arts.

(a) (b)

FIGURE 9.9 Concept in civil architecture. The "Concept" is the departure point for integration of performance and form.   (Source: (a) Funkyfood London-Paul Williams/Alamy (b) Francisco Javier Gil/Fotolia)

In contrast to a conventional product development process, we have used the following "design process with attitude" to draw attention to the most important emergent properties that will determine the architecture's success and longevity.

- Context... The context within which a building exists draws from adjacent existing buildings, cultural norms and associations about the built environment, environmental conditions, and the history and motivations of the organizations and people that the building is designed to serve.
- Content... The content of a building includes both its programmatic requirements and its functional performance. Content describes the purpose the building is meant to serve and the elements through which that purpose is realized.
- Concept... The concept for a building project is the departure point for organization and experience. The concept is the unifying idea by which all design decisions are evaluated (Figure 9.9).
- Circuitry... The connective tissue between functional components in a building. Circuitry, when done well, provides more than a path to get from point A to B but also creates interest and anticipation for how boundaries, thresholds, and edges of spaces are dynamically and progressively revealed.
- Character... As with people, the character of a building can yield long-lasting appeal, short-lived fascination, or very little response at all. The possibility for a building to connect with nature, its site, and a sense of life beyond the aggregations of its materials speaks of it character. Over time, positive associations with the building's use and its visual presence become the substance that lies beyond the façade and the patterns that we all use to evaluate the qualities of our environment.
- Magic... Magic is a largely intangible quality that is defined primarily by the user's experience with a building. Buildings that exhibit magic can have widespread attraction and life beyond the owner's and designer's intent. When working well, the magic of a place fosters an inherent delight and joyful appreciation for experiencing the unexpected. Buildings that exhibit the best qualities of magic often become the most iconic and memorable structures in our designed world (Figure 9.10).

I'd like to focus on two elements, the concept and the magic.

What gets added to a design to make it beyond practical is a challenge that architects deal with every day. A concept is an idea that follows the whole project, yet it is more than a one-liner. The idea has substance, depth, richness, subtlety, and clarity. The idea has to have substance; it is not a coating or a camouflage or a metaphor. The concept has to encompass both the structure of the building and its skin. Concept generation is successful when it integrates and balances the nature of the content and the nature of the context. To this end, architects do deep research to identify this "nature." Concept is the departure point. It is never clean.

Due to the complex nature of building projects, it is helpful in the design process to keep oneself anchored to a bigger, simpler idea. A strong concept provides the framework for the integration of performance and form. The architecture of a building should rationalize the experience of the user. A clear concept coupled with excellent execution will make the idea understandable to the user without any outside information.

(a)   (b)

FIGURE 9.10 Magic in civil architecture. Qualities of "Magic" allow structures to become memorable and iconic to a culture.   (Source: (a) Wim Wiskerke/Alamy (b) EvanTravels/Fotolia)

Magic is very important in architecture. It is the emergence of elegance and character at the building level, the sleight of hand. Magic is a quality that may be difficult to define, but when something is magical, the quality is immediately recognizable. Magic happens when a building engages and delights [so as to leave an indelible image in a person's memory]. Magic is the unusual circumstance that becomes normalized in good architecture. Magic is that moment of surprise that leaves a person in wonder and awe. Eladio Dieste said, "Surprise, as all art and architecture, helps us contemplate. Life wears out our ability for surprise. Surprise is the beginning of a true vision of the world." Architecture exudes a magical quality when its formal elements and functional characteristics combine to create an emotional, timeless, and unanticipated experience.

The performance of a ballerina and that of a gymnast have characteristics in common involving athleticism, dynamic patterning, and nuance of expression, yet one of the activities is considered more in the realm of "art" while the other is more in tune with "athletic expertise." In magical "architecture," we experience the intrigue of artistic expression with functional achievement in a fully integrated way, which imparts a simple and seemingly effortless elegance that elevates our spirit beyond common expectations (Figure 9.11).

(a)   (b)

FIGURE 9.11 An architect's opportunity and responsibility is to overlay performance and poetry and to understand how those ingredients interact.   (Source: (a) Cheese78/Fotolia (b) Gerard Rancinan, Jean Guichard/Sygma/Corbis)

The mystery behind the magic provides another layer of surprise that is released over time. Real magic makes architecture symbolic and iconic, not just at first glance, but in a sustained manner that reaches beyond the original intent. A magical building has universal appeal because it communicates viscerally with its users in a way that touches the innate emotional tendrils of the human experience. A magical place incorporates the strongest ideas with qualities and ingredients that are mutually reinforcing. Magic is not an easy quality to design, but when it is created, it has the power to shift perceptions and change the world.

Architecture is risky business.

References

[1] Benjamin S. Blanchard and Wolter J. Fabrycky, *Systems Engineering and Analysis* (Englewood Cliffs, NJ, 1990), Vol. 4.

[2] Edward B. Roberts, "A Simple Model of R & D Project Dynamics," *R&D Management* 5, no. 1 (1974): 1–15.

[3] Sterman, John D. Business Dynamics: Systems Thinking and Modeling for a Complex World. Vol. 19. Boston: Irwin/McGraw-Hill, 2000.

[4] James M.Lyneis and David N. Ford, "System Dynamics Applied to Project Management: A survey, Assessment, and Directions for Future Research," *System Dynamics Review* 23, no. 2–3 (2007): 157–189.

[5] Qingyu Zhang and William J. Doll, "The Fuzzy Front End and Success of New Product Development: A Causal Model," *European Journal of Innovation Management* 4, no. 2 (2001): 95–112.

[6] http://www.c2es.org/federal/executive/vehicle-standards/fuel-economy-comparison

[7] Project Management of GFE Flight Hardware Projects. EA-WI-023. Revision C, January 2002. Retrieved August 12, 2014 from: http://snebulos.mit.edu/projects/reference/International-Space-Station/EAWI023RC.pdf

[8] Robert Shishkoand Robert Aster, "NASA Systems Engineering Handbook," *NASA Special Publication* 6105 (1995).

[9] B.J. Sauser, R.R. Reilly, and A.J. Shenhar, "Why Projects Fail? How Contingency Theory Can Provide New Insights–A Comparative Analysis of NASA's Mars Climate Orbiter Loss," *International Journal of Project Management* 27, no. 7 (2009), 665–679.

[10] Klaus Ehrlenspiel and Harald Meerkamm, *Integrierte produktentwicklung: Denkabläufe, methodeneinsatz, zusammenarbeit* (Carl Hanser Verlag GmbH Co KG, 2013).

[11] D.T. Jones and D. Roos, *Machine That Changed the World* (New York: Simon and Schuster, 1990).

[12] K. Yang and B. El-Haik, *Design for Six Sigma* (New York: McGraw-Hill, 2003), pp. 184–186.

[13] J. Freire and L.F. Alarcón, "Achieving Lean Design Process: Improvement Methodology," *Journal of Construction Engineering and Management* 128, no. 3 (2002), 248-256.

Chapter 10

Upstream and Downstream Influences on System Architecture

10.1 Introduction

Having identified a wide variety of upstream and downstream influences in Chapter 9, in this chapter we select several to explore in more detail. Our intent is not to develop a framework for each, but to stimulate the reader to consider the relevant coupling between these influences and the architecture. Which influences will meaningfully reduce ambiguity? Which influences will help us select among candidate architectures? In this context, it is helpful to think about influences on *architectural decisions* (Box 10.1).

> **Box 10.1 Principle of Architectural Decisions**
>
> *"We have no reason to expect the quality of intuition to improve with the importance of the problem. Perhaps the contrary: high-stake problems are likely to involve powerful emotions and strong impulses to action."*
>
> Daniel Kahneman
>
> Architectural decisions are the subset of design decisions that are most impactful. They relate to form–function mapping, they determine the performance envelope, they encode the key tradeoffs in the eventual product, and they often strongly determine cost. Separate these architectural decisions from other decisions, and take the time to carefully decide them up front, because they will be very expensive to change later on.
>
> - Architectural decisions lead to architectures that are fundamentally different from each other—the number of driven wheels on a car, whether or not an aircraft has a tail, whether an algorithm runs in real time or not, and the like.
> - Architectural decisions are different from design decisions. The choice between leather and cloth for the seats in a car is not an architectural decision but a design decision. It does not materially impact technical parameters or important metrics. The choice between leather and cloth could affect the cost, but it will do so predictably. It is less likely to have emergent consequences.
> - Identify architectural decisions early and study them carefully, because making the wrong choices can irrevocably hinder realization of the eventual product, where no amount of detail design or component optimization will fix the fundamental issues.

We will revisit architectural decisions in detail in Part 4, but it is a helpful concept to carry while discussing upstream and downstream influences, because we will have to answer questions about the magnitude of the impact of each of the influences. (For example, will marketing affect whether the car is two-wheel or four-wheel drive?)

With these questions in mind, we begin our discussion with upstream influences, where we pose the question *"What upstream influences impact architectural decisions?"* We ask it in four focus areas, beginning with corporate strategy in Section 10.2, followed by marketing, regulation, and technology infusion in Sections 10.3 through 10.5. The second half of the chapter focuses on downstream influences. Whether it be capturing relevant information about the manufacturing capabilities of the enterprise or planning a series of derivative products, it is important to consider the downstream influences on architectural decisions. This is a generalization of the idea that the product design community calls Design for X (DFX), [1] where X is whatever is downstream of design (manufacturing, service, upgrades, and so on). Implementation, operations, Design for X, and product evolution are addressed in Sections 10.6 through 10.9. We then summarize these upstream and downstream influences in an iterative model called the "Product Case."

10.2 Upstream Influence: Corporate Strategy

Among the most fundamental upstream influences on architecture is the strategy of the enterprise. In a corporate context, the firm's corporate strategy describes the means by which the firm achieves corporate goals, such as increasing profitability and building competitive shareholder value. In a government context, the enterprise's strategy similarly defines the means by which the agency achieves its mission. Strategy should explicitly differentiate how the firm or enterprise will do this; stating corporate strategy as "grow the revenue of the firm while maintaining the product margins" conveys little of what *decisions* the firm made. Strategy defines the specific activities of the organization: the mission of the enterprise, the scope of activities, the long-term and intermediate-term goals, resource allocation decisions, and planned initiatives.

An effective corporate strategy is most crucially a way to communicate vision and direction to the stakeholders and team, and a guide to the investment of scarce resources. It should provide a coherent, unifying, and integrative pattern of decisions. The strategy should reflect what markets, features, and customers are considered, and which are excluded from consideration. Strategy involves calculated gambles. If success were assured, strategy would be called investing in risk-free bonds!

In many technology firms, the architecture of the system has an enormous impact on the firm's corporate strategy, and vice versa. IBM's corporate strategy decision to outsource the operating system development to Microsoft, combined with its decision to create a stable architecture for the PC, shaped the desktop computer market for 25 years. Airbus's decision to define a common glass cockpit for the A319, A320, and A321 had a strong positive impact on the firm's market strategy, enabling it to sell to customers the benefits of commonality through training and maintenance savings.

Many engineers unfamiliar with strategy are initially frustrated by the lack of useful information in what in a for-profit firm would be called the *shareholder annual report corporate strategy*. This level identifies the long-term objectives (such as "be the largest automotive

OEM by units sold in 2018" [2]), action programs ("drive cost synergies across business units"), and resource allocation priorities ("continue our commitment to R&D in energy storage"). By definition, information that the firm makes publicly available to its competitors does little to define what is excluded from consideration and what is truly being prioritized. For example, which of the largest five automotive firms doesn't want to be the largest? This statement alone does not provide much information about what actions the firm is taking to realize that goal.

Strategy assesses the opportunities and threats in the competitive environment, and the strengths and weaknesses in the enterprise. It defines initiatives and action programs around which the enterprise will mobilize. Financial analysts consume information at the level of the *shareholder annual report* and produce summaries of their positions on the firm's strategy, which can provide another mechanism for understanding the firm's strategy. Analysts' reports can include assessments of the opportunities and threats in the competitive environment (such as which competitors are losing structural advantages due to deregulation) and of the strengths and weaknesses of the organization (such as its core competency in designing paper-routing mechanisms and its potential weakness in assessing and crafting service plans).

The shareholder annual report corporate strategy is distinctly different from the *executive corporate strategy*. The executive management strategy level defines what businesses the firm is in or should be in, and then establishes a means of investing selectively to develop the capabilities for sustainable competitive advantage. This level involves clear decisions on priorities, primarily executed through investment decisions, and, even more important, what to exclude. Executive management strategy defines corporate philosophy, sets shareholder expectations, segments the business, creates horizontal/vertical strategies, and considers global and macroeconomic trends.

For example, BMW decided in 2009 to exit Formula 1 (F1) racing at a savings of several hundred million U.S. dollars per year. This top management team weighed the benefits, in terms of BMW brand awareness, brand association with F1, and F1 technology trickle-down to production vehicles, against the opportunity cost of other marketing and R&D investments. BMW indicated in its press release that "Premium will increasingly be defined in terms of sustainability and environmental compatibility. This is an area in which we want to remain in the lead." [3] Thus BMW's forecast of macroeconomic trends resulted in a new business segment, *BMW i*, focusing on small, efficient, city-driven vehicles. [4]

In larger companies, there exists a *business unit strategy* below the executive level corporate strategy. Business unit strategy is specialized to the sector, market, geography, or technology that defines the business unit. In the BMW case, the business unit strategy for the BMW i sub-brand asks these questions: Given that BMW is targeting small, efficient, city-driven vehicles, how will the BMW i sub-brand segment its customers, which threats will it actively mitigate with investments, and which opportunities will it exclude from consideration? To be effective, the architect must be intimately familiar with the business unit strategy, in order to understand the context, the priorities, the decisions, and the investment process.

Within a business unit (and sometimes across business units) there are *functional strategies* that aggregate the goals with regard to the firm functions—marketing, R&D, sourcing, and so on. These functional strategies can express how the function serves a corporate goal ("What actions does sourcing need to take, given that the firm wants to be first to market with technology in the industry?") or they can express improvement plans that are only indirectly tied to the firm's profitability. Here are some of the key strategy questions that often appear in these functional strategies:

- **Marketing.** Where a company competes (markets, segments, competitors, geography) and how it competes (the "attack plan"—cost, quality, services, feature innovator/follower).
- **R&D/Technology.** How important new technology is to the business, how technology will be developed/acquired, at what level of investment, with what type of multiproduct planning.
- **Manufacturing/Sourcing.** How important low-cost manufacturing is to the business, whether internal manufacturing or outsourcing will be used, and how suppliers will be integrated.
- **Product Development.** How new products will be developed, whether to employ any platforming, global vs. regional products, and whether the development will be internal, outsourced, or an alliance effort.

The architect must be deeply engaged with these functional strategies.

At all four levels of corporate strategy (*shareholder annual report, executive, business unit, functional*), goals are frequently expressed in corporate finance terms. If the architect is unable to interpret a compound annual growth rate (CAGR) target of 20% for a market entry strategy in Asia in terms of its implications for product development timelines, she or he will fail in the role of communicating goals for the project. Similarly, an architect must be able to define whether a R&D level of 3% of revenues will be sufficient to develop the lead variant in a product family, whether the second and third variants will contribute to the amortization of that investment, and what implications this has for the return on investment (ROI) of the variants individually as compared with the product platform as a whole.

How does corporate strategy impact architecture? How does architecture impact corporate strategy? These are questions that the architect must thoughtfully consider. Some of the many impactful relationships between aspects of strategy and architecture are noted below.

MISSION AND SCOPE   The architecture must directly address the mission and stakeholder needs of the enterprise. The architecture must reflect the scope defined by corporate strategy—in particular, activities that will be excluded. In turn, the architecture can constrain the scope available to corporate strategy, which must be appropriately communicated.

ENTERPRISE GOALS   The architecture must meet or contribute to the financial goals for revenue, margin, and return on investment, and it must support the growth of the firm. The new system should be positioned to align with a market opportunity identified or to defend against a competitive threat. The architecture may have to address other goals on brand, technology direction, and the like. The architect must plan to build the product based on the core competencies of the enterprise, advocate for expanding these competencies, or have a plan to go outside of the enterprise to engage others.

RESOURCE ALLOCATION DECISIONS   The development of the architecture must fit within the resource allocation guidelines, and the architect must be aware of the decision process and how to defend the allocation of resources.

INITIATIVES AND ACTION PLANS The architecture should leverage corporate initiatives and functional strategies where applicable. The architecture can play a central role in enabling new functional strategies, but it must sometimes communicate constraints that complicate initiatives and action plans.

10.3 Upstream Influence: Marketing

The essence of marketing from the engineer's perspective is aptly communicated in the film *Office Space*, where the marketing representative states, "I deal with the...customers so the engineers don't have to." In some organizations, marketing and engineering sit in tension, each taking credit for the creation of products, and each arguing that it has a more accurate view of customer needs based (according to the former) on sales and marketing data or (according to the latter) on customer feedback, warranty reports, and product tests. A consumer goods firm we worked with expressed this tension: Engineering developed a product roadmap completely independent of marketing's product roadmap. Although it is plausible to have independent marketing and engineering product ideation for simple systems (for toothpaste, a new tube filled with the same toothpaste may be a new product), separating them is not so simple for complex systems.

We will argue that the architect plays a critical role in bridging the divide between the marketing and product development communities, and that successful firms link these two functions closely. We begin with a generic definition of marketing: "Creating, communicating and delivering value to customers, and...managing customer relationships in ways that benefit the organization and its stakeholders." [5] This definition links upstream and downstream functions in the sense that it covers the creation of value, as well as the downstream delivery. We find it useful to distinguish between *inbound marketing*, the process of discovering user needs, and *outbound marketing*, the process of satisfying consumer needs. In modern practice, inbound marketing is sometimes called "product development" in the marketing community.

Outbound marketing (Figure 10.1) is what most people understand by the term marketing. It is enshrined in Edmund McCarthy's Four P's: product, price, promotion (communications), and place (distribution). It is a fallacy, in most cases, to think that marketing occurs downstream of product development without any necessary interaction upstream. The history of engineered products is rife with examples of individual features renamed for outbound marketing campaigns. The Cessna Aircraft Company in the 1960s and 1970s was known for originating terms

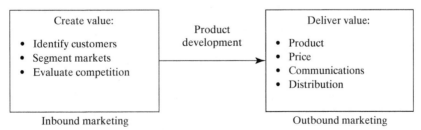

FIGURE 10.1 Inbound and outbound marketing as seen by the architect.

Office Space, 20th Century Fox, 1999.

like Land-o-Matic (tri-cycle landing gear), Stabila-Tip (wing tip fuel tanks), and Omni-Vision (rear windows). However, rather than conceiving outbound marketing as an oversimplification of the product value proposition, we argue that successful architects often intertwine the creation of the architecture with a market strategy. This is not to say that all aspects of the Four P's play a role in architectural decision making, but rather that the architect should consider which couplings to outbound marketing will be important to consider.

Inbound Marketing

Inbound marketing is the aspect of marketing that most concerns the architect. The traditional role of inbound marketing comprises customer identification and needs analysis, market segmentation, and competitive analysis, as shown in Figure 10.1. An expanded view of inbound marketing could also include identifying possible new products, testing value propositions, helping to define product requirements, and even defining a business case for the product. It is important to recognize that this list originally arises from the reference case of a consumer seeking a replacement product: "Everyone knows what a red polo shirt is, and we're creating a new red polo shirt that is better." The job of the architect of complex technical systems is to generalize this reference model to non-consumer cases as necessary.

Of the various tasks in inbound marketing, far and away the most critical are identifying stakeholders (including customers), understanding their needs, and converting needs to system requirements. We have devoted Chapter 11 to this process, which may follow, or may run parallel to, the market segmentation and competitive analysis of inbound marketing, as described below.

In an established market, the purpose of segmentation is to group customers who have similar needs and who will respond in a similar fashion to outbound marketing, so that the firm can allocate its efforts among them. The description of the customers in these market segments can be used to create use cases for the product or system, to structure upstream customer needs surveys or interviews, or to perform downstream prototype evaluations or test advertising messages. In the consumer context, market segments were traditionally defined geographically and demographically (perhaps based on the availability of census data). Today there is increased opportunity to define them with much more granularity, in terms of attitudes, values, personality, and interests. Indeed, online advertising is rapidly changing the information available to marketers.

An important attribute of market segments is the size (current and anticipated growth) of the segment, because this is often used in the business case to weigh against investment. Specifically, the business case takes into consideration the size of the total addressable market, defined as yearly sales of the product assuming that 100% of the market is captured. Based on the market, the firm can speculate on the initial and steady-state penetration rates to arrive at sales volumes, against which the costs of the product development can be weighed in a business case.

Generalizing traditional consumer segmentation to complex systems is difficult. The traditional consumer context is rarely a good model for complex systems, but it is frequently used nonetheless. Estimating the size of the total addressable market is an excellent case in point. For some complex systems, the problem is somewhat easier because there are very few potential customers who have the need and can afford the purchase price (this would be true of military attack helicopters or 200-MW gas turbines, for example). However, most complex systems compete in markets that are more difficult to predict. History is full of examples where the market was misestimated by two orders of magnitude (up and down!). Especially in capital-intensive industries, a factor of 2 can represent the difference between product success and failure.

Consider the challenge of predicting the market for optical communications equipment in the satellite market. Optical communications has been cited [6] as the next technological step beyond radio for in-space communications between satellites, as well as between the satellites and ground stations, offering much higher data rates with lower equipment mass. The underlying markets served are governments purchasing communications satellites predominantly for military use, and commercial satellite operators providing consumer services (such as satellite television and satellite phones). Although the technology has recently been demonstrated in space, [7] significant uncertainty remains with respect to technological factors (such as the pointing accuracy and stability required for laser beams) and operational factors (such as how the market will respond to the constraint that optical signals cannot be transmitted between a satellite and Earth when the weather is cloudy). Furthermore, the market for the underlying services is difficult to predict; witness the challenges encountered by Iridium when the market rejected the size of its satellite phone handset.

Conducting competitive analysis is the final step of inbound marketing. One of the most common mistakes in technology-based enterprises is to assume that the technology of competitors will not improve from their current market offerings. It is therefore important to characterize the performance and features of competitive products and predict their future trends. More generally, the marketing approaches of competitors should be understood so that barriers to their success can be built into the product or plans of the enterprise. As a final note, one should not assume that competitors are bad: Their presence helps inform and educate customers and create a growing market.

As we have noted, the interaction between marketing and system architecture is critical, especially for inbound marketing. Below, we have summarized some of the important interactions between marketing and architecture.

STAKEHOLDERS AND THEIR NEEDS   The architecture is shaped by an understanding of stakeholder needs. The chosen architecture can help satisfy some needs, and it may place other needs of stakeholders out of reach.

SEGMENTATION OF MARKETS, MARKET SIZING, PENETRATION   The architect must deeply understand the segmentation in order to properly develop system requirements that reflect a real customer community. The estimated size of the market for the product greatly impacts the estimates of the financial return that are part of the business case. If the estimated sales do not provide robust margin, the firm will probably not provide financial support for the product development. The timing of market penetration can vastly change how stable the architecture will be in the market.

COMPETITORS AND COMPETING PRODUCTS   All customers will compare the function and performance of the enterprise-developed product with that of the competitors. Therefore, the architect must make the best possible estimate of competitors' systems in order to build the basis of successful competition.

OUTBOUND MARKETING   Outbound marketing will define the revenues for the product. The architecture must enable outbound marketing, regardless of whether it reflects the central core of the architecture or minor features. The architect should also be generally aware of plans for product definition by marketing, communications plans, and channels of distribution (which affect service provision).

10.4 Upstream Influence: Regulation and Pseudo-Regulatory Influences

Regulation can be both an enabler of competitive advantage and a barrier to entry. Regulation can act to preserve an existing system architecture well past its useful life or generate an opportunity for new architectures to disrupt the market. For example, the introduction of Tier 4 Final emissions regulations for off-highway trucks, which essentially upheld stricter emissions standards, required firms to update their entire engine portfolio, but it simultaneously offered an opportunity in the market to compete anew on fuel efficiency.

The obvious impact of regulations is their implications for architecture and design, such as the mandatory inclusion of backup cameras on U.S. cars currently expected in 2015, or regulation on materials used in medical devices.

There is a second pathway by which regulations can affect architecture: through their impact on downstream factors. For example, regulations can have an impact on the desired location of manufacturing facilities, the rate at which team members can be hired and fired, and the markets in which the product can be sold. All of these would affect the economic viability of the product.

Which elements of regulation are important to consider when choosing among architectures, and which elements can be addressed in detailed design and testing? We will discuss regulations as well as pseudo-regulations (standards, anticipated regulation, and litigation), all of which can have implications for the architecture (Figure 10.2). For an overview of regulatory economics, see Viscusi (2005). [8]

Sources of Regulation

What are the sources of regulation? Regulations occur at the federal, state, and city/county levels in the United States, in addition to international regulations. We will use the United States as a reference case for discussion, but there are some regulatory similarities among major industrialized nations. In the United States, federal regulation is incorporated in the U.S. code and overseen by agencies such as the Food and Drug Administration (FDA), the U.S. Department of Agriculture (USDA), the Federal Aviation Administration (FAA), and the Environmental Protection Agency (EPA). Additionally, some states have product and systems regulations, such as those imposed by the California Air Resources Board (CARB) and state motor vehicle safety equipment laws.

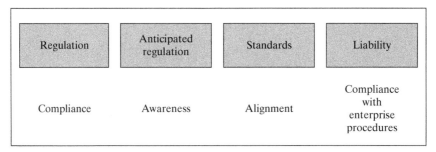

FIGURE 10.2 Regulation and pseudo-regulatory influences that the architect encounters, and the level of engagement with the architecture.

City and county regulations can be enshrined in local codes, but they are typically too diffuse for complex systems to cover exhaustively. Regulations are not usually hard to find for regulatory environments requiring certification (such as the issuance of aircraft type certificates), but in other cases they can be challenging to unearth. It is the responsibility of the architect, perhaps working with legal functions in the enterprise, to be aware of all relevant regulations.

In terms of architecture, we believe the first step in *complying* with any regulation is to develop an understanding the *regulatory intent*. Regulations exist for a variety of purposes:

1. Protect the consumer. For example, under recent regulation from the Consumer Financial Protection Bureau, credit card bills must display the payback period that will result if only the minimum payment is made each month.
2. Set standards to help prevent workplace injury to individuals. For example, the Occupational Safety and Health Administration (OSHA) sets standards on ventilation systems and protections required for individuals working with toxic materials.
3. Protect the environment, such as by restricting the production of pollutants. For example, the Corporate Average Fuel Economy (CAFE) standards set fleet average emissions for car manufacturers.
4. Implement industrial policy by providing incentives to producers (such as the plug-in electric drive vehicle linked to U.S. tax credits) or by dis-incentives (such as implicit tariffs and rules of operation).
5. Prevent dissemination of sensitive or defense technology. For example, the International Traffic in Arms Regulations (ITAR) regulate disclosure of munitions-related data to non-U.S. Citizens.

Although it is not exhaustive, our list of purposes for regulation reveals that there are a variety of regulatory intents. It is important to understand that by our definition, regulations are issues of *law,* so compliance is mandatory. The reasons for understanding the regulatory intent are, first, to understand how compliance will be implemented and, second, to help predict and manage anticipated regulation.

Compliance with regulation is overseen in a variety of ways: A certificate is required for operating the product or system; plans are submitted for review and approval; designated examiners must approve the product; or the firm is subject to regular or surprise inspections. In the extreme case, the firm is responsible for demonstrating to an internal team that compliance with a standard has been met, and it is thereafter responsible for maintaining records. These reviews function without external oversight until a complaint or lawsuit is launched, at which point the firm must produce the documentation.

The implementation of compliance needs to be judged in order to establish how rigid the terms of the regulation are. Many leading firms exceed regulations, either because doing so provides value to the customer or because the firms derive a marketing benefit from it. As a general rule, regulations lag the leading edge of the market, which means that leading firms exceed the regulations. However, regulation is sometimes used to push industries to promote the larger social good. As a result, firms engineer their products to come as close to the limit as possible, because this allows them to deliver additional operating value and reduce the cost of the product. It is therefore important to determine the rigidity of the regulations as a function of the regulatory intent.

Pseudo-Regulation: Anticipated Regulation, Standards, Liability

The second reason for understanding regulatory intent is to forecast and manage anticipated regulation. In the short term (up to several years, depending on the industry), potential regulations can be tracked as they make their way through government. Notices of proposed rulemaking are issued, and often industry players are invited to comment. For example, automotive firms lobbied against mandatory backup cameras, including supplying estimates of increases in car prices and commented on proposed deadlines for implementation as a function of development lead times. Where regulations are immersed in the political process, outcomes are uncertain, but they can often be tracked and to a certain extent can be anticipated by the architect.

For longer-term anticipation of regulations, the leading global jurisdiction can be a helpful indicator. For example, Switzerland issues some of the most stringent noise regulations for domestic white goods (appliances). Although some standards may never be replicated in other markets, white goods have become significantly quieter over the last 20 years. Similarly, Sweden set some of the most stringent automotive safety regulations, and the State of California almost acts as a federal agency in setting fleet emissions regulations. By watching developments in these leading global jurisdictions, architects can to some extent forecast the future evolution of regulations.

In addition to regulations and anticipated regulations, standards and sometimes litigation can act as pseudo-regulations. Examples of standards include IS14001, an environmental management certification, and MIL-STD 1553, which defines parameters for operating a serial data bus. The principal intent for standards is often interoperability, which would suggest that firms can decide whether to comply with the standard. However, meeting standards is often mandated as a condition of membership in industry associations, where the industry as a whole intends to self-regulate to stave off external government regulation. The architect must judge the benefits and penalties associated with compliance with standards. Furthermore, standards are sometimes used as a tool in competitive strategy, where a firm seeks to set and control an industry standard in order to create demand for its products. For example, Intel has long played a role of standard setting, such as with wafer-size transitions for semiconductor manufacturing. Non-compliance with standards can result in litigation; Sun Microsystems, the owner of Java, sued Microsoft for improperly implementing the Java 1.1 standard. Other sources of litigation include product liability cases brought by individual class actions, and actions associated with lack of contractual compliance.

The architect's role is to understand the firm's corporate strategy and legal strategy as they relate to regulation and pseudo-regulation, to comply where mandatory, and otherwise to implement to the extent necessary or desirable. It is important to note that the legal department in a firm has a substantially different role. Its role is to be conservative—to protect the firm from risk and liability. Legal counsel is not incentivized based on product success in the market. Additionally, legal counsel is based on existing law; legal departments are rarely tasked with anticipating regulation. Therefore, it is the role of the architect to bridge the gap with legal counsel and to provide comprehensive guidance for the product based on regulation and pseudo-regulatory influences.

10.5 Upstream Influence: Technology Infusion

A new technology is often at the heart of a new product, and a shift in technology is often a primary motivation for developing a new architecture. Especially in the entrepreneurial community, a proprietary piece of technology is the central feature of the company's first product. One of the pivotal roles of the architect is deciding whether and how to infuse a new technology into an architecture. An enormous amount has been written about this process, and we will therefore provide only a brief overview.

The first question that should be addressed by the architect is: *Will the technology be ready for infusion?* Here we are focused primarily on technologies early in the "S" curve of their development. Technology development is a process that proceeds at its own pace, independent of the timing of a new product or system. A new technology may be ready in time for your system, it may be late, or it may have lain fallow so long that the development team has dispersed. Advocates sometimes overstate the readiness of the technology. A useful tool for assessing the readiness of a technology is the Technology Readiness Level (TRL) scale shown in Figure 10.3. Despite the impression left by the TRL rating, technology development is not a linear process, and setbacks and iterations before final infusion readiness are common. It is the responsibility of the architect to understand the potential technologies that might be infused into the product.

Another key question for the architect is: *Will this technology actually create additional value from the perspective of the customer and other stakeholders?* This is best determined by comparative analysis in conceptual design (Chapter 12) of the architectures that would be possible with the new technology versus those without the technology. The associated risks of timely delivery and long-term supplier base of the technology must be included in such an analysis.

If it is judged that the technology is ready and will create customer value, the final question is: *How will the technology be effectively transferred into the product or system?** This discussion often jumps immediately to ownership of intellectual property (IP). Although consideration of IP is necessary, it is not sufficient for effective knowledge transfer. IP ownership or licensing covers the "know what" of transfer, but the real problem is the transfer of "know how."

People are by far the most effective means of knowledge transfer. Where there are corporate research labs, the people who develop a technology often move with it to the product development team. In modern practice, the approach of many firms is to de-invest in technology and use a supply chain approach to acquiring technology from suppliers, universities, and government laboratories and through the acquisition of smaller companies.

As a marketplace of technologies, this market is very inefficient. The developers of the technology often do not know of the potential value it may bring, they almost certainly do not know of the architect's intended application, and they almost never have enough money and time to bring the technology to the level of readiness that the architect would prefer before making a final decision on infusion. It is the role of the architect, working with the technology team of the enterprise, to devise a mechanism for effective knowledge transfer and successful infusion of new technologies into the product.

10.6 Downstream Influence: Implementation—Coding, Manufacturing, and Supply Chain Management

The downstream development of the system is an equally significant source of ambiguity for the architect. An important role of the system architecture is to compress and extract the relevant downstream details and inject them back into the decision-making process up front. We begin with implementation as downstream influence.

---
*With thanks to Prof. Rebecca Henderson.

Technology Readiness Level	Description
1. Basic principles observed and reported	Lowest level of technology readiness. Scientific research begins to be translated into applied research and development (R&D). Examples might include paper studies of a technology's basic properties.
2. Technology concept and/or application formulated	Invention begins. Once basic principles are observed, practical applications can be invented. Applications are speculative, and there may be no proof or detailed analysis to support the assumptions. Examples are limited to analytic studies.
3. Analytical and experimental critical function and/or characteristic proof of concept	Active R&D is initiated. This includes analytical studies and laboratory studies to physically validate the analytical predictions of separate elements of the technology. Examples include components that are not yet integrated or representative.
4. Component and/or breadboard validation in laboratory environment	Basic technological components are integrated to establish that they will work together. This is relatively "low-fidelity" compared with the eventual system. Examples include integration of "ad hoc" hardware in the laboratory.
5. Component and/or breadboard validation in relevant environment	Fidelity of breadboard technology increases significantly. The basic technological components are integrated with reasonably realistic supporting elements so they can be tested in a simulated environment. Examples include "high-fidelity" laboratory integration of components.
6. System/subsystem model or prototype demonstration in a relevant environment	Representative model or prototype system, which is well beyond that of TRL 5, is tested in a relevant environment. Represents a major step up in a technology's demonstrated readiness. Examples include testing a prototype in a high-fidelity laboratory environment or in a simulated operational environment.
7. System prototype demonstration in an operational environment	Prototype near or at planned operational system. Represents a major step up from TRL 6 by requiring demonstration of an actual system prototype in an operational environment (e.g., in an air-craft, in a vehicle, or in space).
8. Actual system completed and qualified through test and demonstration	Technology has been proven to work in its final form and under expected conditions. In almost all cases, this TRL represents the end of true system development. Examples include developmental test and evaluation (DT&E) of the system in its intended weapon system to determine if it meets design specifications.
9. Actual system proven through successful mission operations	Actual application of the technology in its final form and under mission conditions, such as those encountered in operational test and evaluation (OT&E). Examples include using the system under operational mission conditions.

FIGURE 10.3 Technology Readiness Levels (TRL) as defined by the U.S. Department of Defense. [9] The TRL scale allows the architect and technology developers to set common expectations for readiness. (Source: Department of Defense, Technology Readiness Assessment (TRA) Guidelines, April 2011.)

All systems must be implemented. In physical systems, the distinction between design and manufacturing is sharp (although it may be iterative). In software systems, the boundary is blurred, because both the design and the product are information objects. We use the term "implementation" rather than "manufacturing" (a traditional hardware notion) to cover both hardware and software. Designing software consists of choosing algorithms and abstractions, designing data structures and program flow, writing pseudo code, and so on. Implementing is

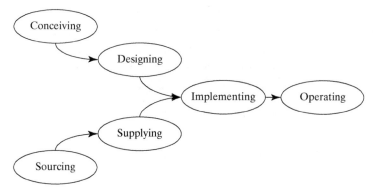

FIGURE 10.4 The flow of conceiving the product architecture and its design merging with sourcing and supplying at the implementation stage. These two flows must be carefully coordinated by the architect and the implementation team.

writing the actual code and testing. There is a great deal of contemporary work on coding, manufacturing, and supply chain management, so this discussion will focus on the interrelationships with architecture.

Like technology development, implementation is an ongoing business process. The product under the control of the architect will fit in a queue of other projects that the implementation system produces. Therefore, it is critical that the architect understand the implementation process that will be used for the product. The architect must design for implementation and must work with the implementers from early in development.

Implementation is a broad set of activities that includes sourcing (strategic partnerships, supplier selection, supply contracts); implementation capitalization, set up and ramp-up; software module development and testing; hardware element manufacturing and quality control; supply chain management; system integration; system-level testing; certification; and ongoing implementation operations.

The activities of conceiving the architecture and designing must merge with sourcing and supply chain management at the actual implementation process (Figure 10.4).

From the perspective of the architect, the key implementation decision is whether to make or buy: Will the implementation of the product be done within the enterprise or at a supplier? The issue hinges on what is within the enterprise's defined core implementation competence. Does the firm have (or want to develop) the implementation capabilities to build the entire system? A secondary question is capacity: Is there "space" in the implementation process to realize this product?

If key suppliers are to be involved, the architect faces two additional questions: How early to involve suppliers and how to deal with their impact on the architectural decomposition? Early involvement of suppliers has benefits (access to more technology, implementation expertise), but it also has drawbacks (the suppliers will start influencing the architecture for their own good). Once a supplier is slated to provide a component, that component becomes a hard boundary and will influence decomposition. *Suppliers can define decomposition or defy decomposition.* The architect should involve the implementation team early, should participate in make/buy discussions, and should carefully time the introduction of suppliers into the architecting process.

A final implementation issue for the architect is the dynamics of the supply chain. The implementation system is large, and there are many points of coordination (often with incomplete information) and lags. Modern approaches to supply chain management work to harmonize these issues, but the architect should be prepared to understand them, because they affect many timing decisions, such as time to market of the first product.

10.7 Downstream Influence: Operations

System operation, the act of deploying and using the system, is an essential downstream influence because the system delivers value during operations. *Operations* is centrally concerned with how people use the product or system. When a product reaches its operational environment, it will be operated by operators as they see fit and will encounter operational environments that are often difficult to predict. In this section, we provide a framework for thinking holistically through the different types of operations that the system may encounter in its service lifetime, as depicted in Table 10.1 for an air transport service, a hybrid car, and a refrigerator. It is important for the architect to understand this full range of operational issues that the system will confront, because many of these issues will drive the choice of architecture.

In Chapter 6, we learned that systems also have supporting and interfacing functions, and that when these functions execute in sequence, there is an emergent operational behavior of the system. But all of this still focuses only on the system nominally delivering primary and secondary value, when everything is up and running, and running perfectly. Unfortunately, systems also have to be commissioned and decommissioned. And there are unusual operating modes such as contingencies and emergencies. The system must be architected to accommodate all of these off-nominal modes.

Commissioning and Decommissioning

Systems do not magically appear at their final place of operation ready to execute their functions. They must be moved, prepared, primed, started, oriented, warmed up, and so on. We will call this *commissioning*. After operations, there is *decommissioning*. It is often easy to forget the commissioning and decommissioning processes and represent only the steady-state operations. Think of a dinner party: Commissioning is buying food, cooking, and setting tables. Primary value operation is the dinner itself. Decommissioning is cleaning up and storing everything that was used.

Get Ready, Get Set, Go is a simple framework to stimulate creativity in imagining potential necessary processes. We use *Get Ready* to refer to getting the system installed: transporting, retrieving, connecting, powering up, initializing, and so on. One can also trace the system's distribution and sales channels as a part of getting ready (Table 10.1). *Get Set* refers to processes that are executed every time the system executes, compared with some *Get Ready* processes that may be executed only once. This might include things such as loading data or consumables and positioning the system in the right physical or logical place. Then there is *Go*, which refers to nominal operations.

We can mirror *Get Ready, Get Set* after the value processes as well. *Getting Unset* involves routine post-operations steps: unloading, archiving, regular inspection. *Getting Unready* involves

TABLE 10.1 | Operations framework as applied to three examples: an air transport service, a hybrid car, and a refrigerator

	Air Transport Service	**Hybrid Car**	**Refrigerator**
Get Ready	Ticket purchasing, IT supporting, HR supporting, staff training, airport supporting, aircraft scheduling	Moving car to dealership, transferring ownership, registering car, moving to home or base of operation	Moving to point of sale, transferring ownership, moving to home, installing
Get Set	Traveler online check-in, traveler baggage tagging, traveler loading, catering loading, flight plan planning	Fueling car, removing from garage or short-term storage, starting car, adjusting seats and mirrors, loading baggage, seating passengers, car backing	Initial cool-down, loading water, adjusting shelves, setting intial temperature, loading initial load of food
Go - Nominal - Primary Value	Traveler transporting	Cargo and passenger transporting	Food preserving
Go - Nominal - Secondary Value	Traveler nourishing, traveler entertaining	Climate control, passenger entertaining, navigational information providing	Cold water and ice providing, freezing food, place for posting notices
Go - Contingent	Informing traveler of delay information, informing traveler of missed connection procedure	Flat tire repairing, skid preventing, car protecting in low-velocity impact	Easy clean-up of spills
Go - Emergency	Positioning traveler in crash landing position, traveler exiting aircraft via slides	Protecting passengers in crash	Preventing people from being locked into operating or abandoned refrigerator
Go - Stand-Alone	Deadheading (moving aircraft without passengers)	Auto starting car without driver inside, door auto lock and window auto opening	NA
Get Unset	Unloading baggage from overhead compartment, unloading checked baggage, deplaning traveler	Unloading passengers and baggage, parking car	Cleaning out refrigerator
Get Unready	Cleaning aircraft, storing aircraft overnight	Cleaning car, storing car for extended periods	Un-installing, disposing of refrigerator
Fix	Routine maintenance, system upgrades	Regular and event-driven maintenance and repair	Event driven maintenance and repair

the less frequent terminating, disconnecting, depowering, and storing steps that are usually associated with the long storage of the system or its retirement. Finally there is *Fixing,* which is associated with regular or event-driven maintenance: inspecting, calibrating, repairing, overhauling, updating, and so on.

Contingencies, Emergencies, and Stand-Alone Operations

During operations of the system, things do not always go as planned; parts fail, schedules are missed, communication links go down, and sometimes lightning actually strikes. A system must be resilient to respond to all of these expectable situations. We will distinguish two classes of such events: contingencies and emergencies.

Contingency operations are operations, other than normal, that might reasonably be encountered, and from which recovery is necessary without loss of primary function and without personal or property damage. Performance may or may not degrade gracefully, but no loss of life or property is expected. Examples of contingencies include a spill in a refrigerator, a skid on a slippery surface by a car, or a missed packet in TCP/IP (the basic communication language or protocol of the Internet). The operator expects to be able to gracefully recover from these contingencies, and provisions for them are built into the system: easily cleanable surfaces that don't let fluid leak into the rest of the refrigerator; ABS braking; or resend of packet if it is not received within a certain time. The architect must anticipate contingencies and provide for them.

Emergency operations are operations, other than normal, that are outside the contingency window. All hope of primary value function is abandoned, and emphasis shifts to saving life and minimizing property damage. Emergencies include death of a child inside a refrigerator (prevented by not allowing latches on the outside of the door); the crash of a car (in which the occupants are protected by crash-worthy design, seat belts, and airbags); and loss of a communication link in TCP/IP (in which the protocol waits to retransmit or times out). Loss of more than one engine in a commercial aircraft is an emergency, and literally hundreds of the design features of the aircraft (from doors and seats to emergency power sources and safety briefings) are designed to minimize losses in such a situation. The architect has a serious responsibility to address foreseeable emergency operations.

A slightly different off-nominal mode is stand-alone operations. These occur when the system must operate without connection to its normal supporting systems. Such situations often occur during system test, but they can occur in operations as well, when the system temporarily is separated or goes off its grid. The ability of a refrigerator to keep food cold for some period of time when there is a loss of power is a stand-alone operation (no power supply). Testing of a piece of code, without its being embedded in a larger context, is a stand-alone operation, as is the ability of an aircraft to navigate if it loses radio communications.

10.8 Downstream Influence: Design for X

Design for X (DfX) is a term used to describe a series of design guidelines, such as Design for Manufacturing, Design for Six Sigma, Design to Cost, and Design for Test. [10] For example, Design for Test can include inserting probe points on a circuit board. Each of these captures downstream constraints or considerations and propagates them back upstream. Having covered implementation and operations broadly, let's see what other considerations have been captured under Design for X.

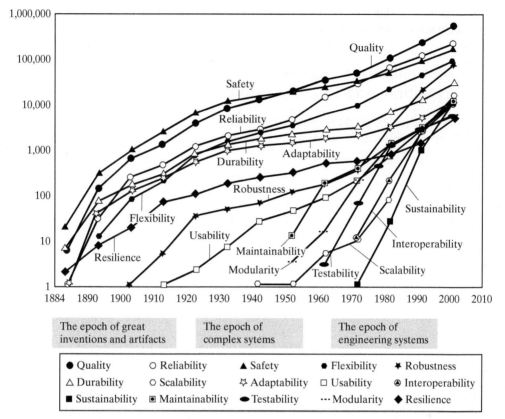

FIGURE 10.5 Graph showing the prevalence of "ilities" over time. (Source: de Weck, Olivier L., Daniel Roos, and Christopher L. Magee. *Engineering Systems: Meeting Human Needs in a Complex Technological World,* © 2011 Massachusetts Institute of Technology, by permission of The MIT Press.)

One way to think about the universe of Design for X is in terms of the following list of "ilities," or attributes of the system that are related to its implementation and operations. The list of 15 "ilities" shown in Figure 10.5 was compiled over time by tracking mentions in journal articles; note the log scale on the vertical axis. Each of these attributes could act as a downstream influence.

However, not all "ilities" are architecturally distinguishing, and no one "ility" is always architecturally distinguishing. Although reliability may help distinguish between potential architectures for a pump, it may not distinguish between vehicle architectures if the architectural decisions don't define the attributes that relate to reliability; these attribute may arise only from detailed design decisions or choice of assembly method. Maintainability may be a function of whether a tablet's architecture has an integral case design or a modular case design, but it may also be a function of an individual component that has no bearing on the architecture, such as how easy it is to clean the screen as a function of the coating on the glass.

It is also worth noting that the "ilities" are not independent. For example, the durability of the architecture may be correlated with the emphasis placed on quality in manufacturing, the modularity of the architecture may impact the adaptability of the architecture, and the robustness of the architecture could be inversely proportional to its flexibility.

10.9 Downstream Influence: Product and System Evolution, and Product Families

In modern practice, few products or systems of any complexity are built without some consideration of how they are related to other products. Terms often used to describe this planning include:

- Reuse, legacy, product extensions
- Product lines, preplanned product improvements
- Product platforms, modularity, commonality, standardization

So far in this text, we have focused on the creation of architecture from a blank slate. In our experience, questioning the first principles of an architecture leads to the most holistic evaluation of opportunity, even if some of the options considered are later eliminated as infeasible due to constraints from the architectural legacy (Box 10.2).

In the majority of architectures, starting with a blank slate is not possible. A printing company once conducted an exercise to evaluate the gains and costs of recoding the software for its mid-size printers. Legacy software had been passed from one generation to the next, with code

---

Box 10.2   Principle of Reuse of Legacy Elements

"We learn in this life by remembering knowledge originally acquired in a previous life."

Plato

"Poor is the pupil who does not surpass his master."

Leonardo da Vinci

"To make an apple pie from scratch, you must first invent the universe."

Carl Sagan

All built systems involve reuse of legacy elements, either in the conceptual sense or in the sense of physical/informational elements. Understand the legacy system and its emergent properties thoroughly, and include the necessary elements in the new architecture.

- Legacy has attributes that one may not completely understand or appreciate, which might be lost in a complete redesign.
- Legacy systems are built on an architecture that is often incrementally modified over time. The architect should assess whether the previous architecture is best positioned to meet new system goals, and whether that architecture reflects assumptions that have since been modified.
- Legacy may involve substantial retesting to verify that the elements can be used in a system for which they were not originally intended.
- Document design elements to facilitate their subsequent reuse where possible.

accreting over 15 years, thus motivating a blank-slate recoding of the software base. A back-of-the-envelope calculation suggested that it would cost the company its entire operating profit for a year just to redevelop the code for the mid-size printers.

Reuse and Legacy Elements

Reuse captures the idea that the part or module that was not intentionally designed to be used beyond the original design (Box 10.2 Principle of Reuse of Legacy Elements). There is very little uncertainty associated with reuse, in the sense that the part, component, or code already exists; its performance and functionality are known (at least in the context of the legacy system in which it was implemented); and it has passed tests and validation and may have been subject to market testing or product improvements. For reuse, the part, component, or code is a known quantity. The unknown element that arises is how it will integrate in a new context.

For example, NASA has considered reusing components from legacy designs of the Space Shuttle on several future launch vehicles, such as using the Space Shuttle's solid rocket motors on the (since-canceled) Ares-1 crew launch vehicle, and in the Space Shuttle main engines on the Space Launch System. Doing so involved not only repurposing and requalifying legacy spares (identical reuse) but also potentially redesigning pumps, valves, and other parts that are 40 years old.

The benefits of reuse are that the development cost has already been expensed (although integration or modification may be necessary), the tooling has been paid for, and the performance characteristics are well known. The downside of reuse is that the legacy elements impose constraints. For example, the reuse of a network bus imposes a limit on data rates achievable.

Reuse is different from platforming. Reuse represents sharing that was not intended in the original design, whereas platforming represents intended sharing.

Product Lines

Product lines or product families are linked groups of products, such as the GE Healthcare LOGIQ family of ultrasound machines shown in Figure 10.6. Product families do not necessarily share underlying components or architectures. They can be arranged under similar names and stratified for feature content for marketing purposes alone, separate from objectives (such as reducing development time) that are explicitly linked to sharing components or architecture. Conceiving product families explicitly requires the architect to consider how much variety is demanded. (Will one ultrasound machine serve the whole market, or will ultrasound machines with differentiated offerings be required?) Conceiving a product family also hints at product evolution. (Will the family be launched in parallel, or will it be phased into the market?)

Analogous to the case of reuse, much of product family evolution (Box 10.3 Principle of Product Evolution) is unforeseeable. Stable markets such as rail are more likely to see component-level innovation within a stable architecture, whereas mobile phones move at a significantly higher clockspeed, limiting the architect's ability to forecast the dimensions of the market that will demand variety and innovation. The architect may invest in real options during the design, such as choosing to create additional interfaces to enable future modularity. The best practice we have observed is to explicitly set interfaces between slow- and fast-clockspeed components, on the presumption that unplanned evolution will occur on the high-clockspeed side of the interface.

LOGIQ E9 with XDclear
Use expert tools to help achieve extraordinary images and easy workflow with the flagship LOGIQ* E9.

LOGIQ S7 Ultrasound
Amazing versatility across multiple clinical specialties and applications.

LOGIQ S8
The balance of price and performance you've always wanted in a lightweight, portable design.

LOGIQ P6
An ultrasound system that combines performance, ergonomic design and value.

LOGIQ P5
This portable system helps provide quality ultrasound capabilities across a broad spectrum of care areas.

LOGIQ P3
Deep penetration, high resolution and great color sensitivity for advanced imaging.

LOGIQ A5
A lightweight, portable ultrasound system offers capabilities to help meet your imaging needs.

LOGIQ *e* Ultrasound
Get advanced imaging in a compact platform that will grow with a clinician's practice.

FIGURE 10.6 GE Healthcare's LOGIQ product family of ultrasound machines. (Source: Used with permission of GE Healthcare.)

Platforming and Architecture

Platforming is the intentional sharing of parts and process across or within products. Platforms require that the architect invest once in a common part or software module, predicting which models that part or module will be used in. Unlike reuse, platforming allows the architect to choose which constraints to impose, but it must deal with the uncertainty of future applications, whereas in reuse, the parts or module performance is known, and the current use case is known as well.

Platforming has become an important strategy for delivering product variety. Platforming makes it possible for firms to deliver similar products to multiple markets and multiple vertical

> **Box 10.3   Principle of Product Evolution**
>
> *"Learning is like rowing upstream: Not to advance is to drop back."*
> Traditional Chinese Proverb
>
> *"In the struggle for survival, the fittest win out at the expense of their rivals because they succeed in adapting themselves best to their environment."*
> Charles Darwin
>
> *"Panta Rhei" [Everything flows]*
> Heraclitus 540–480 BCE
>
> Systems will evolve or lose competitive advantage. When architecting, define the interfaces as the more stable parts of the system so that the elements can evolve. Plan for evolution with a master vision and sufficient reserve of resources.

segments, enables them to dynamically enter niche markets more quickly as a result of identified needs, and can be coupled with strategies of mass customization.

One of the chief advantages of platforming is cost sharing across products. Examples include Volkswagen's A platform (including the VW Jetta, Audi TT, and Seat Toledo), the Joint Strike Fighter program (with variants for the Air Force, Marines, and Navy), and Black & Decker's electric hand tools. Development costs can be shared, learning curves lower touch labor hours in manufacturing, and demand aggregation can reduce safety stock levels in inventory. Platforming strategies enable firms to deliver more product variety to the marketplace on a smaller cost base. Figure 10.7 summarizes several of the benefits associated with platforming.

However, platforms can substantially increase the complexity of product development. Whereas previously decentralized products could optimize to a market segment, platforming requires judging the variety of offerings demanded, determining where over-performance can be carried for the sake of shared components, and defining interfaces that can meet the needs of multiple components. Component-level commonality can be non-architectural (standardizing on M8 and M10 bolts is unlikely to impact the architecture). However, many commonality strategies at either the subsystem level (sharing engines across trucks) or the system level (sharing a rolling chassis, drivetrain, and geometry across trucks) are enabled by the architecture. The architecture can define which modules can be paired with which other modules (which axles can be paired with which transmissions).

Platforming for the purpose of saving development cost often involves making linchpin parts or modules common—parts or modules that are at the heart of the architecture. For example, Black & Decker designed a common electric motor to be shared across a variety of portable hand tools. Changes to these parts or modules often cascade down through many other parts as a result of system-level tuning.

The results of the MIT Commonality Study [11] suggest that many firms invest substantially in platforms (development spending on platforms can be up to 50% higher than the cost of stand-alone development), as shown in Figure 10.8. But many firms realize lower levels of sharing than originally envisioned, a phenomena called divergence. Some divergence is defined early in product families, when large offsets are planned between variant development schedules (often months, and sometimes years, between variants), and where the goals set for the variants fail to articulate the opportunities for carrying margin and over-performance for the sake of common components.

In summary, product families, product evolution, and platforming are a substantial downstream influence for the architect. The architecture chosen shapes the product's longevity and its modularity for new functionality. As seen in the case study that concludes this chapter, the choice of architecture locks in substantial downstream performance and cost consequences.

10.10 The Product Case: Architecture Business Case Decision (ABCD)

We have discussed a number of upstream and downstream influences on architectures. It is the responsibility of the architect to prioritize among these influences and to determine which require careful attention. In thinking about influences as a list, one can become too *linear*. It is

Phase	Commonality Benefit	Explanation
Conception	Quicker variant time to market	Only the unique portion has to be designed
	Revenue in niche markets	The firm can recognize and enter markets as they appear with less effort if the common parts can be reused
	Field new technologies	Effort to deploy technologies is lower where interfaces to the platform are the same
	Reduced technology risk	Technology risk more concentrated in fewer components
Development	Reduced development cost	Less engineering effort needed for later derivatives
	Reduced testing and/or commissioning time	Learning curves in test labor for later variants
	Shared test equipment across variants	Testing equipment can be distributed across several variants
	Shared external testing/certification	Type certificates can be reused or amended, regulatory approval granted for family
Production	Shared tooling across variants	Tooling cost can be distributed over several variants
	Reduced touch labor	Fewer hours/unit required due to learning curve
	Manufacturing economies of scale	Lower per unit cost as a result of capital investment
	Volume purchasing	Lower price from suppliers for larger orders of same part
	Lower inventory	Reduced safety stock levels from demand aggregation
	Flexibility in variant volumes (for a fixed platform extent)	Ability to adjust to variant demand changes
Operations	Lower sustaining engineering	Fewer parts to be sustained
	Shared fixed costs	Sharing of facility cost across more variants
	Reduced operator training	Operators can transfer skills across products
	Economies of scale in operations	Move to higher volume operating procedures as a result of capital investment
	Volume purchasing of consumables	Lower price from suppliers for larger orders of same parts
	Lower inventory	Decreased variable costs due to more efficient logistics and sparing
	Slower replacement rate for spares	Fewer spares must be purchased due to higher quality
	Flexibility in operations	Ability to switch staff between variants
	Shared inspections/recurring regulatory compliance effort	Less effort required for regulatory compliance

FIGURE 10.7 Platforming involves many possible benefits. Reproduced from Cameron (2013) in Simpson (2013). [12]

Phase	Costs and Drawbacks
Conception	Constraints on future variants to use shared components
	Schedule risk is more concentrated in shared components
	Brand risk to firm from lack of variant differentiation
	Risk of sales cannibalization from high margin to lower margin products
	Risk of price escalation from single suppliers
Development	Cost of designing to multiple use cases
	Cost of creating more flexible test equipment
Production	Cost to lower performance products of higher performance shared parts
	Additional configuration management on the manufacturing line
	Additional rate capacity of production equipment
Operations	Risk of common part failure affecting many variants
	Possible performance shortfall relative to unique optimized design

FIGURE 10.8 Platforming always involves additional investment, and it often involves other drawbacks or compromises. (Reproduced from Cameron (2013) in Simpson (2013)).

not possible just to tackle the influences in sequence, enumerate their goals and constraints, and then sum them up to produce the concept. This section explores a simple iterative model that combines influences explicitly.

Innovation in products and systems is traditionally thought to derive from one of two sources: The firm either has identified a new customer need or has identified and commercialized a new technology that delivers a higher performance/cost ratio in satisfying an existing need. Adding an icemaker to a refrigerator is responding to customer need with existing technology, whereas adding automatic tracking of stored food depends on the development of new RFID (radio-frequency identification) technology. This is a very simplified perspective on product strategy. It does not express strategies that capture new value in the value chain by backwards or forward integration, strategies that leverage home markets against regional or global expansion, or strategies that leverage the firm's installed base of hardware to deliver high-margin service strategies.

Figure 10.9 shows a simple model that includes technology and customer need in an iterative process and provides a simple anchor around which we can enumerate the highest-level considerations when introducing an architectural innovation in a corporation. We call this the Product Case.

The diagram is intended to convey that there is a strong interaction between two processes: the development of a business case and the choice of system architecture. When the product innovation has identified a customer need, this need sets *goals* for the system architecture. The architect must then evaluate whether the solution is feasible and help determine whether the business case "closes" (stands up to financial scrutiny, see below) based on the cost of the solution. When product innovation derives from a new technology, the architect must help evaluate whether the

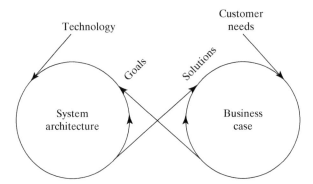

FIGURE 10.9 The initial aspects of the ABCD (Architecture Business Case Decision) framework, showing the two most common insights leading to new products: the availability of new technology and a new understanding of customer needs.

solution would address customer needs. The challenge is broader than this, of course, because there are additional external and internal factors.

Figure 10.10 shows the complete ABCD framework. First, two external drivers are added to the framework: the competitive environment and regulations. The competitive environment primarily impacts the business plan, through barriers to entry, competitor product offerings, and market timing. Regulation primarily acts to constrain the available technical solutions, although creative use of the regulatory environment can create strong advantages in the business case. For

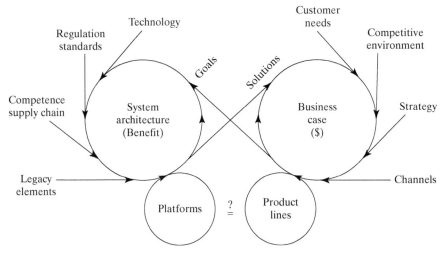

FIGURE 10.10 The full ABCD (Architecture Business Case Decision) framework, showing the principal influences on the two iterative cycles that interact, converging on a decision to proceed with the product or not to proceed.

example, the introduction of Tier 4 Final emissions regulations for off-highway trucks primarily set emissions targets that had to be met by new designs. But it also had an impact on the business case and created market opportunity by resetting the products on offer. Thus, it both challenged firms to sell off finished-goods inventory prior to the deadline and invited them to compete on new dimensions after the deadline.

Two internal drivers are then added: the firm's strategy and its competence in supply chain execution. As stated earlier, the architecture must absorb goals set by strategy. The intent is for investments made at the firm level to contribute positively to the business case for the architecture, rather than imposing constraints on it. By analogy, the architecture must absorb goals set by the supply chain environment. The "currency" of each analysis is also shown—the financial analysis of the business case and the benefit analysis of the architecture. The two combine to inform the value proposition, which is defined as benefit at cost.

In the completed ABCD framework we have also represented the channels or distribution system and the product lines sold through them, as well as the legacy elements and the platforms they adhere to. It may surprise the reader to see channels listed as an architectural consideration. From the perspective of stakeholder needs, however, it becomes clear that architecture can inhibit channel distribution, or take advantage of it. The relationship between platforms and product lines is uncertain. In some businesses they are synonymous; in others the platform is purely a technical artifact, while the product line is channel focused. For example, recent research [13] revealed a case where the retail channels through which a heavy-equipment product was sold were responsible for the addition of five new variants in a product line of four variants. Matching product competitor price points was the originating dynamic, but this would not necessarily have motivated the firm to produce unplanned variants. The channel customer not only identified the opportunity but lobbied hard for it as well.

The ABCD framework is intended to make it clear that creation of the architecture is an inherent process that sits alongside the better-known "business case" deliverable. In some enterprises, the combined document is called the "product case." It is common to speak of the business case "closing"—that is, the projected revenues are earning sufficient margin to merit the investment. Similarly, the technical architecture has to "close;" the goals are met by solutions, the required technology is present or developed, and the architecture can be delivered by the planned supply chain. These parallel cycles end when the technical architecture closes *and* the business case closes. This point is known as the *Decision,* implying that the enterprise makes a decision to move ahead and commit itself to the product.

10.11 Summary

This chapter has focused on the upstream and downstream influences on architecture, which the architect must interpret and condense in order to reduce ambiguity. The architect must always be prepared to declare a current valid information set so that others can work, understanding that it might be corrected and updated. The influences presented in this chapter include:

- An understanding of corporate strategy, so that the guidance on investment of resources is known to the architect, and the development can be aligned with major initiatives of the enterprise.

- Close cooperation with marketing, so that stakeholders can be mutually appreciated, markets defined and segmented, and competition understood.
- An understanding of the current regulation, a prediction of the anticipated regulation, and an approach to other pseudo-regulatory sources from standards and liability.
- A thorough knowledge not only of the technology available and its readiness, but also of the process of infusing it into the system and the value it will create.
- A working partnership with the implementation experts in the enterprise, so that they can be drawn in early, help formulate make/buy decisions, and engage suppliers when appropriate.
- An understanding of the operational environment, including the nominal, off-nominal, commissioning, and decommissioning operation of the system.
- A clear view of the evolution of the product, and of the relationship of the product to the platforms or other products with common sub-systems.
- Application of the iterative Architecture Business Case Decision (ABCD) framework, a tool to integrate many of the key ambiguous upstream flows of information into one decision process, the Product Case.

The other major approach to reducing ambiguity is engagement with stakeholders, which is the subject of Chapter 11.

Box 10.4 *Case Study: Architecture and Platforms: B-52 and B-2*

As an example of architectural decisions related to legacy, product evolution, and product platforms, we discuss the architectural decisions in Figure 10.11, which shows a B-52 bomber and a B-2 bomber. There are many differences between the two aircraft, but which differences were decided early in their development? Which differences have the greatest impact on their performance envelopes?

A major difference is the presence, on the B-52, of a tail, specifically horizontal and vertical control surfaces separated by a significant distance from the main lifting surface, the wing. The B-2 does not have any horizontal or vertical control surfaces other than those on its main wing; this has significant consequences for its stability in pitch (nose up and down) and yaw (nose left and right).

A second difference between these two aircraft is in engine placement. The B-2's design uses concealed engines, which are helpful for reducing the radar signature of the aircraft, whereas the B-52 has under-wing engines. The B-2 engine placement is much more integral, whereas the B-52 visually suggests modularity. In fact, the B-52 has received several engine upgrades, because the clockspeed of engine technology has progressed faster than the structure of the aircraft.

The B-52 architecture is called a "tube-and-wing" architecture, whereas the B-2 architecture has been referred to as a "flying wing." How do these two architectures compare in

CHAPTER 10 • UPSTREAM AND DOWNSTREAM INFLUENCES ON SYSTEM ARCHITECTURE   223

FIGURE 10.11   Left: A B-52 bomber. Right: A B-2 bomber. What are the architectural differences?   (Source: U.S. Air Force photo)

terms of their ability to scale to different sizes? Pictured in Figure 10.12 are two civil aircraft families, based on the two architectures.

It is a common practice for tube-and-wing aircraft to be designed in product families with a variety of seating capacities, often with the explicit intent of platforming (sharing components). The concept is that the wing is sized for the longest (and heaviest) variant; the wing produces the lift and therefore supports the weight of the aircraft in flight. The fuselage (the "tube") is lengthened or shortened as needed. Because wings are expensive

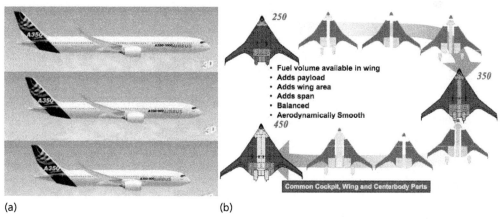

(a)   (b)

FIGURE 10.12   Left: The A350 product family. Right: A proposed "flying wing" civil transport family, more specifically referred to as a Blended Wing Body   (Source: (a) Copyright Airbus (b) Liebeck, Robert H., "Design of the Blended Wing Body Subsonic Transport," Journal of Aircraft 41, no. 1 (2004): 10–25, reprinted by permission of the American Institute of Aeronautics and Astronautics, Inc. [14])

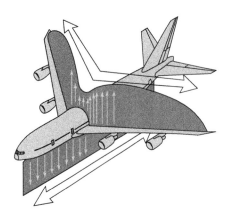 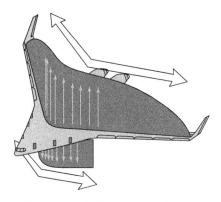

FIGURE 10.13 Lift distribution dark gray on a tube-and-wing architecture and a flying-wing architecture, showing that the wings of tube-and-wing aircraft must be sized for the heaviest variant, whereas for a flying-wing architecture, the lift scales with the size of the aircraft variant. (Source: Robert H. Liebeck, "Design of the Blended Wing Body Subsonic Transport," Journal of Aircraft 41, no. 1 (2004): 10–25)

to optimize, they are frequently shared across the family with minor changes, the net result being that the wing is "overdesigned" for the smallest family member; it can produce more lift than that aircraft needs.

On the other hand, the family concept shown on the right in Figure 10.12 proposes a different scaling scheme. The flying wing is split in half along the centerline, with middle sections (shown in light gray) added to increase the seating capacity.

What, then, were the downstream influences on the product family in these two architectures? The architectural implications are significant. Figure 10.13 shows the lift distribution for the two aircraft, in dark gray. For the tube-and-wing architecture, the fuselage does not produce much lift. To lengthen the fuselage and add seating capacity, the wing must be enlarged to support the additional weight, which is why tube-and-wing aircraft families size the wing for the largest variant. To add seating capacity to the flying wing, a center section is added, but (crucially) that center section is also part of the wing, so it produces lift. The flying-wing architecture can more closely optimize the wing to the desired performance for each of its variants, whereas the tube-and-wing architecture will always struggle with overdesigned wings that are less than optimal for the smallest variant of the product family.

References

[1] T.C. Kuo, S.H. Huang, and H.C. Zhang, H. C., "Design for manufacture and design for 'X': Concepts, applications, and perspectives," *Computers & Industrial Engineering* 41, no. 3 (2001): 241–260.

[2] http://www.volkswagenag.com/content/vwcorp/content/en/the_group/strategy.html

[3] https://www.press.bmwgroup.com/global/pressDetail.html?title=bmw-to-exit-formula-one-at-end-of-2009-season&outputChannelId=6&id=T0037934EN&left_menu_item=node__803

[4] http://www.bmwgroup.com/e/0_0_www_bmwgroup_com/investor_relations/corporate_news/news/2011/Neue_BMW_Submarke_BMW_i.html

[5] "The American Marketing Association Releases New Definition for Marketing," American Marketing Association, Jan. 14, 2008. http://www.marketingpower.com/aboutama/documents/american%20marketing%20association%20releases%20new%20definition%20for%20marketing.pdf

[6] S. Arnon and N.S. Kopeika, "Performance Limitations of Free-space Optical Communication Satellite Networks Due to Vibrations—Analog Case," *Optical Engineering* 36, no. 1 (1997): 175–182.

[7] http://esc.gsfc.nasa.gov/267/278/279/487.html

[8] W.K. Viscusi, J.E. Harrington, and J.M Vernon, *Economics of regulation and antitrust* (Cambridge, MA: MIT Press, 2005).

[9] http://www.acq.osd.mil/chieftechnologist/publications/docs/TRA2011.pdf

[10] G. Pahl and W. Beitz, *Engineering Design: A Systematic Approach* (New York: Springer, 1995).

[11] R.C. Boas, B.G. Cameron, and E.F. Crawley, "Divergence and Lifecycle Offsets in Product Families with Commonality," *Journal of Systems Engineering,* 2012 (accepted, doi: 10.1002/sys.21223).

[12] B.G. Cameron and E.F. Crawley, "Crafting Platform Strategy Based on Anticipated Benefits and Costs," in T.W. Simpson et al.*, Advances in Product Family and Product Platform Design: Methods & Applications* (New York: Springer, 2014).

[13] Bruce G. Cameron, "Costing Commonality: Evaluating the Impact of Platform Divergence on Internal Investment Returns," PhD thesis, Massachusetts Institute of Technology Engineering Systems Division, 2011.

[14] Robert H. Liebeck, "Design of the Blended Wing Body Subsonic Transport," *Journal of Aircraft* 41, no. 1 (2004): 10–25.

Chapter 11
Translating Needs into Goals

11.1 Introduction

In Part 2, we discussed the identification of value for existing systems, where we have to reverse-engineer the function from the form and determine the design intent from possible use cases. In this chapter, we will find that identifying value for new systems is significantly more challenging.

Many transformative architectures arise out of new value propositions or serve as-yet-unidentified needs. Who knew they "needed" a Podcast of the *Economist* on their morning commute before Podcasts were invented? Who knew they "needed" key card readers for hotel rooms when keys were the dominant choice? The identification of value is difficult, because it blends the challenge of identifying the underlying needs of the user with the mental cage of previous designs.

Complex products (goods and services) serve stakeholders who often have differing needs. A new power station might focus on the power utility as the primary stakeholder, but the state government could be shepharding its development as an opportunity to demonstrate greener power, or the federal government could be funding it with the intent of increasing the grid's redundancy. Although these three stakeholders all want to see the project implemented, their needs and priorities might conflict. Separately, the project may have stakeholders who were originally opposed and whose needs reflect potential hazards or mitigation strategies.

In addition to new value propositions and prioritizing multiple stakeholders, a hallmark challenge of complex systems is the indirect delivery of value. The power station does not interact directly with end users of the power, but it does interact with them through one or many utilities and intermediaries. The federal government may issue funding through intermediating agencies or the state government. Therefore, there are challenges in understanding not just which stakeholders are important but also how value is delivered to them. In this chapter, we focus on developing principles and methods for identifying and prioritizing these stakeholders and their associated needs in the context of complex systems.

We will introduce a running example in this chapter: the Hybrid Car, which captures many of the challenges associated with the development of complex systems. In Chapter 11 we will focus on identifying the needs and goals for the Hybrid Car, and in Chapter 12 we will develop several concepts to meet these goals.

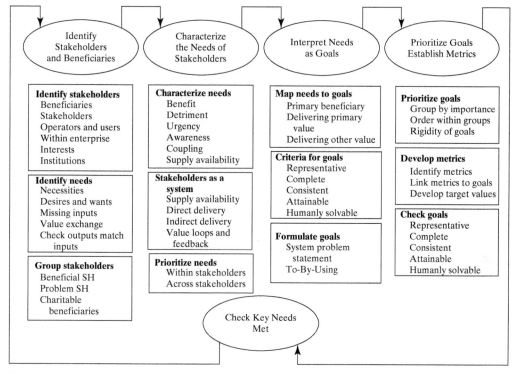

FIGURE 11.1 Needs-to-goals framework, working from the identification of stakeholder needs through to prioritized goals.

We begin with the identification of the stakeholders for the system, then work toward a prioritization of their needs, and subsequently address the establishment of goals. The progression of logic in this chapter is summarized in the needs-to-goals framework shown in Figure 11.1. We first identify stakeholders and beneficiaries (Section 11.2) and then characterize their needs (Section 11.3). These needs are next interpreted as goals (Section 11.4) and then prioritized (Section 11.5). The chapter concludes with a case study involving stakeholder management.

11.2 Identifying Beneficiaries and Stakeholders

Beneficiaries and Stakeholders

The term "stakeholder" has gained broader usage since Edward Freeman's 1984 publication of *Stakeholder Management*, which firmly established the term and the importance of stakeholder management as an active task. However, over the years it has come to mean any and all parties touched by the system, with the net result that "managing stakeholders" is often interpreted as a downstream public relations activity rather than an upstream process identifying and serving potential customers.

We distinguish beneficiaries from stakeholders to help resolve this challenge. *Beneficiaries* are those who benefit from your actions. Your architecture produces an outcome or output that

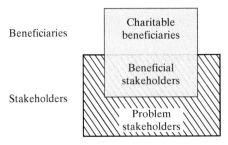

FIGURE 11.2 Stakeholders and beneficiaries.

addresses their needs. You are important to them. It is beneficiaries whom we must examine in order to list the needs of the system.

By contrast, *stakeholders* are those who have a stake in your product or enterprise. They have an outcome or output that addresses *your* needs. They are important to you. This is much closer to the original concept of stockholders, who supply cash (which the firm needs), in return for a stake in the firm, notably a share in the firm's profits.

These two concepts are distinct but they overlap, as shown in the Figure 11.2. At the center of the diagram we have *Beneficial Stakeholders,* who both receive valued outputs from us and provide valued inputs to us. Beneficiaries who aren't stakeholders we call *Charitable Beneficiaries,* in that the firm receives no return (however indirect) from the outputs provided to them. Stakeholders who aren't beneficiaries are *Problem Stakeholders,* in the sense that you need something from them, but there is nothing they need that you can provide in return.

There are many beneficiaries and stakeholders of a product/system, both internal and external to the organization. Internal beneficiaries and stakeholders include technology developers, design teams, implementation, operations, sales, service, management/strategy, marketing, and so on. External beneficiaries and stakeholders include regulators, customers, operators, suppliers, investors, and potentially competitors.

There is a further issue that influences the identification of beneficiaries and stakeholders: There may be an *operator* who is neither the primary beneficiary nor the primary stakeholder. In a taxi, the driver is not necessarily the primary stakeholder if the driver's cab is owned by a firm. The driver is also not the primary beneficiary, because that role is filled by the passenger. Therefore, we distinguish among operators, stakeholder, and beneficiaries. In a consumer context, these are often the same actor; the consumer buys the car, operates the car, and benefits from the car. However, in most corporate and government settings, these are distinct roles with incentives that skew their behavior. This division of roles causes us to realize that to deliver value, we must focus on the stakeholder who is the primary beneficiary.

The first step of any stakeholder analysis is identifying stakeholders and beneficiaries, as shown in Figure 11.1. Stakeholders are identified by asking, "*Who provides inputs that are required to make this project successful?*" Beneficiaries are identified by asking, "*Who benefits from the outputs of this project?*"

Typically, beginning to list potential stakeholders and beneficiaries is not difficult. Starting with existing products on the market, existing customers, existing shareholders, and existing suppliers can quickly give rise to a long list. Additionally, as we saw in Chapter 10, we can source possible stakeholders and beneficiaries from upstream and downstream influences on the architecture.

The main difficulties in identifying stakeholders and beneficiaries are setting the boundaries of the analysis and setting the granularity of the analysis.

First, we must set *bounds*—that is, decide how far from our system or enterprise the analysis should extend. For example, the economic concept of the multiplier effect of government spending states that for any direct payments to contractors, there are reverberations in the broader economy as those contractors engage suppliers, and as the employees of the contractor spend discretionary income at local businesses. Where should we draw the line? Later in this chapter, we will illustrate a quantitative analysis by which we can prune downstream stakeholders if the impact diminishes, but the reality is that this depth of analysis is challenging to undertake (indeed, there is substantial discussion on measuring the multiplier effect). [1], [2]

Second, it can be challenging to set the right *granularity*—in other words, to choose the right abstractions for stakeholders. Should we consider individual suppliers, or should we abstract them as "Suppliers"? Are individuals important, or just their department or their firm as a whole? Do individual suppliers have different needs and different influence, or are there many suppliers with similar needs and influence? Is the purpose of the stakeholder analysis to understand the market context at a 30,000-ft level, or is it to develop product specifications?

Our guidance for setting the bounds of the analysis and choosing abstractions is to test for architectural impact. Testing for architectural impact involves understanding whether the choice among architectures *under consideration* could be important to the stakeholder or beneficiary. If all architectures will deliver the same benefit to them, they will still need to be managed as stakeholders, but they may not merit consideration in making the architectural decisions. Further guidance on stakeholder identification is given in Box 11.1. In Part 4 we discuss quantitative methods for conducting this test.

> **Box 11.1   Principle of the Beginning**
>
> *"The beginning of the work is most important."*
>
> Plato, The Republic
>
> *"Great warriors position themselves where they will surely win, prevailing over those who have already lost."*
>
> Sun Tzu, The Art of War
>
> The list of stakeholders (internal and external to the enterprise) that are included in the early stages of product definition will have an outsized impact on the architecture. Be very careful about who is involved at the beginning and how they shape the architecture.
>
> - The initial list should consider which stakeholders have a strong need for your project outputs, as well as stakeholders whose outputs are important to the project.
> - The architect should balance the need to engage stakeholders for their views against the decision rights of the stakeholders, such as whether they are informed, consulted, or voting members and whether or not they have a veto.

The output of this procedure is a first-pass list of stakeholders and beneficiaries. Subsequently, we will prioritize stakeholders, which offers us an opportunity to revisit the list.

- Driver/owner (primary beneficiary)
- Regulator
- Producing enterprise or company
- Investors in the enterprise
- Suppliers
- Local community
- Environment NGOs
- Oil companies
- The Hybrid Car project

Using this procedure, we can create the list of stakeholders and beneficiaries for the Hybrid Car. Note that we could also have included other car companies in the analysis, either to reflect market share questions, or because we believe they play a positive role in growing the market or boosting consumer comfort with hybrid systems. However, in this analysis we have omitted other car companies. We have also included the Hybrid Car project itself as the reference, because we will want to illustrate the interactions between the stakeholders and the project.

Identifying the Needs of Beneficiaries and Stakeholders

Beneficiaries and stakeholders have *needs*. Needs are a product/system attribute in that you build a system to meet needs. Therefore, an important part of the first step of our stakeholder analysis (Figure 11.1) is identifying the needs of our beneficiaries and stakeholders.

Box 11.2 defines needs. Needs can be expressed—you know you want something—or unexpressed or even unrecognized.

Box 11.2 Definition of Needs

A need may be defined as:

- A necessity
- An overall desire or want
- A wish for something that is lacking

Needs exist in the mind (or heart) of the beneficiary, and they are often expressed in fuzzy or general (ambiguous) terms. Needs can be unexpressed or even unrecognized by the beneficiary. Needs are primarily *outside* of the producing enterprise; they are owned by a beneficiary.

Needs are a product/system attribute. They are interpreted (in part) by the architect, in order to identify goals for the product or system.

In many cases, the identification of requirements is levied from the customer-specified technical requirements or past technical systems, without examining where needs arise. Furthermore, there is often a selection bias that tends to highlight technical needs because they can be more easily quantified as requirements. Requirements analysis has not always mated well with stakeholder analysis in the past, because there are difficulties translating between the outputs of stakeholder analysis and the inputs for requirements analysis. For example, the *NASA Systems Engineering Handbook* of 1995 makes no reference to stakeholders. A recent version of the "NASA Systems Engineering Processes and Requirements" document (2006) now requires stakeholder analysis as the first step in the procedure for defining requirements, in order to "elicit and define use cases, scenarios, operational concepts, and stakeholder expectations."

There is no structure to the way that people talk about needs. It is part of the role of the architect, sometimes aided by business development or marketing functions, to understand the needs of the beneficiary.

When setting out to identify needs, consider each stakeholder and beneficiary, and ask what needs the stakeholder has that might be met by the system under consideration.

The most important guidance we can offer is to start with the primary beneficiary and the primary need of that beneficiary, whose needs are satisfied by changes in the principal value-related operand, as discussed in Chapter 7. The test is that if the primary needs of the primary beneficiary are not met, the system will never be a success. In the Hybrid Car example, the driver derives the primary benefit, transportation. The operator may have needs as well, and the stakeholder may play an important role in interpreting the needs of the beneficiary and the operator. We will treat the Hybrid Car's beneficiary, stakeholders, and operator as the same individual, the driver/owner, even though we recognize that they could be separate. Therefore, the driver/owner is our primary beneficiary.

The driver/owner wants transportation, specifically fuel-efficient transportation. Figure 11.3 illustrates the needs of the stakeholders and beneficiaries we identify for the Hybrid Car. Notice that these needs are not complete at this stage; that is in order to keep Figure 11.3 manageable. Notice also that not all of the needs of the driver/owner are satisfied by the

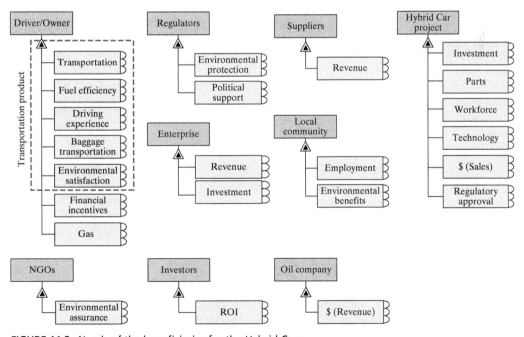

FIGURE 11.3 Needs of the beneficiaries for the Hybrid Car.

actual Hybrid Car. For example, the need for financial incentives, such as Hybrid Car federal tax deductions, cannot be met by the car-producing firm.

We identified three challenges for stakeholder analysis of complex systems in the introduction:

1. Complex systems, particularly new architectures, often provide new value propositions or address latent needs.
2. Complex systems serve many stakeholders, often with differing needs. It is often difficult to prioritize among these stakeholders.
3. Complex products (goods and services) often deliver value indirectly, which may make it difficult to identify all the needs in the system.

Our discussion of beneficiaries and stakeholders has touched on serving many stakeholders and on indirect value delivery, but we have not discussed latent needs. Although listing two or three needs for each beneficiary might be easy, the difficulty does not lie in identifying which needs existing products serve. The difficulty arises in identifying latent needs—those needs that are not yet served by current products. This is a challenge in synthesis of architectures. How did Whirlpool know customers "needed" an ice dispenser on the exterior of the refrigerator before this feature was on the market? How did GM know that customers who already had cell phones would "need" the OnStar™ system to call a central service in the event of an emergency?

The term "latent needs" is used to convey the idea that the needs are fundamental to the beneficiary, before the product that meets those needs comes into being. Needs are not chosen by the firm. We deliberately stay away from the idea that the firm shapes the needs of the market. The idea that advertising changes the customer's value function is common in marketing. But in our experience, this idea is dangerously overused in an engineering context. The history of product development is rife with products that contained engineered features not valued by the market.

How, then, does one identify latent needs? One of the great breakthroughs in product needs was the idea of observing beneficiaries and the user using the product. Ernest Dichter pioneered in-depth interviews with consumers in the 1950s, searching for the underlying qualities that users perceived in the product. The design firm IDEO uses what is termed empathic design to observe potential customers in context, on occasion spending hours following shoppers through grocery stores. By definition, there is no set practice for identifying latent needs; a procedure that identified needs with accuracy through analysis would contravene the idea of *latency*. The only answers are deep scrutiny of users, including customer feedback on previous products, iteration on product concepts, and hard work.

As a caution against the hubris of asserting the latent needs of the customer, consider the litany of failed architectural predictions. Thomas Edison said in 1922, "The radio craze will die out in time." Ken Olson, the founder of Digital Equipment Corp, said in 1977, "There is no reason anyone would want a computer in their home." Imagine all the predictions made by overconfident architects whose product did not outlive their failed understanding of latent needs!

Identifying Stakeholders and Needs from an Exchange

A central tenet of stakeholder theory is that stakeholder value results from an *exchange*. We are identifying the needs of the beneficiary so that we can architect our system to satisfy these needs. This is the first half of the exchange; our outputs or outcomes satisfy the beneficiary's needs. The other half of this exchange is that stakeholders have outputs or outcomes that meet our input

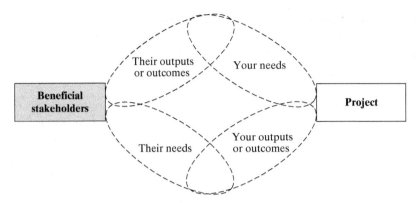

FIGURE 11.4 Value delivery as an exchange.

needs, as shown in Figure 11.4. This exchange is at the heart of stakeholder analysis. This observation will help us identify other potential stakeholders and their needs.

When we were identifying beneficiaries, we asked, "*Who benefits from the outputs of this project?*" Now we can see that the root of that question has two parts: the outputs or outcomes of the project, and the needs of the beneficiary. It is often worthwhile to explicitly list the outputs of the project and identify whether all of the project's outputs are received by a beneficiary. For example, if the project produces regulatory approval for a new hybrid technology, might other projects within the enterprise benefit or have a stake? Might suppliers benefit? This question may lead to the identification of additional needs.

The complementary question we asked in identifying stakeholders was "*Who provides inputs that are required to make this project successful?*" We can now see that this question has the same exchange quality. We have already listed the needs from the perspective of the project, but we could similarly list the outputs of the stakeholders and beneficiaries and determine whether any of these outputs might be important to the project. In essence, we are changing the frame of reference between stakeholders, asking about the inputs and outputs of each. This can be helpful in stimulating creativity about needs.

For example, the driver/owner will produce driving behavior—she or he will operate the car. Is the driving behavior important to the project? It could be very helpful to understand the needs of drivers. An electric vehicle with a gasoline engine charging the battery might cause range anxiety with a range of only 80 km, and if we collected data on a prototype to show that a segment of buyers take trips of greater than 80 km only 5% of the time, it might impact our perspective on the importance of range. Therefore, there might be classes of hybrid vehicles where returning driving behavior is an important component of the exchange, such as with BMW's MiniE and ActiveE demonstrator programs.

This fundamental exchange nature of beneficial stakeholder relationships acts as a *completeness* criterion to enumerate stakeholders and beneficiaries. The crux of this method is that the input needs of each stakeholder and beneficiary are listed (traditionally, it is easiest to argue what you need and hardest to indicate to whom you provide valued inputs), then the method works to identify which stakeholders provide those inputs. After iterating around a network, this method provides a complete list of stakeholders and their outputs at a given level of detail. This is shown as the *check outputs match inputs* step in Figure 11.1.

234   PART 3 • CREATING SYSTEM ARCHITECTURE

This is the critical link that will enable us to compare the needs we identify for each beneficiary with the outputs or outcomes produced by our architecture. Therefore, we can restate "stakeholders who must be considered and satisfied" as stakeholders whose outputs or outcomes are important to fulfilling our needs. This is the core principle by which we will later prioritize stakeholders according to what needs the project has.

This is fundamentally different from the consumer context. In consumer product markets, the only traditional output of the consumer is money, given in exchange for purchasing products. For complex systems, we will argue that there are important differences. The business-to-business markets and nonmarket environments that prevail for many complex systems cannot be treated with the logic of consumer markets, which typically assumes that beneficiaries are static entities whose only feedback is acceptance or rejection in the market. Stakeholders and beneficiaries are often more actively engaged in product definition. These stakeholders and beneficiaries can and must be deeply engaged in project definition and criteria setting.

Grouping Stakeholders

Having listed the potential needs that the Hybrid Car could satisfy, we proceed to the final task of Step 1 in Figure 11.1: taking a first pass at prioritizing stakeholders and beneficiaries by grouping them.

In Figure 11.5, we have classified our stakeholders according to whether we believe they are charitable beneficiaries, beneficial stakeholders, or problem stakeholders. For example, we receive regulatory approval from the regulator, but it is not clear that we provide anything of value to the regulator. By contrast, we need the parts the supplier produces, and the supplier needs our volume in order to fill its capacity. Similarly, we need gas from oil companies, but an aim of our project is to consume less gas by producing a Hybrid Car, so we have classified them as problem stakeholders. This classification might change if, for example, we partnered with the retail arm of an oil company to install fast charging stations at their gas stations. In that context, the oil company would need our cars to earn revenue on the electricity sold, and we would need its network of charging stations to create a use case for consumers. In this situation, we might reclassify our oil company partner as a beneficial stakeholder.

It would seem logical to prioritize problem stakeholders above charitable beneficiaries, but at this stage it is unclear which beneficial or problem stakeholders should take first priority. Similarly, we haven't discussed how to prioritize within each group. Which of the problem stakeholders is most important?

Having made a distinction between beneficiaries and stakeholders, and having identified beneficial stakeholders, we will now condense to using the more traditional term "stakeholder" for brevity unless otherwise noted.

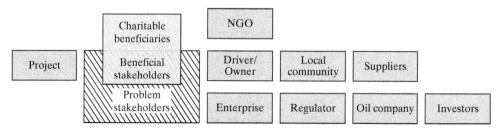

FIGURE 11.5 Charitable beneficiaries, beneficial stakeholders, and problem stakeholders for the Hybrid Car.

There are many other frameworks for grouping stakeholders. Another useful approach is to group stakeholders on the basis of how involved we want them to be in decision making: stakeholders who have veto power over decisions, stakeholders who have a vote in decisions, stakeholders whose opinions are important but who do not receive a vote, stakeholders who should be consulted about decisions, and (finally) stakeholders who should be informed of decisions.

The simplest method we have found is to group stakeholders in three buckets:

- Stakeholders who must be considered and satisfied
- Stakeholders who must be considered and should be satisfied
- Stakeholders who should be considered and might be satisfied

We've applied this taxonomy to the stakeholders in Figure 11.6, but we do so at this point without an explicit method. This is our first guess. Our experience suggests that writing down the expectations of stakeholder importance before conducting an in-depth analysis is a useful way to compare expectations against analysis. In Step 2 of Figure 11.1, we will prioritize the needs, and

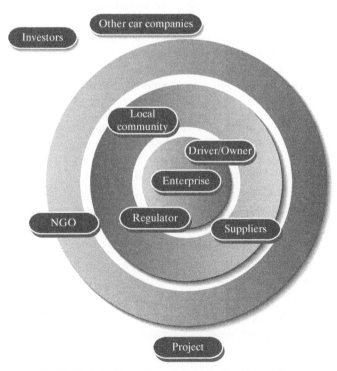

FIGURE 11.6 Stakeholder prioritization for the Hybrid Car.

we will begin to consider our stakeholders in a system, both of which procedures will inform our prioritization of stakeholders.

In summary:

- Beneficiaries have needs; you are important to them.
- Stakeholders produce outputs that you need; they are important to you.
- Value delivery occurs in an exchange, when your outputs meet the needs of beneficial stakeholders, and their outputs meet your needs.
- Think carefully about which stakeholders and beneficiaries are included in the early stages of product definition, because they will have an outsized impact on the architecture.

11.3 Characterizing Needs

Dimensions of Stakeholder Needs

> *"Find the appropriate balance of competing claims by various groups of stakeholders. All claims deserve consideration but some claims are more important than others."*
>
> **Warren G. Bennis**

We have now enumerated our stakeholders and identified their needs, so we have completed column 1 of Figure 11.1. Our next step is to characterize these needs, with a view to prioritizing among them.

There are many dimensions along which we could characterize needs.

- **Benefit intensity:** how much utility, worth, or benefit the fulfillment of the need will bring
- **Detriment:** the adverse reaction that will occur if the need is unmet
- **Urgency:** how quickly the need must be fulfilled
- **Awareness:** how aware stakeholders are of their need
- **Coupling:** the degree to which the fulfillment of one need relieves or intensifies another
- **Supply availability:** whether other sources of supply exist that could fulfill the need

There are many methods available for characterizing needs. We could survey each stakeholder and ask them to rank their needs, either by directly assigning values from an anchored scale, or by answering questions about their preferences in pairwise comparisons, as described in conjoint analysis and the analytic hierarchy process (AHP). [3]

One useful conceptual model for integrating several of these dimensions is Kano analysis. [4] The Kano methodology was originally used to classify product attributes from the consumer's perspective. For each need, the Kano method asks how you would characterize the presence/absence of this need in terms of the following categories:

- **Must have:** Its presence is absolutely essential, and I would regret its absence.
- **Should have:** I would be satisfied by its presence, and I would regret its absence.
- **Might have:** I would be satisfied by its presence, but I would not regret its absence.

The three Kano categories are shown in Figure 11.7. Braking is a "Must have" attribute. If the car has good brakes, the stakeholder will not be particularly satisfied, but if the car does not have good brakes, the stakeholder will be upset! Fuel efficiency is generally a "Should have"

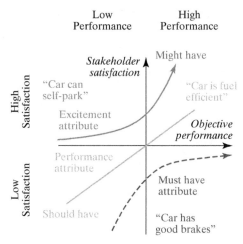

FIGURE 11.7 Kano analysis showing three types of needs on two dimensions: objective performance and stakeholder satisfaction.

attribute; the better the fuel economy, the more satisfaction it will produce. Self-parking is currently a technology feature; it is not expected, but its inclusion can be a selling feature. It is a "Might have" now but may transition to a "Must have" in time!

We make two important observations about characterizing needs. In many markets, formal methods (such as Kano analysis) for prioritizing needs are outmatched by the judgment of seasoned individuals. Where large data sets are available to perform analysis on consumer buying patterns, or where mature markets revolve around technical buying criteria, so much the better when analysis is present, but analysis should never supplant judgment. Second, there are many methods for comparing the needs of one stakeholder but few methods for comparing needs across multiple stakeholders. This is a problem that tests the limits of human reasoning as a consequence of the sheer number of needs possible. In the next section, we introduce a method for comparing them.

Stakeholders as a System; Indirect Value Delivery and Stakeholder Maps

We made a fundamental observation that stakeholder relationships are an *exchange*. Now we will take that observation a step further: Taken as a group, the set of stakeholder exchanges can form a system. The first step in understanding this system is to illustrate the individual exchanges.

When we were identifying needs, we checked for completeness by examining whether beneficiaries produced outputs that we needed, and whether stakeholders needed inputs provided by our Hybrid Car project. Up until now, we haven't explicitly represented the outputs of beneficial stakeholders. We will do so here.

If we take the existing list of stakeholders and their needs (Figure 11.3), we can observe that they are interrelated. For example, the Hybrid Car project's need for parts is satisfied by the suppliers. The suppliers need for revenue is satisfied by the project. Illustrating these relationships in Figure 11.8, we can now see the principle of needs existing in equilibrium with the outputs of the stakeholders. For example, the Hybrid Car project provides a transportation product to the driver/owner in exchange for sales revenue.

238 PART 3 • CREATING SYSTEM ARCHITECTURE

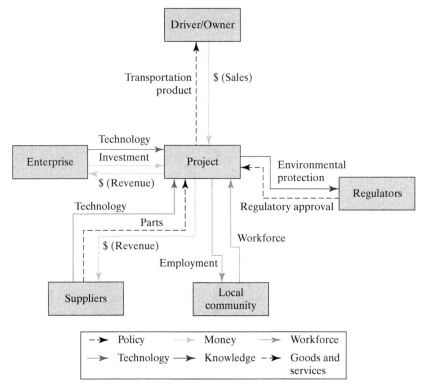

FIGURE 11.8 Hub-and-spoke stakeholder map for the Hybrid Car.

We are moving toward constructing a system out of our stakeholders. To build this network, we will use three steps.

First, we ask, *"Who could potentially satisfy the needs of each of the stakeholders?"* For example, this leads us to the supplier as the source of parts for the project.

Second, we ask, *"What are the outputs of the project, and to whom are they provided?"* For example, the project produces employment, which suggests that the project has a corresponding need for a workforce. However, it is important that needs not be created simply to match outputs. Those outputs that do not link to true needs do not deliver value. Unused outputs are an important product of the analysis and will be treated separately.

These first two steps could produce a diagram like Figure 11.8. Notice that we have represented the needs of the stakeholders as a potential flow. This does not imply that we are specifying which stakeholders the project has to satisfy, nor does it imply that we have chosen the concept for how the needs of the stakeholders will be satisfied. The language of these links is deliberately written in a solution-neutral format. We use the term *"value flow"* to mean the connection of an output to an input in the model; it is the provision of value from one stakeholder to another. An individual value flow is unidirectional and does not necessarily imply a return transaction. We categorize the types of flows into six categories, as shown in Figure 11.8.

We now have a basis for estimating which stakeholders are not in equilibrium exchanges (they are those where we do not supply any of their needs), and we might well re-categorize them

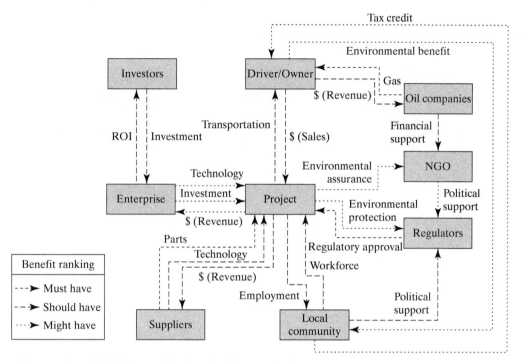

FIGURE 11.9 Stakeholder map for the Hybrid Car, with characterization of the needs illustrated.

as problem stakeholders. We can also identify charitable beneficiaries on this diagram as those who do not return a flow back to the project.

Notice that several stakeholders are missing from Figure 11.8: the NGOs, the oil companies, and the investors. They are not present because they do not produce anything, or receive anything directly from the Hybrid Car project.

The third step is that we *combinatorially pair all stakeholders and ask whether there are relevant transactions that play out between them*. Understanding which transactions are relevant will become clearer as we progress, but at this stage, it is best to err on the side of discovering more needs. Figure 11.9 completes this third step. Notice the flow of tax credit from the local community to the driver/owner of the hybrid vehicle.

What we have done here is interpret the stakeholders for the project as a *system* (Figure 11.9). We've identified the entities and the relationships between them. We have not defined a *system boundary*, other than imposing the completeness criterion when we required that all potential stakeholders be listed. This excludes other possible entities that will not be impacted by the project. As with all systems perspectives, choosing which entities to represent is a decision with consequences for the utility of the analysis and the complexity of the representation.

We use the term "value loop" to denote a series of value flows that return to the starting stakeholder—the system behavior in our stakeholder system. Previously we had only a hub-and-spoke network (Figure 11.8); now we have a broader stakeholder system with feedback loops. Value loops are at the heart of this system in that they illustrate which stakeholder needs are satisfied by strong feedback loops, and which needs are not well satisfied.

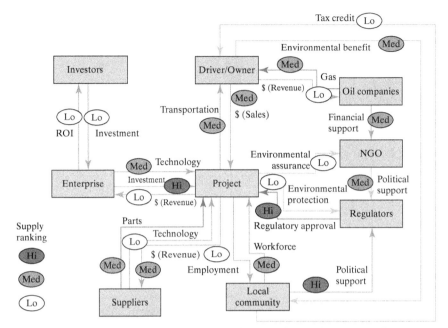

FIGURE 11.10 Stakeholder map for the Hybrid Car, with details on availability of alternative suppliers.

It is possible to overlay other dimensions of the need on top of this diagram. In Figure 11.10 we show the availability of alternative suppliers for each flow. In theory, if the products of the alternative suppliers are substitutes, the presence of alternatives will dilute the bargaining power of suppliers. [5] Our project is one of many suppliers of environmental protection to the regulator, but the regulator is the only possible supplier of regulatory approval.

Strictly speaking, competition in supply is not an attribute of our needs but an attribute of the value flow. We may choose to incorporate this knowledge into our prioritization of needs; high supply availability would effectively reduce the priority of a given need.

It is sometimes helpful to combine the influence of many dimensions of the needs. Recall that both of these can be composites of many dimensions; we listed six for needs alone, one of which we later noted is more a property of the flow. In Figure 11.11, we define a meshing between them, producing an aggregate score that can be used for comparing all the needs that are inputs to the project.

Prioritizing Needs across Stakeholders

We now have a representation of the stakeholder context in which the project exists. We have illustrated several qualitative analyses that are possible: checking for unmet needs, checking for unused project outputs, and categorizing problem stakeholders and charitable beneficiaries.

We have already prioritized the needs within each stakeholder. Regardless of whether we used the Kano analysis or another method, this is the easy step, because we can ask a single organization or actor.

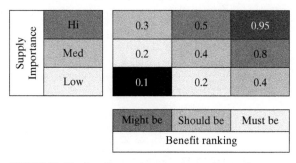

FIGURE 11.11 Creating a single measure for each flow, a combination of the flow characterization (here represented as the supply availability), and the need characterization (from the Kano analysis).

The next step is to prioritize needs across all the stakeholders to determine how we should prioritize the outputs of the project. This is the harder of the two prioritizations.

We have ranked needs for each stakeholder, so one solution is to meet all of the most important needs first, regardless of which stakeholder they come from. This can be difficult in some cases, because many organizations manage at the stakeholders level. Important clients are often accorded a higher service level, even for their minor requests.

Another solution is to define weights for different stakeholders in order to arrive at a weighted set of needs among many stakeholders. However, this is basically putting our thumb on the scale—it doesn't use the analysis we conducted at the needs level.

We break this problem into two conceptual pieces [6]:

- Prioritizing stakeholders who provide a single good to the firm
- Tracing indirect transactions with stakeholders

Prioritizing stakeholders is not difficult if they return only a single revenue stream to the firm. In this case, we can prioritize stakeholders by revenue (see the left side of Figure 11.12). However, this will not work for non-monetary flows, and it also does not capture supplier relationships well. How do we prioritize a relatively inexpensive component that is a critical part of the system (see the right side of Figure 11.12)?

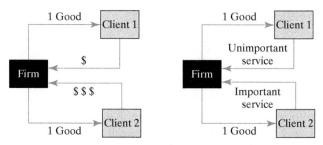

FIGURE 11.12 Prioritization of outputs in direct interactions with two clients.

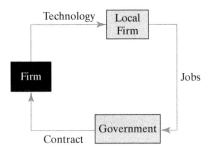

FIGURE 11.13 Example of an indirect transaction.

The important underlying principle here is that stakeholders should be prioritized by the inputs that they provide to the project! The comparison can be made with monetary flows, but if we have already categorized the needs of the project using the Kano analysis, we should use that information to prioritize stakeholders.

This direct exchange framing of the problem enables us to examine a more complex version of the problem, the stakeholder system, where transactions are no longer direct, two-party exchanges, but instead involve indirect flows of goods and services in value loops (Figure 11.13). Indirect transactions—value loops—are defined as those transactions with one or more intermediaries between the firm and the end stakeholder. In an indirect transaction, we do not require that the price paid for a benefit come from the party that receives the benefit.

This indirect behavior arises in many transactions for complex systems—for example, in securing foreign contracts with the help of a local partner. The firm provides technology to the local partner, the local partner provides employment to the government, and the government provides a contract to the firm, as illustrated in Figure 11.13.

Despite the fact that the relationship between the firm's outputs and its resulting inputs is no longer direct, it is still possible to prioritize the firm's outputs. This is accomplished by tracing backwards from the firm's important inputs to the stakeholders who provide them, and then further to the outputs of the firm that satisfy the needs of those important stakeholders. We might well prioritize an output to a small, local firm that we derive relatively little revenue from if that firm has a strong influence on the government and thus on our likelihood of winning a contract.

Cameron describes an analytical method for computing the relative importance of each of the needs, based on these principles. Where there are multiple paths to a given project input, we prioritize the stronger value loop. Other things being equal, a value loop is likely to be stronger if it satisfies important needs for all stakeholders along the path. For example, if two local suppliers are equal in their lobbying ability, we might give higher priority to the firm for whom our contracts represent a greater fraction of revenue. That firm has a stronger incentive to complete the value loop. We're not prioritizing it because of the size of our contract with it, but because of the importance to us of the government contract.

A simplistic analytical procedure for determining the strength of the value loop is therefore to multiply the importance of the needs on each of the links in a value loop together, to produce a metric by which we compare all value loops. There are many more sophisticated modeling techniques, but Cameron demonstrates that this approach can be effective in some situations.

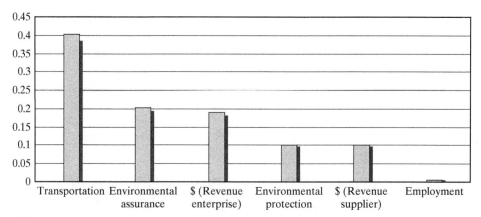

FIGURE 11.14 Prioritization of stakeholder needs for the outputs of the Hybrid Car project normalized to sum to one.

Using the metric for the loops, we could weight the outputs of the project according to the loops in which they participate.

Using the principles above, we can compute the priority among the six needs of the stakeholders that are directly fulfilled by the outputs of the Hybrid Car project, shown in Figure 11.9. The input data necessary to produce this output are shown at the end of this chapter. The results are shown in Figure 11.14.

This method produces a view of the most important stakeholders and their most important needs, but it does so without the firm intervening to judge which stakeholders are more important than others. The only inputs to this model are the priority of needs, which are judged by each stakeholder individually, and the structure of the stakeholder network. The model did not force us to say *a priori* that the enterprise was more or less important than the regulator.

If we had disregarded indirect transactions, we could have simplified this method. We would have listed the needs of the Hybrid Car project in order, ordered the stakeholders according to how many of our important needs they met, and then prioritized the strongest needs of those stakeholders. The only thing wrong with this analysis is that it would disregard the third-party transactions, such as those with the NGO, that are unimportant to the firm in its bilateral exchanges but important as a lever on the much-sought-after regulatory approval. The architect must exercise judgment to determine in which complex systems it is important to consider indirect transactions, and in which cases only bilateral stakeholder relationships are important.

Summary of Needs Prioritization

In summary, in this section we set out to characterize the needs of stakeholders. We defined a number of dimensions along which we could characterize needs. We used these dimensions to prioritize the needs of individual stakeholders. We recognized that the real challenge lies in prioritizing needs across stakeholders. To help prioritize stakeholders, we illustrated the needs of each stakeholder as inputs, and we built a stakeholder system diagram showing the dependencies among them.

This stakeholder system contains a number of bilateral exchanges most easily understood as transactions—money exchanged for a Hybrid Car. However, this system representation of stakeholders forced us to consider another type of exchange, an indirect exchange or feedback loop.

Analysis of direct exchange more than suffices in many product markets, particularly in consumer markets. The primary stakeholder is the consumer, and we interact directly with consumers. Networks at the level of individual consumers are diffuse enough that we cannot reasonably analyze their interactions, so we simplify to concepts such as "word of mouth" and "lead adopters."

By contrast, many complex systems are beholden to several different stakeholders, whose order of priority is not always clear (Box 11.3 Principle of Balance). It is for this reason that we presented the stakeholder as a system discussion. Our intent is primarily to enable the reader's mental model. Although we have presented a method for reasoning through the stakeholder system, it is the underlying principles that are more useful to system architects on a daily basis.

Box 11.3 Principle of Balance

"Bring balance to the Force, not leave it in darkness...."

Obi-Wan Kenobi

"No complex system can be optimized to all parties concerned."

Eberhardt Rechtin

Many factors influence and act on the conception, design, implementation, and operation of a system. One must find a balance among the factors that satisfies the most important stakeholders.

- In a built system, one may never be able to even enumerate, let alone quantify, all of the factors. One must make tradeoffs and compromise among the recognized factors to find solutions that come close to satisfying the identifiable influences, recognizing that formal optimality may never be achieved or may be meaningless in view of the un-enumerated factors.
- The system must be balanced as a whole, and not just the sum of balanced elements.
- In particular, an architect is always balancing optimality with flexibility. In general, the more optimized the architecture is for a given application, the less flexible it will be to respond to new use cases and new technologies. In general, the more flexible the architecture, the more margin will be carried to enable flexibility, making it less optimized. [7]

11.4 Interpreting Needs as Goals

For any product/system there will be many stakeholders, each potentially with many needs. It is hardly ever possible to meet all the goals of all the stakeholders, but there should be a well-defined subset of all of these needs that will be represented by the goals adopted by the enterprise. In this section we move to the third column of Figure 11.1, Interpret Needs as Goals.

We have structured this analysis to proceed from needs to goals, but we have yet to define goals. We define *goals* (Box 11.4) as what is planned to be accomplished and what the designer hopes to achieve or obtain. The purpose of this definition is to explicitly recognize that goals are decisions made by the architect, whereas needs are fundamental to the beneficiary.

> **Box 11.4 Definition of Goals**
>
> Goals are defined as:
> - What is planned to be accomplished
> - What the producing enterprise hopes to achieve
>
> Goals are a product/system attribute.

Needs exist in the mind and heart of the stakeholders, but goals are defined by the producing enterprise with the intent of meeting them. They define how the system that the enterprise builds will meet the needs of the beneficiary. Goals are under the control of the producing enterprise. They are not static and are often traded off against other product/system attributes in design. Therefore, they should be considered independent attributes of the system. Goals are defined, at least in part, by the architect.

We will create goals by choosing which needs of the beneficiary we believe the product or system should address, in addition to which needs of other stakeholders the project should address, such as corporate strategy and regulation. This raises concerns for many. We have suggested that corporate strategy needs are an active choice, rather than an imperative, and similarly for regulation. It is not our intent to suggest that architects should build systems that violate mandatory regulations. Rather, our point is that the firm has a choice of markets in which to compete, in whether or not to deploy products well in advance of regulatory deadlines, and so on.

Various industries speak in terms of goals, specifications, requirements, intent statements, and constraints and use these terms to mean different things. We will simply call them all goals, which incorporate various statements of intent. In the spirit of expressing decisions rather than imperatives, we explicitly avoid the term "requirements." It is our observation that many engineering organizations do not use the prioritization 'shall, should' for goals, and, further, do not view requirements as tradable. System architecture fundamentally represents tradeoffs among competing objectives. Paradigms based on the search for a single design that meets all constraints often fail to reflect the potential value delivery, because value is not well formulated as a constraint. These issues are treated informally in Part 4, but here we present the scalable, human process.

Criteria for Goals

If goals are decisions made by the architect, we should establish the criteria by which those decisions can be evaluated. We advance five criteria for goals:

Representative Goals are representative of stakeholder needs, so that a system that meets the goals will in turn meet those needs.

Complete Satisfying all the goals will satisfy all *prioritized* stakeholder needs.

Humanly Solvable Goals are comprehensible and enhance the problem solver's ability to find a solution.

Consistent Goals do not conflict with each other.

Attainable Goals can be accomplished with the available resources.

For comparison, the *INCOSE Systems Engineering Handbook* defines eight criteria for requirements (Necessary, Implementation-independent, Clear and Concise, Complete, Consistent, Achievable, Traceable, Verifiable). In view of the fact that Necessary, Traceable, and Verifiable are all part of Representative, and that Implementation-independent and Clear and Concise are part of Humanly Solvable, these are essentially the same criteria. Our intent is to express goals, which others might call high-level requirements, in such a way that we can place more emphasis on trading between goals.

Although we did not state it as such, the purpose of the detailed stakeholder needs analysis is to build the relevant knowledge to write *representative* and *complete* goals. Representative and complete goals are analogous to validation of requirements, where the fundamental assertion is that satisfying the requirements will satisfy the customer. Included in this notion is the idea of completeness—that all customers have been satisfied. A systems perspective on the stakeholder network allowed us to make all of the stakeholders an endogenous part of the system, and by closing the loops, we have a (partial) check on completeness. The other key completeness check that can be performed is to compare the goals to the list of upstream influences to see whether each has been captured or rationally excluded. The criteria Consistent and Attainable are discussed in Section 11.5.

Humanly Solvable can be the most difficult criterion to capture. One of the greatest levers the architect has available is crafting the problem statement used in the design of the system. How will the human designers and builders interpret the goals? We offer several ideas about human solvability.

First, the goals should be clear and minimalist. The information contained in the problem statement predisposes readers to consider them. Take, for example, "Find the voltage drop across a 5-ohm resistor to which a 12-A current is applied in a 120-V DC circuit." The information "120-V DC circuit" is extraneous at best, but also potentially distracting or disruptive. In the context of simple problems with known physics, this information can be eliminated with confidence, but in complex systems, additional information cannot be so easily eliminated. In crafting a problem statement, architects should be judicious in their use of information.

Second, information not contained in the problem statement may be ignored or, worse, used to reason about priorities. For example, we could reason that the exclusion of temperature data implies that there is no resistance–temperature dependency.

The principle of solution-neutral function (Box 7.1) influences how the problem statement is crafted. INCOSE calls this implementation-independent. Solution-neutral statements enhance the creativity of the process and leave a larger design space open. We will build a procedure for goals on this principle below.

To paraphrase Herb Simon, problems are more complex than our brains can handle, so we oversimplify and ignore parts of the problem statement, or fill in the unknown with our mental models of the way we think it should be, sometimes with dramatic effects on the results.

Creating a useful problem statement is therefore a balancing act, a matter of providing enough solution-neutral information to capture important needs, without providing extraneous or misleading information.

Maier and Rechtin [8] note that the original problem statement for the F-16 was to build a Mach2+ fighter. This performance goal was set by the Air Force, but it was based on an underlying need to be able to exit a dogfight quickly. The problem statement for the Mach2+ fighter focuses the mind on the maximum speed of the fighter, rather than on its acceleration capability.

A revised problem statement might have addressed building a Mach1.4 fighter with a high thrust-to-weight ratio, enabling it to out-accelerate its opponent when necessary.

Humanly Solvable Goals; System Problem Statements

We focus our discussion of creating humanly solvable goals around the idea of a *System Problem Statement (SPS)*. Much like the mission statement for an organization, the system problem statement is the single assertion of what the system is intended to accomplish to deliver value, which is representative of real success (Box 11.5 Principle of the System Problem Statement). A current challenge with requirements engineering is that the long documents produced are not quickly comprehensible; they fail to articulate the forest through the trees. In this frame, the de facto problem statement often reverts to a previous product with a modifier, such as "cheaper electric toaster" or "more efficient roadster." Maier and Rechtin note that problem statements are often ill-formed, incomplete, or over-constrained. Further, they are often a mixture of needs, goals, functions, and forms.

> **Box 11.5　Principle of the System Problem Statement**
>
> The statement of the problem defines the high-level goal and establishes the boundaries of the system. It divides content from context. Challenge and refine the statement until you are satisfied that it is correct.
>
> - The statement of the problem has enormous impact on the eventual design. It provides the origin for all other statements of goals, and it is a consistent and pervasive message to communicate to the team.
> - The statement of the problem is rarely stated properly upon its first formulation. Challenge the scope and statement of the problem early and often.

We will begin with a typical problem statement for the Hybrid Car and work to modify it until we have crafted a humanly solvable goal statement.

ITERATION 1: SYSTEM PROBLEM STATEMENT　Manufacture and sell a successful hybrid gas/electric car to environmentally conscientious consumers.

What is wrong with this statement? It tells us something—"manufacture" implies that we will not outsource production, and "consumers" indicates that the primary customer is not government. However, it is also ambiguous. How is "successful" an attribute of the car? Who would want to architect an "unsuccessful" product? What type of car is implied (sedan, city car, coupe?) and, more important, which subset of consumers?

Phrased more specifically, what is the solution-neutral operand implied in Iteration 1 of the SPS? What is the solution-specific process? Clearly it shouldn't be "selling" or "manufacturing," because neither is related to the primary beneficiary.

The canonical framework for problem statements is shown below. We call it "To-By-Using" but it is by no means our invention.

To...[the statement of (solution-neutral functional) intent]
By verb-ing [the statement (solution-specific) of function]
Using [the statement of form]

An example appears in the U.S. Constitution:

... to form a more perfect union,... promote the general welfare,....
(By)... laying and collecting taxes, paying debts,...
(Using)... the powers of... The Congress

The most important purpose of the To-By-Using framework is to focus the problem statement on the intent on value delivery to the primary beneficiary. For example, we focused our Hybrid Car on the driver/owner as the primary beneficiary. If we had targeted our Hybrid Car at police vehicle fleet buyers rather than consumers at large, the SPS would have changed, and this change would have had a substantial impact on the architecture of the system, in terms of prioritization of crash safety, modularity for equipment additions, vehicle idle time, and the like. The value delivery to the primary beneficiary needs to be presented first.

Note that the full problem statement contains solution-neutral function (To...) and the solution-specific function (By...) and form (Using...). Yet we cautioned before that crafting the statement of specific function and form too early can prejudice the eventual design and limit creativity. In the discussion that follows, we will present the complete SPS, from which a detailed design could be executed, but the architect needs to recognize that the SPS is developed during the architecting phase. Early stage meetings should make use of only the solution-neutral functional intent. Later, the solution-specific function and form will be added.

Figure 11.15 shows a detailed view of the To-By-Using framework and its relation to the now-familiar Object Process Methodology (OPM) representation of the functional intent and concept of a system (Figure 7.4)—this should not come as a surprise!

For the hybrid gas/electric car, we identify the driver/owner as the primary beneficiary and "changing the location of people and their possessions" as the solution-neutral transformation, which could also be called transporting people and their possessions. Below, we have translated

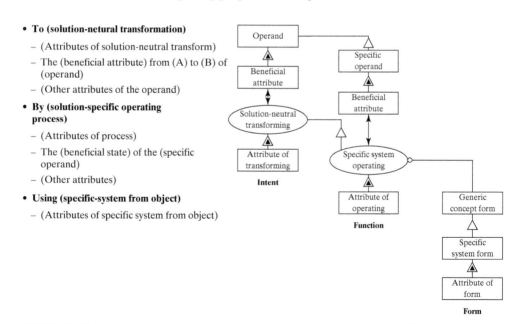

FIGURE 11.15 To-By-Using framework for formulating a good System Problem Statement.

- **To (change)**
  - (Inexpensively and environmentally)
  - The (location) from (A) to (B) of (people and possessions)
  - (Other attributes of the operand)
- **By (driving)**
  - (Fuel efficiency and good handling)
  - The (location) of the (driver, passenger and light cargo) (other attributes)
- **Using (car)**
  - (Hybrid gas/electric)

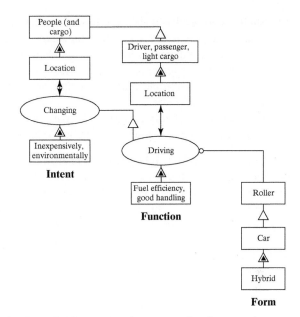

FIGURE 11.16 System Problem Statement for the Hybrid Car using the To-By-Using framework.

this into the second iteration of the problem statement. This statement of the problem indicates which attributes in the goal are related to intent, function, and form.

ITERATION 2: SYSTEM PROBLEM STATEMENT Provide our customers a product:

- To transport them and their possessions inexpensively and in an environmentally sound manner
- By allowing them to drive themselves, their passengers, and light cargo fuel-efficiently and with good handling characteristics
- Using a hybrid gas/electric car

Enterprise goals such as "provide our customers..." often appear at the beginning of a goal statement. They are indicative of internal organizational intent, not external value delivery.

Figure 11.16 represents the system problem statement in the full template. Given this SPS, we can now proceed from item to item in the template and begin to enumerate a more detailed set of descriptive goals. For now, we will enumerate goals for each of the needs that the Hybrid Car project intends to satisfy, regardless of their relative priority.

Descriptive goals:

- Shall provide transportation performance (range, speed, acceleration, etc.)
- Shall be inexpensive
- Shall have an environmental satisfaction to the driver
- Shall accommodate a driver (size)
- Shall carry passenger (size and number)
- Shall carry cargo (mass, dimension, and volume)
- Shall have good fuel efficiency
- Shall have desirable handling characteristics

- Shall require modest investment from corporate, shall sell in volume, and shall provide good contributions and return, *with the intent of securing investment and technology*
- Shall engage suppliers in long-term, stable relationships with good revenue streams to them, *with the intent of an uninterrupted supply of quality parts*
- Shall provide stable and rewarding employment, with the intent of securing a dedicated workforce
- Shall satisfy regulatory requirements, particularly regarding environmental protection, and environmental assurance to the NGO

Notice that the first eight goals reflect attributes of the vehicle produced to satisfy the transportation need and should be evaluated against the beneficiary's value function. The remainder of the goals reflect desired value delivery to other stakeholders, as illustrated in the stakeholder system (Figure 11.14). This implies that we will have to interpret desired flows in the network as goals, sometimes concatenating flows together into a desired indirect value exchange. For example, take the value delivery loop whereby the project provides environmental assurance to the NGO, which in turn provides political support to the regulator in support of regulatory approval for the project. To capture this as a goal, we need to write a goal on the delivery of the environmental assurance, but it is also necessary to record the intent that the NGO pass this on to the regulatory agency. Note that at this stage, all goals are stated as "shall." They have not yet been prioritized.

In summary, we use the term goals rather than requirements to emphasize that the architect must frequently make tradeoffs among goals. We developed the To-By-Using framework for creating a System Problem Statement, the purpose of which is to succinctly communicate the goal of the product. We advanced five criteria for goals and developed a list of detailed goals that follow from the SPS.

11.5 Prioritizing Goals

Our desired end state includes a prioritization of goals, which is the final column in Figure 11.1. To these ends, we will advance three categories: critical goals, important goals, and desirable goals. Critical goals contain the three to seven absolute necessities for product success; they are sometimes called "live or die goals." Important goals contain major goals that contribute to, but are not absolutely necessary for, success. Desirable goals contain objectives that it would be desirable, but not important, to meet.

This framework suggests how the project manager should allocate the limited project resources, spending resources on critical goals, conducting trades to determine resource allocation among important goals, and applying any leftover resources to desirable goals.

In practice, many organizations confuse the importance of goals with the rigidity of a need, as stated by a customer. Needs can be absolutely constraining ("the car must meet regulatory requirements"), constraining ("the car must accommodate a driver of dimensions XYZ"), or unconstrained ("the car must have good fuel efficiency"). The architect has to decide which constraining needs to satisfy and which to ignore (with the consequence of potentially shrinking the target market).

Determining how important each need is to each stakeholder was the easy part. But how should goals across stakeholders be compared? As we described in the stakeholder analysis, the fundamental principle is that stakeholders should be prioritized according to the importance of the inputs *they provide to the firm.*

Based on the prioritization among stakeholders, we've provided a possible categorization of system goals. Note that the rigidity of goals (absolutely constraining or constraining) is indicated where applicable in brackets after the goal. An analysis of the rigidity would show that it is not necessarily correlated with the importance assigned to the goals.

Descriptive goals:

Critically

- Must have an environmental satisfaction to the driver
- Must engage suppliers in long-term, stable relationships with good revenue streams to them (mildly constraining)
- Must accommodate a driver (size) (constraining)

Importantly

- Should provide transport range
- Should satisfy regulatory requirements (constraining)
- Should carry passenger (size and number)
- Should provide stable and rewarding employment
- Should have good fuel efficiency

Desirably

- Might require modest investment from corporate, might sell in volume, and might provide good contributions and return (mildly constraining)
- Might have desirable handling characteristics
- Might carry cargo (mass, dimension, and volume)
- Might be inexpensive

What is missing from this list of goals? There are no target values, and the metrics are implicit rather than explicit at this stage. For example, we can guess that fuel efficiency would be measured in miles per gallon (MPG), but this was not stated, nor have we set a target value to define good fuel efficiency. A goal should have a metric and a target value. The prioritized list of goals shown below has been annotated to reflect these metrics, and targets would be supplied.

Critically

- Must have an environmental satisfaction to the driver [as measured by EPA standards]
- Must engage suppliers in long-term, stable relationships with good revenue streams to them (mildly constraining) [as measured by supplier survey]
- Must accommodate a driver (size) (constraining) [as measured by % U.S. male and female]

Importantly

- Should provide transport range [miles]
- Should satisfy regulatory requirements (constraining) [certificate of compliance]
- Should carry passenger (size and number) [how many]

- Should provide stable and rewarding employment [turnover rate]
- Should have good fuel efficiency [MPG or MPGe (miles per gallon equivalent)]

Desirably

- Might require modest investment from corporate, might sell in volume, and might provide good contributions and return (mildly constraining) [ROI, revenue $]
- Might have desirable handling characteristics [skidpad acceleration in g's]
- Might carry cargo (mass, dimension, and volume) [cubic feet]
- Might be inexpensive [sales price not to exceed]

Although this analysis is simplistic, we quickly developed a large quantity of goals. Critically, the quantity here has been managed by prioritization, and the complexity is partially managed by clear indication of the chosen metric and target value. Regardless of the method chosen for prioritizing the goals, the sheer act of prioritizing goals is more important. Modern systems-engineering culture includes some prioritization in the difference between *shall* requirements and *should* requirements, but our experience with complex systems is that requirements are often recorded as equally important, with the offline task of prioritizing left to the system architect or the marketing representative. Managing this complexity is key to the idea of creating humanly solvable goals, as discussed in Box 11.6 Principle of Ambiguity and Goals.

Box 11.6   Principle of Ambiguity and Goals

"Always aim higher than you want to climb, for one usually stops below the goal one sets."

Traditional Chinese Saying

The early phase of a system design is characterized by great ambiguity. The architect must resolve this ambiguity to produce (and continuously update) a small set of representative, complete and consistent, challenging yet attainable, and humanly solvable goals.

- Ambiguity contains known unknowns and unknown unknowns, as well as conflicting and false assumptions.
- In general, no one designs or rigorously controls the process upstream of architecting, so there should be no expectation of unambiguity. There will be uncoordinated, incomplete, and conflicting inputs.
- The goals set must be representative of the needs of the beneficiaries. For a system to be successful, it must meet these real needs.
- The goals must be set so as to be mindful of competitive pressures, consistent with enterprise strategy, challenging yet reachable with available resources and available technology, and respectful of regulatory and pseudo-regulatory influences.
- Because the context and customer needs will evolve, there must be a process to continuously reexamine and update the goals.

Consistent and Attainable Goals

The consistency of goals is a very subtle property that is at the heart of systems engineering. By definition, many goals will conflict; a Hybrid Car may sacrifice range for reduced battery weight, for example. However, understanding whether the tradeoff between range and weight is shallow, steep, or otherwise shaped is not obvious.

Sometimes consistency can be checked through logical examination. For example, two goals with the same metric and opposing target values would be a clear indication of inconsistency. Some system metrics add linearly; every additional hardware feature on the Hybrid Car adds weight, and the weight of the Hybrid Car can be calculated as the sum of the weights of the parts. However, some metrics are emergent, which makes logical evaluation difficult.

Sometimes consistency can be checked analytically, using a model, as we will do in Part 4. Any model, low- or high-fidelity, could be used to compare the impact of different requirements on the metrics of interest—for example, minimizing the weight of the car compared with maximizing acceleration. As with all models, we would have to evaluate whether the model is accurate enough for us to base this decision on it.

The architect may choose to set goals that deliberately "stretch" the organization or get ahead of competition. Attainability implies that all product and strategic goals can be reached within technology limitations, and also that they meet development schedule, resource, and risk objectives. This is much easier for derivative products, much more difficult for clean-slate designs. When goals are not attainable, there are two options: reduce the scope of the goals or add resources. In Part 4, we will use explicit cost models in the loop with performance models to evaluate the attainability of models. Our focus here has been on the synthetic process, as unassisted by models, to speak directly to current industry practice.

The reader will note that we have not covered validation and verification in depth. These topics are well covered in systems engineering texts. [9], [10] As discussed under downstream influences in Chapter 10, sometimes these concerns create significant constraints or opportunities that rise to the level of architectural decisions.

Figure 11.17 shows the inner loop of verification and the outer loop of validation. We use it to summarize approximately where the five criteria for goals come into play: representative, complete, humanly solvable, consistent, and attainable. For comparison, the reader might compare the goal criteria represented by the acronym S.M.A.R.T. Advanced by George Doran in *Management Review* and frequently used in project management education, S.M.A.R.T stands for Specific, Measurable, Attainable, Relevant, Time Bound. Doran's framework focuses on an individual requirement, not a system of requirements, so it omits Complete and Consistent, but it does place added emphasis on time for a project management context.

11.6 Summary

Complex systems frequently face challenges managing stakeholders. There are many stakeholders, and their relative priority can be unclear, especially if they provide a non-monetary input such as regulatory approval or if they do not participate in a bilateral exchange with the firm. Additionally, complex systems with new architectures face the challenges of all new products in identifying latent needs, as well as the needs served by existing products in the market.

In this chapter, we have focused on deriving prioritized goals for the architecture, based on stakeholder needs. Several important ideas have emerged in this discussion. First is the idea of value as an exchange, where your outcomes meet their needs, and their outcomes meet your

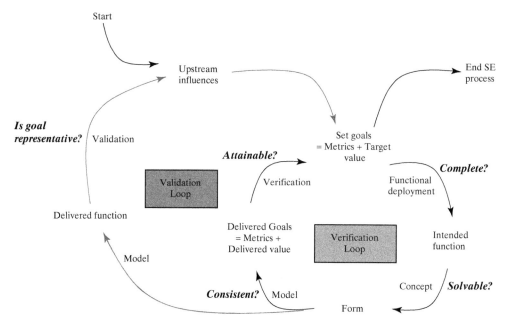

FIGURE 11.17 Summary of framework for checking whether goals meet criteria.

needs. Second is the idea of prioritizing stakeholders according to their importance to you. Third, represent the goals in a System Problem Statement, and pay careful attention to its creation, because small differences in the problem statement can shape the problem-solving activity.

11.7 Box *Case Study: 1998 Nagano Winter Olympics: Stakeholders in IT Architecture*

By Dr. Victor Tang. Dr. Tang was the Director of IT Client Relations for IBM's 1998 Olympic Winter Games, with responsibility for architecture and systems management.

Introduction

IBM was one of 11 top sponsors for the 1998 Olympics Winter Games in Nagano, Japan. Each sponsorship has been reported to cost $60 to $80 million (Humphreys 2008, Djurdjevic 1996). As the IT sponsor, IBM was contractually obligated to provide and operate the entire IT system for the games, in return for which, IBM was allowed use the Olympic logo for marketing.

The Challenge

IBM's scope was daunting. It had to provide, build, operate, and maintain a customized IT system just for the Games. This system had to compute and deliver competitors' results for stakeholders, media, and the general public; update and archive athletes' records; accredit athletes and staff; record drug testing results, and so on (Figure 11.18). The built system consisted of 25 million lines of software to support 10 mainframes, 5,000 PCs, 1,500

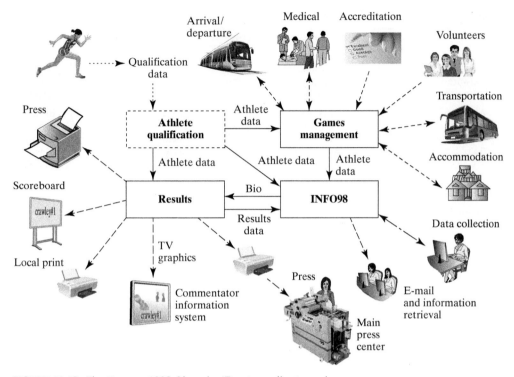

FIGURE 11.18 The Nagano 1998 Olympics IT system, clients, and users.

printers, and 160 servers networked through 2,000 routers that served about 150,000 athletes, officials, and volunteers. Key stakeholders included the IOC, the media, games administration, sports federations, the Nagano Organizing Committee (NAOC), and dozens of others. The Winter Olympics IT system is a *system-of-systems (SoS)* in which each constituent system has a different stakeholder with distinct priorities, interests, and biases.

Beyond the daunting scope, IBM faced a loss of confidence from the IOC and its stakeholders as a result of poor execution at the 1996 Olympic Summer Games in Atlanta. The IOC told IBM that they "didn't find any part which worked well."

Rethinking IBM's Strategy

For the Atlanta Summer Games, IBM had the wrong strategy. It chose to showcase products, technology, and systems that were not ready for showcasing. This lack of readiness caused widely visible disruptions in Games operations and business management. To avoid another fiasco, IBM changed its myopic product-technology-centric strategy to one in which products and technology were *the means* to deliver a service, *not the end* itself. IBM communicated this strategic shift as: "IT as a service...helping those running the Games, run the Games...[and] enhancing the Olympic experience." In other words, *IBM explicitly moved the system boundary.*

The social architecture was implemented using the organizational pattern of *X-teams* (Ancona et al. 2009). X-teams are formed by enabling domain-experts to work across organizational boundaries with their counterparts. IBM formalized this approach by

Without X-teams

	IBM	Xerox	Sports	News	NAOC	IOC
Requirements	do	review				review
Specifications	do	review				review
Software	do	review				review
Documentation	do	review				review
Write test cases	do	do some	review	review	review	review
Run test cases	do	do, review	review	review	review	review
Test performance	do	concur	concur	concur	concur	concur
Training	concur	do, concur	concur	concur	concur	concur

With X-teams

	IBM	Xerox	Sports	News	NAOC	IOC
Requirements	*do*	*do, review*	*review*	*review*	*review*	*review*
Specifications	*do*	*do, review*	*review*	*review*	*review*	*review*
Software	*do*	*review*	*review*	*review*	*review*	*review*
Documentation	*do*	*review*	*review*	*review*	*review*	*review*
Write test cases	*do*	*review*	*do, review*	*do, review*	*do, review*	*review*
Run test cases	*do*	*review*	*do, review*	*do, review*	*do, review*	*review*
Test performance	*do*	*concur*	*concur*	*concur*	*concur*	*concur*
Training	*concur*	*do, concur*	*do, concur*	*do, concur*	*do, concur*	*concur*

FIGURE 11.19 Co-development with and without X-teams.

announcing that (i) IBM was inviting stakeholders to participate in its IT system development, and that (ii) IBM was embedding itself in the Olympic stakeholders' processes. There was a sharp contrast in how development work was completed before and after the formation of X-teams (Figure 11.19). This approach to stakeholders' interactions enabled IBM engineers to develop the Games IT solution with a substantially more comprehensive understanding of the technical requirements and reciprocal organizational dependencies. Stakeholders became competent spokespersons and also credible advocates for IBM's work. Stakeholders were no longer just critics; they were part of the IT development team.

The Requirements

From the Atlanta debacle, four "IBM Must Do" goals emerged: (i) meet all technical and management system requirements, (ii) ensure no surprises during the games, (iii) provide 24/7 operations, and (iv) ensure highly secure and usable systems. IBM had its work cut out for itself.

These imperatives were easier to impose than to satisfy. The technical and sports-specific requirements were exhaustively documented in a voluminous library compiled by the IOC and the sports federations. A less formal, but equally formidable, set of technical requirements were those dealing with timing equipment from Seiko; printing and distribution requirements from Xerox; accreditation and security requirements for those providing press, TV, and radio coverage, medical services, transportation, and accommodations; technical training and education for volunteers; and so on. IBM pondered how it could convincingly satisfy this enormous set of requirements.

IBM found the answer in the International Competitions Prior to the Games (ICPG). The practice of the sports federations is to hold competitions, before the actual Olympic

Games, using new or modified Olympic rules and IT systems. Many World Cup competitions are ICPGs. The goal is to make sure everything works. Instead of the ad hoc debriefings of the past, IBM held a plenary convocation to discuss IBM's performance at the ICPGs. IBM invited the winter-sports federations, IOC representatives, and key stakeholders to this meeting in Nagano. They came well armed with papers documenting IBM's errors and shortcomings. After protracted negotiations, it was decided that solving the documented errors and shortcomings would be interpreted as IBM satisfying all requirements. Thus management-by-exception became the strategy for the "IBM Must Do" work.

Architecture

Louis Sullivan, inventor of the skyscraper, famously said, "Form follows function"; that is, the architecture is the output of a *mapping* from a function-space into a form-space. Mappings begin with ideas. Meaningful ideas require workable *concepts* that describe how an idea can be formed into an artifact that will function and operate. This is known as the embodiment of an idea, and it is the most creative and challenging part of architecting. Brilliant architectures require brilliant embodiments.

IBM re-conceptualized the Olympics IT system as a *utility* (highly secure and plug-&-play usability) to provide data and information to run the games. Networked servers would be the equivalent of a utility's power plant and distribution grid (with 24/7 availability), applications would be the equivalent of generators, and structured data would be the output of the plant to serve the stakeholders and the public as clients and consumers (Figure 11.20). These *concepts* were initially received with reserve and skepticism.

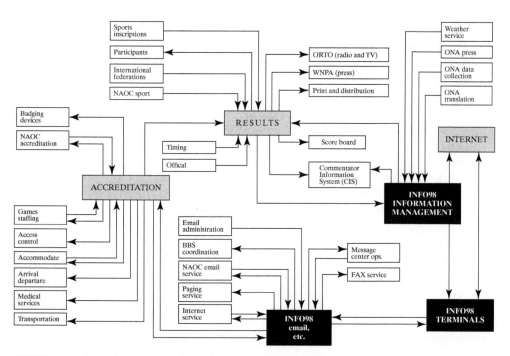

FIGURE 11.20 Data flow view of the architecture of the Nagano 1998 Olympics IT system.

258   PART 3 • CREATING SYSTEM ARCHITECTURE

Architecting a complex system is seldom a green field endeavor; it must consider the legacy subsystems and components that need not be reinvented but can be reused, modified, or protected. For the Nagano IT System, legacy was concentrated in the Athlete Qualification, Results, Games Management, and INFO98 (a Web-based system for Olympic information) systems. They *formed* the core of the Nagano IT SoS (Figure 11.18). Clearly, the most important subsystem is the Results System that must process athletes' performance. These systems were formed into a coherent whole using principles of data and function flow (Pahl and Beitz 1995). In the remainder of this case study, we will focus on the Results System as a concrete example of the Nagano IT architecture.

Architecture is the product of a mapping from function to form. These mappings, except for the simplest systems, are seldom completed in a single step. Rather, they require sequences of deliberate steps. For the Nagano IT system, IBM used the classical pattern of stepwise system decomposition and refinement. For example, the overall Results System is hierarchically decomposed into Results Subsystems for each competition venue. The first-level decomposition of the Results Subsystem is based on the principles of data-function flow within the constraint of using Seiko timing for data entry (input) and the score board (output) for displaying results. IBM provided the IT hardware and software systems that join Seiko's inputs to Seiko's output. Figure 11.21 shows how the architectural pattern (Gamma 1995) "reliability through redundancy" was used. The idea is simple, though

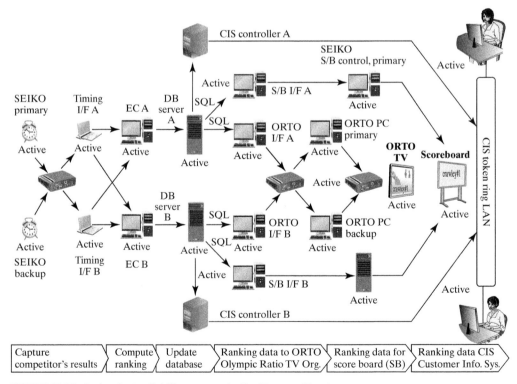

FIGURE 11.21 Redundant reliability pattern in the Nagano IT system.

challenging to implement. In this pattern, key functional components and subsystems are duplicated. Should any of them fail, its duplicate takes over so that operations remain uninterrupted. This pattern was used repeatedly throughout the IBM IT system, ranging from timing interfaces, PCs, LANs, WANs, and databases to medium and mainframe servers—all designed to prevent any single failure rendering the entire IT system inoperable.

Key Lessons Learned

From these Olympic Games, IBM learned two important lessons about managing stakeholders and IT architecture.

1. To paraphrase Clemenceau's famous dictum, "IT architecture is too important to be left to technical experts." IT is a technical expression of a business's policy. IT architecture has to be positioned within the context of a business policy for the architecture to be meaningful to decision makers, stakeholders, and IT users (Henderson and Venkatraman 1983). As a corollary, IT is no longer just about products; it is also about services. Far from being satisfied with just "break/fix and maintenance" services, customers increasingly need knowledge-intensive services to deal effectively with the cradle-to-grave complexities of multivendor and multi-stakeholder social technical systems. These are now known as product-service systems (PSS) (Tang and Zhou 2009).

2. The architect is always accountable for the integrity of the system but is not responsible for its implementation, which must necessarily be delegated. This is particularly true for large-scale social and technical systems. The architect is accountable for the specification of the mappings, descriptions of form, and specification of intended emergent system outcomes. Responsibility for implementation must be delegated and diffuse.

IBM began its efforts for the Nagano Olympic Winter Games with no confidence from the IOC and its stakeholders. But after the games, IOC's press release said, "IBM wins the gold in Nagano." Francois Carrad, IOC director general, a man not known for compliments, stated, "Technology at the Winter Games in Nagano has been outstanding." IBM's strategy of conceiving of IT as a service—and IBM's technical and organizational architecture—were validated.

References

Ancona D., H. Bresman, and D.Caldwell. 2009. Six Steps to Leading High-Performing X-Teams, *Organizational Dynamics*, Vol. 38, No. 3, 217–224.

Djurdjevic, B. 1996. *A 5-Ring Circus*. http://www.truthinmedia.org/Columns/Atlanta96.html.

Gamma E., R. Helm, R. Johnson, and J. Vlissides. 1995. *Design Patterns: Elements of Reusable Object-Oriented Software*. Reading, MA: Addison-Wesley.

Global Employment Trends January 2008. International Labor Office. ISBN 978-92-2-120911-9. Geneva.

Henderson, J.C., and N. Venkatraman. 1983. Strategic Alignment: Leveraging Information Technology for Transformaing Organizations, *IBM Systems Journal* 32(1): 4–16.

Hubka, V., and W.E. Eder. 1988. *Theory of Technical Systems: A Total Concept Theory for Engineering Design.* Berlin: Springer-Verlag.

Humphreys, Brad R. 2008. Rings of Gold. *Foreign Policy*: 30–31.

Pahl, G., and W. Beitz. 1995. *Engineering Design.* Edited by K. Wallace. London: Springer.

Tang, V., and R.Y. Zhou. 2009. First Principles for Product Service Systems. *International Conference On Engineering Design, ICED '09.* 24–27 August, Stanford University, Stanford, CA.

Tang, V., and V. Salminen. 2001. *Towards a Theory of Complicatedness: Framework for Complex System Analysis and Design.* 13th International Conference on Engineering Design, ICED'03. August, Glasgow, Scotland, UK.

Wirth, N. 1971. Program Development by Stepwise Refinement. *Communications of the ACM*: 221–227.

[1] Brian Snowdon and Howard R. Vane, *Modern Macroeconomics: Its Origins, Development and Current State* (Edward Elgar, 2005), p. 61. ISBN 978-1-84542-208-0.

[2] E. Ilzetzki, E.G. Mendoza, and C.A. Vegh, "How Big (Small?) Are Fiscal Multipliers?" *Journal of Monetary Economics* 60, no. 2 (2013): 239–254.

[3] Thomas L. Saaty and Kirti Peniwati, *Group Decision Making: Drawing Out and Reconciling Differences* (Pittsburgh, PA: RWS Publications, 2008). ISBN 978-1-888603-08-8.

[4] Noriaki Kano et al., "Attractive Quality and Must-Be Quality," *Journal of the Japanese Society for Quality Control* (in Japanese) 14, no. 2 (April 1984), 39–48. ISSN 0386-8230.

[5] Porter, M. E. *"How Competitive Forces Shape Strategy,"* Harvard Business Review 57, no. 2 (March–April 1979): 137–145.

[6] B.G. Cameron et al., "Strategic Decisions in Complex Stakeholder Environments: A Theory of Generalized Exchange," *Engineering Management Journal* 23 (2011): 37.

[7] A.P. Schulz and E. Fricke, Incorporating Flexibility, Agility, Robustness, and Adaptability within the Design of Integrated Systems—Key to Success? Proc IEEE/AIAA 18th Digital Avionics Syst Conf, St. Louis, 1999.

[8] M.W. Maier and E. Rechtin, *The Art of System Architecting* (CRC Press, 2002).

[9] *INCOSE System Engineering Handbook* V3.2.2. INCOSE, October 2011.

[10] Dennis M. Buede, *The Engineering Design of Systems: Models and Methods,* Vol. 55 (New York: John Wiley, 2011).

Addendum – Stakeholder System Needs Characterization

To	Needs	From	Demand Ranking			Supply Importance			Weight
			Might Be	Should Be	Must Be	Low	Medium	High	
Project	Regulatory Approval	Regulator			x			x	0.95
	$ (Sales)	Driver/Owner		x		x			0.2
	Workforce	Local Community		x			x		0.4
	Parts	Suppliers			x		x		0.8
	Investment	Enterprise		x		x			0.2
	Technology	Enterprise		x			x		0.4
Enterprise	$ (Revenue)	Project		x		x			0.2
	Investment	Investors		x		x			0.2
Local Community	Employment	Project		x		x			0.2
	Environmental Benefit	Driver/Owner	x				x		0.2
Regulator	Environmental Protection	Project	x			x			0.1
	Political Support	NGOs	x				x		0.2
	Political Support	Local Community	x				x		0.2
Investors	ROI	Enterprise		x		x			0.2
Driver/Owner	Transportation	Project		x			x		0.4
	Fuel	Oil Companies			x		x		0.8
	Tax Credit	Local Community	x			x			0.1
Suppliers	$ (Revenue)	Project		x			x		0.4
NGOs	Environmental Assurance	Project	x			x			0.1
	Financial Support	Oil Companies		x			x		0.4
Oil Company	$ (Revenue)	Driver/Owner		x			x		0.4

Chapter 12
Applying Creativity to Generating a Concept

The authors would like to acknowledge the significant contribution of Dr. Carlos Gorbea to this chapter. C. Gorbea, "Vehicle Architecture and Lifecycle Cost Analysis in a New Age of Architectural Competition." Verlag Dr. Hut, Dissertation, TU Munich, 2012

12.1 Introduction

Whereas the purpose of stakeholder analysis and goal writing is to reduce the ambiguity of the system, developing the system concept is fundamentally a creative process. We have already framed the problem statement as solution-neutral (Chapter 11), at the appropriate level of abstraction to help generate novel ideas without venturing beyond the firm's core competence or desired scope. Clearly, the concept is not defined only through analysis. Creativity is involved in producing concepts that excite, resolve design tensions, and bring in new functionality.

Although many consider the concept to be the central deliverable that the architect produces, it is our experience that the architect must be prepared to architect up and down. That is to say, the available levers on the concept lie in the upstream *and* the downstream processes, and they will shape the extent to which we can apply structure as well as creativity to the concept.

The funnel in Figure 12.1 illustrates the idea that the concept is the simplest representation of the system. It is honed from a litany of possible stakeholder needs, and it is succeeded by the complexity and depth of the eventual system implementation. If we reexamine the deliverables of the architect from Chapter 9, we can categorize them according to the three themes in the diagram shown in Figure 12.1. Upstream influences and stakeholder analysis relate to reducing the ambiguity in the definition of the system. Downstream of concept, the primary task of the architect is managing the complexity of the system, so that it remains manageable and understandable to those involved, which is the subject of Chapter 13.

In Chapter 7, we analyzed concepts and identified their constituent parts. In this chapter, we use the results of that analysis to help structure the creation of a new concept for the Hybrid Car, with emphasis on applying creativity.

We begin by discussing different modes of creativity. We review our framework for generating the concept, but we explicitly scope the activity without using Object Process Methodology (OPM) as a representation (Section 12.2). The process of generating the concept is then applied in detail to concepts for Hybrid Cars (Sections 12.3 to 12.5). We conclude with a case study on architectural competition in the automotive industry.

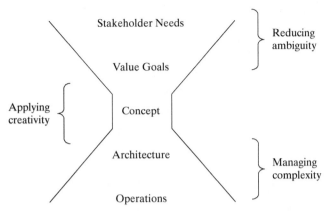

FIGURE 12.1 Three themes in architecting complex systems, suggesting that the architect must be prepared to architect "up" and "down" from the concept.

12.2 Applying Creativity to Concept

Creativity

What is creativity? Many would agree that creativity must result in a novel output, in the sense of not previously known. However, there is wide disagreement on two other potential dimensions of creativity: whether it must be intentional, and whether it must have influence or impact. We believe that *creativity must be intentional*. Products that are created unintentionally and that remain unrecognized, such as accidentally spilling paint on a canvas and then discarding the canvas in the trash, are not fundamentally creative. [1] Intention does not imply a linear or even a known process, only that effort is being deliberately applied. We also believe that *creativity should not be defined by influence or impact;* this is the role of the metrics applied after ideation. Experience would suggest that a focus on impact can restrict creativity. For example, brainstorm participants are encouraged not to criticize the ideas of others during the ideation phase.

The key observation we make here is that in an ideal creative process, the number of concepts under consideration should balloon (Figure 12.2). This idea was first articulated by Alex Osborn, the originator of the word "brainstorming," who hypothesized that "quantity breeds quality." [2] Perhaps quantity alone is not sufficient, but we will take this idea as a starting point here and refine it in Part 4.

There are two broad schools of thought on how to ideate many concepts: unstructured creativity and structured creativity.

Unstructured Creativity

The approach of *unstructured creativity* is far more prevalent. It includes brainstorming, blue sky ideas, free association, and related techniques. This school focuses on ideating without prejudices or biases from previous experiences. Edward de Bono [3] asserted that the mind is physically constrained by channels of thought and that the purpose of creativity is therefore to form new pathways through the concept-space. This approach is well illustrated using one of

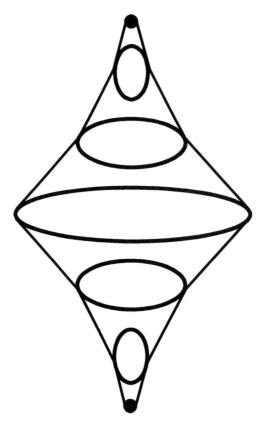

FIGURE 12.2 Applying intentional creativity to concept implies expanding the number of concepts under consideration and then winnowing the list according to "fit against goals."

his creativity techniques: opening a book to a random page, blindly pointing to a word, and then attempting to link the word to the problem at hand.

Unstructured creativity is rooted in a notion of unconscious processing—an "Aha!" moment that arises seemingly out of nowhere, providing the creative thinker with a solution. Some would argue that creativity is by definition unconscious and *must* occur without warning. If the roots of the idea are known, or the process is defined, then it is not creative. Although this is a feasible definition, we believe it is somewhat counterproductive. It holds creativity to a standard at which it cannot be encouraged. We will try to encourage use of both conscious and unconscious processing.

History has a long list of "creative thinkers" (Einstein, Picasso, da Vinci, Maya Angelou, and so on), who in characterizations are imbued with divergent thinking, the ability to come up with many novel ideas. The counterpoint to this concept of unconscious, native creativity is structured creativity, which holds that creativity can be stimulated by analysis and that creative thinking is not fundamentally different from problem solving and ordinary thought.

Structured Creativity

Structured creativity asserts that problem analysis can be helpful in solution synthesis. Recall the corkscrew example in Chapter 7, where "cork removing" generated a narrower list of possibilities than the solution-neutral statements of "wine accessing." We will build on this example of structured creativity with component recombination and completeness frameworks, described below.

A frequent theme in structured creativity is component recombination, the idea that the problem is decomposed into pieces, where more than one choice per "piece" is available, and then a solution is synthesized by selecting one choice for each piece. We have been deliberately vague with the pieces of the decomposition, because they can be form–function assignments (as in Figure 12.3, where a form is chosen for each of the three functions), or they can be specific functions for a given solution-neutral function. This same idea is encoded in morphological matrices (introduced in Chapter 7) and other decision support tools presented in Chapter 14.

Completeness frameworks, another form of structured creativity, use lists to stimulate ideation. For example, we could develop a list of forms of energy: linear kinetic energy, rotational kinetic energy, potential energy, chemical energy, and so on. When faced with a problem, we would ask ourselves what a concept employing each form of energy would look like. For example, many modern hybrid cars use batteries to store energy for use in propelling the vehicle. However, we could also use rotational energy, such as the flywheel used in Porsche's GT3RS hybrid vehicle and in the 2009 Williams F1 car.

A more abstract version of a completeness framework is de Bono's Six Hats. [3] This team-based method defines six roles for team members to play in problem solving, The structure of this approach is based on a theory that challenging different modes of reasoning leads to a more holistic evaluation of the problem and potentially spurs new solutions. Although the six hats (Managing, Information, Emotions, Discernment, Optimistic Response, and Creativity) are not proven orthogonal directions from a "brain and cognitive science" perspective, they illustrate a possible decomposition of team roles for working groups.

One of the most famous structured creativity approaches is known as TRIZ (Theory of Inventive Problem Solving) and was developed by Genrich Altshuller in the Soviet Navy, who

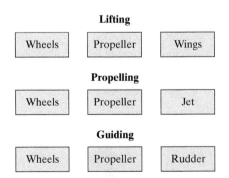

FIGURE 12.3 Stimulating creativity by component recombination: choosing one form (shaded box) from each row yields a variety of different concepts, as seen in Chapter 7.

reviewed 40,000 patent abstracts to define 40 principles of invention. Altshuller begins with seeming contradictions, such as "a faster train will require a more powerful engine, but a more powerful engine will be heavier... (thus reducing the acceleration gains to be had from the higher power)—we wish to go faster for the same weight." These contradictions are resolved using the set of 40 inventive principles, such as Mechanics Substitution: Replace a mechanical means with a sensory means (optical, acoustic, or the like), or use electric, magnetic, or electromagnetic fields to interact with the object. This might lead from a diesel locomotive to an electric locomotive or a linear induction motor. Note that the principles do not ensure that the resulting system is the same weight, but they provide a concept for the new system.

Both unstructured and structured methods are necessary in the application of creativity. It is difficult to prescribe unstructured creativity, so it may appear that our view of system architecture more closely reflects structured creativity. Indeed, our contention in the introduction to this text was that we would prefer to succeed with an architecture chosen well rather than with one chosen by luck. Simply put, it is our experience that many forms of creativity are at the heart of elegant architectures (see Box 12.1 Principle of Creativity).

Identifying Concept

We defined the system *concept* in Chapter 7 as a vision, idea, notion, or mental image that maps function to form. Necessarily, the concept embodies a principle of function and operation and includes an abstraction of form. The architect creates the concept for the system. This is a time of peak *creativity*, because the selection of concept will have a deep and far-reaching impact on the system. The concept should establish the solution-specific vocabulary—it is the beginning of the architecture. The civil architect Steve Imrich notes that "the concept rationalizes the structure of the architecture." The concept is *not* a product attribute; it is *a mapping from one attribute (function) to another (form)*.

The concept is separate from the architecture; it is a partial answer. We explicitly separate these two ideas to recognize that concepts are the working language of ideas, whereas architecture is the working language of implementation.

In Chapter 11, we described a method for analyzing concepts in OPM, based on completing the To-By-Using framework with information gathered from the solution-neutral function, the solution-specific function, and form. In this chapter, we move to complex systems where representation in Object Process Methodology may be possible at a first level, but where domain-specific language and methods quickly become more important.

We will structure the concept ideation phase according to the four steps outlined below. In Part 2, we built a framework of questions (Questions 4a to 8b) to first represent the concept and

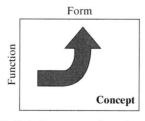

FIGURE 12.4 Representation of concept.

> **Box 12.1 Principle of Creativity**
>
> *"Imagination is more important than knowledge."*
> Albert Einstein
>
> *"Change takes place through the struggle of opposites."*
> Vladimir Lenin, *Philosophical Notebooks*, 1912–1914
>
> *"When you are face to face with a difficulty, you are up against a discovery."*
> Lord Kelvin
>
> *"Creativity is bred by creating a gap between current reality and the vision for the system."*
> Peter Senge
>
> For any interesting and real problem, there will be essential tensions among the goals set for the system. Creativity in architecture is the process of resolving the tensions in the pursuit of good architecture. To maximize the possibility of creativity, treat all goals as tradable, remove barriers imposed by organizational or cultural environments, and consider the entire space of possibility.
>
> - Because architecting is design at a strategic level, architects must resolve the tension in the high-level goals by creating the overall system concepts.
> - The tensions are resolved by applying creativity to find a concept that allows all goals to be met. Alternatively, if no concept can be found within existing technology, the tension is resolved by revising the goals.
> - Creativity is an iterative process of finding one's way when the destination is unclear. As E.L. Doctorow put it, "Creativity is like driving a car at night. You never see further than your headlights, but you can make the whole trip that way."
> - Generate creativity by provoking the mind to create new associations between existing pieces of information.
> - Think outside the box and fully explore the space of alternatives to discover new and innovative solutions; otherwise, the solution set is confined to incremental improvements on existing designs.

then the architecture. Here we will explicitly focus on generating multiple concepts, essentially an expansion of Questions 5a and 5b of Table 8.1.

1. Develop the Concepts
   - Start with the system problem statement (SPS) and descriptive goals.
   - Analyze (and reinterpret) to identify solution-neutral operands and processes related to value.
   - Apply creativity, and specialize to determine a specific operand/process/instrument concept.
   - Check that each concept meets the system problem statement goals, and reformulate the solution-neutral statement if necessary.
2. Expand the Concepts and Develop the Concept Fragments
   - For rich multifunctional concepts, expand or decompose the concept to reveal principal internal function or expanded operands/processes/instruments.

- For each of these, repeat the steps listed under "Develop the Concept" to identify operand/process/instrument concept fragments.
3. Evolve and Refine the Integrated Concepts
   - Search systematically through the space of concepts and fragments to ensure coverage.
   - Combine fragments combinatorially, with constraints, to identify integrated concepts.
4. Select a Few Integrated Concepts for Further Development
   - Apply backwards considerations: What is most likely to satisfy goals?
   - Apply forward considerations: What is most likely to make a good architecture?
   - Check that selected integrated concepts meet SPS and descriptive goals, and reformulate if necessary.

For the remainder of the chapter, we apply this framework to the generation of concepts for developing an architecture for the hybrid gas/electric car.

12.3 Develop the Concepts

We begin with the first step in our concept framework, developing the concepts. Recalling the stakeholder and goal analysis for the Hybrid Car, we began with *changing the location of the people and their possessions* as the solution-neutral operand and process, modified by the attributes of "inexpensively" and "environmentally."

Our system problem statement has already begun to define the concept, in that we specialized the process to driving with the attributes of fuel efficiency and good handling:

Provide our customers a product:

- To transport them and their possessions inexpensively and in an environmentally sound manner
- By allowing them to drive themselves, their passengers, and light cargo fuel-efficiently and with good handling characteristics
- Using a hybrid gas/electric car

The process of defining goals caused us to expand the list of specific processes. We used a structured approach to applying creativity by enumerating the needs of beneficial stakeholders (Sections 11.2 and 11.3), and then we chose which needs to encode as descriptive goals of the Hybrid Car.

Thus far, our concept of form is simply a hybrid, but we have not elaborated on the form or the function–form mapping of the concept. The term "hybrid" is a short form, where both the instrument to which the adjective "hybrid" refers is suppressed (Hybrid Car), and the two poles of the spectrum are left off. The reader probably assumed that we meant a hybrid gas/electric vehicle, but as the case study at the end of this chapter shows, there was a time when hybrid steam/gas vehicles was the prevalent meaning.

Given the prior development of the system problem statement in Chapter 11, the specialization we began in defining the goals, and the prior development of the transportation processes and instruments from Chapter 7, we will proceed directly to the second step: Expand the Concepts.

Note that at this stage, we are carrying only one concept (hybrid gas/electric car), but we *could* already be carrying several concepts to expand.

12.4 Expand the Concepts and Develop the Concept Fragments

Expanding the Propulsion Function

In Chapter 7, we noted that concepts can often be broken down into constituent ideas, which we call *concept fragments*. Step 2 in creating the concept is to expand the concept and develop fragments. For the Hybrid Car, the specific process we identified is driving—a concept rich in meaning! In order to expand this concept, we are going to decompose the driving process to identify concept fragments. We begin with the concept fragment of propulsion as one of the constituents of driving, and then we will examine seven additional concept fragments.

Specifically, we will categorize hybrids first by how their propulsion systems are combined. Three types of overarching vehicle concept classifications exist based on the dependence of the car's propulsion system on external energy sources: monovalent, bivalent, and multivalent architectures.

- **Monovalent Architectures:** Cars that exhibit a propulsion system dependent on one external energy source. Most cars today are monovalent cars that use an internal combustion engine with one liquid fuel such as gasoline or diesel. Hybrid cars that exhibit a secondary internal fuel source in the form of a high-voltage battery are also considered monovalent, because such cars remain dependent on one external fuel source.
- **Bivalent Architectures**: Cars that exhibit a propulsion system with two external energy sources. An example of a bivalent car, or fuel-flexible car, is a plug-in hybrid electric car wherein two external energy sources are transferred and stored within the vehicle: electricity and fuel.
- **Multivalent Architectures**: Vehicle architectures that exhibit a propulsion system with more than two external energy sources. These cars are designed to obtain and store three or more sources of energy. An example of such a system is the Fiat Siena Tetrafuel, which is designed to run on gasoline, E20 to E25 blends, on pure ethanol (E100), or as a bi-fuel with natural gas (CNG). [4]

This decomposition highlights that our original concept was anchored to the bivalent architectures, but there is a much broader space in multivalent architectures that would still solve the System Problem Statement. Notice, however, that the choice among these types of hybrid propulsion is not a concept fragment in that it does not map function to form.

Decomposing the propulsion function further into energy carrying/storage and vehicle moving (typically referred to as the powertrain) may help ideate additional concepts. Four general energy storage concept pathways are known to date. These include steam, internal combustion engine, battery, and fuel-cell-based concepts, along with multiple combinations thereof. Four powertrain concepts are summarized, together with their input energy sources, in Figure 12.5. Note how the diagram illustrates two primary internal functions: energy carrying and vehicle moving.

Steam-based concepts, which dominated the early automotive market from approximately 1790 to 1906, have not seen successful commercialization at a large scale since. Attempts have been made to combine the early steam concepts with other concepts, such as a steam–hybrid

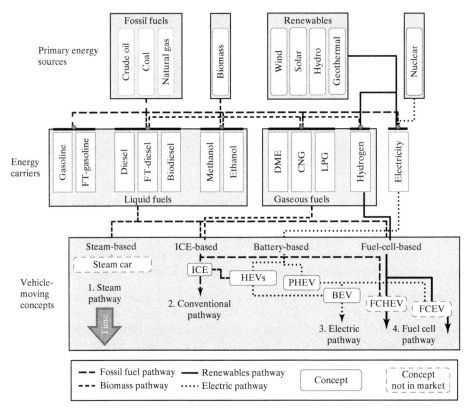

FIGURE 12.5 Four primary vehicle-moving concepts (ICE, HEV, PHEV, and BEV), together with the primary energy sources they rely on.

electric car or the use of internal combustion engine exhaust gases to generate steam. [5] The BMW turbo-steamer project developed a proof of concept that could use the combustion engine exhaust gases to generate steam and use the excess energy to boost the car's torque by 10%, but at a weight increase of 220 pounds. Arguably, the overall vehicle efficiency of steam, steam-electric, and ICE-steam combinations lies below that of traditional cars today.

Gasoline-powered spark ignition (SI) engines and diesel-powered compression ignition (CI) internal combustion engine (ICE) types are well-known alternatives. Other combustion engine alternatives for cars, such as turbine engines, have also been studied but have failed to achieve fuel consumption equivalent to that of SI and CI engines. [6]

Battery electric powertrains are now gaining favor as environmental demands for the reduction of exhaust gases in transportation have become a leading issue. Hybrid electric vehicles (HEVs) represent the first commercially available alternative to the conventional ICE. Initial hybrids feature small electric systems that assist the internal combustion engine in delivering power to the wheels. These first successful hybrid models are expected to lead the way for larger battery electric concepts that feature external battery charging, as in the case of plug-in hybrid electric vehicles (PHEVs). Battery electric vehicles (BEVs) will gain importance for city driving and short commuting customer use cases. [7] The key advantages of the battery-based concepts are the ability to reduce tailpipe emissions and increased flexibility in selecting less CO_2-emission-intensive production of electricity from primary energy sources. The greatest limitation to the battery-based concept is the battery itself. Improvement in battery life, energy density limitations, and costs will be crucial in making the battery-based pathway a success. [8]

Finally, the fuel cell concept is considered to be several decades of development away from market-readiness. [9] The first commercial fuel cell vehicles are expected to combine a large battery electric system and a fuel cell range extender with the ability to chemically convert fuel into electricity to be used in powering electric motors. Fuel cell powertrains may feature hydrogen as a fuel or a variety of liquid fuel carriers, such as methanol.

Concept Fragments for Seven Additional Internal Functions of the Hybrid Vehicle

Just the propulsion system of a car is clearly a rich multifunctional concept (Figure 12.5). In the interest of expanding to view the concept fragments, we will create a more complete list of the internal functions. Seven basic hybrid vehicle internal functions encompass the value added of hybrid systems for the customer: motor start-stop, regenerative braking, power boost, load level increase (battery charging), electric driving, external battery charging, and gliding. These functions are briefly discussed below.

CONCEPT FRAGMENT 1: MOTOR START-STOP The motor start-stop function is a basic function found in all hybrid vehicle concepts. As soon as the hybrid control system senses that the vehicle will come to a complete stop (for example, at a traffic light), the engine will shut off and be prevented from idling. The engine is restarted by means of an electrical motor or starter-generator as soon as there is a power requirement that merits its starting again. In micro hybrid systems that do not offer electric driving, the automatic start-stop feature is able to start the engine and have it available for acceleration in less than a second. The driver's signal to start the engine is normally depressing the clutch (for manual transmission cars) or releasing the brake (for automatic transmission cars).

For concepts that have electric driving capability, the transition from rest to starting the engine can be delayed by using the electric driving mode as a first means of propulsion before starting the engine for additional power.

Figure 12.6 shows a 5 to 7% saving in fuel consumption from a reference conventional car facilitated by the elimination of idling through the motor start-stop function during city driving conditions. [10] Under optimal control strategy, it is estimated that there is a further 5 to 9% saving available by combining the functions of load level increase, boosting, electric driving, and gliding (to be discussed below). Of our seven concept fragments, note that only external battery charging is not present in Figure 12.6; it adds energy to the system outside the nominal operations of driving. The values shown in Figure 12.6 are representative of full hybrid powertrains with limited electric driving and are not representative of plug-in hybrid systems that can essentially replace fuel consumption in greater proportions.

CONCEPT FRAGMENT 2: REGENERATIVE BRAKING The term "regenerative braking" refers to the capturing of braking energy that would normally be lost to friction and heat in conventional car systems. Brake energy recuperation is achieved by setting the electric traction motor in a generative mode that serves as a counterforce to the vehicle's direction of movement. The energy obtained through regenerative braking can be directly stored in the high-voltage battery and later used for

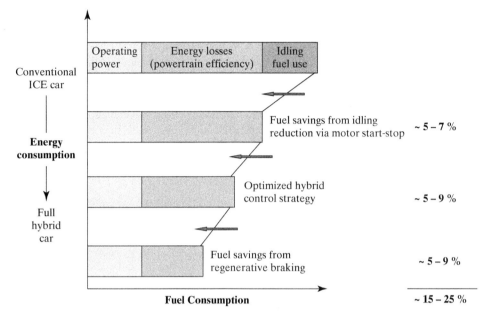

FIGURE 12.6 Fuel consumption savings potential for full hybrid systems. [11]   (Source: M. Ehsani, A. Emadi, and Y. Gao, *Modern Electric, Hybrid Electric, and Fuel Cell Vehicles: Fundamentals, Theory, and Design,* CRC Press, 2009)

boosting or powering the electrical system components. Figure 12.6 indicates an additional 5 to 9% saving from regenerative breaking.

The use of electric motors as brakes could be sufficient for most braking situations. However, redundant friction braking systems are still required for safety purposes. Hybrids with high-voltage battery systems (greater than 42 V) display a regenerative braking capability that can prolong the life of traditional friction brakes as an added benefit to the customer. Regenerative braking is limited by the battery system's ability to allow for impulse power storage in short time scales. Super-capacitors have been proved to be well suited for regenerative braking in the case of micro and mild hybrid systems, where two or three seconds of high power inputs and outputs are used in charging and discharging from the capacitor device. Sustained electric driving is not currently possible with super-capacitors.

CONCEPT FRAGMENT 3: POWER BOOST   When the driver's situation requires excess acceleration power beyond what the combustion engine can deliver, the electric motors provide additional torque to the wheels known as boosting. Power-boosting situations also include driving on inclines and towing. In this mode, the battery charge is depleted, and power is delivered through the electric motors as an additional source of power.

Boosting is particularly effective in improving a car's 0–100 km/h (0–60 mph) acceleration specifications. Figure 12.7 shows that the electric motor delivers the highest moment starting from rest and low RPM values (0 to 900 RPM), whereas the typical Otto-cycle internal combustion engine achieves maximum power at higher RPM values (2000 to 2500 RPM). In a typical Hybrid Car, the resulting system performance is enhanced when accelerating from rest by initially using the torque that the electric motor supplies to the drive train.

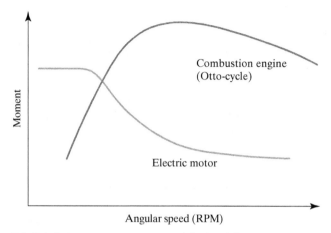

FIGURE 12.7 Moment versus speed (in RPM) for an electric motor and a combustion engine. Boosting in hybrid systems allows for additional torque for acceleration, especially when starting from rest.

CONCEPT FRAGMENT 4: LOAD LEVEL INCREASE   The load level increase or generative mode allows the engine to deliver some of its excess power to generate electricity, together with an electric motor in generative mode. The extra load level can be used to increase the engine torque and RPM to a more efficient operating point when excess power is available. The generated electricity can be used to charge the battery or power other electrical system loads.

CONCEPT FRAGMENT 5: ELECTRIC DRIVING   Electric driving is achieved by using electric energy stored in the high-voltage battery to power the traction motor to power the wheels. During the electric driving mode, the combustion engine is decoupled from the powertrain. It is either shut off or used to generate electric power. Electric driving is limited by the energy availability of the electrical storage system.

CONCEPT FRAGMENT 6: EXTERNAL BATTERY CHARGING   External battery charging differentiates plug-in hybrid concepts from all other hybrid vehicle concepts. In addition to the typical hybrid components, a battery charging unit can be added to the car, offering the possibility of plugging into an external electrical grid. Otherwise, an electrical charging station is required. Both charging strategies impose a limitation on the PHEV market, because customers are forced to have access to plug in their cars at home or at a charging station.

The option of connecting hybrid and electric cars to the electrical grid opens up possibilities for nighttime charging when electricity is cheapest and the electric load capacities of local power stations are at their lowest. Vehicle-to-grid studies within electric mobility research are complementary areas of study that have garnered recent attention. [12]

CONCEPT FRAGMENT 7: GLIDING   The last hybrid function of "gliding" is somewhat trivial but nevertheless useful in optimizing a hybrid control strategy. Gliding consists of decoupling both the engine and the electrical system from the wheels and using the force of gravity to propel the vehicle without friction losses of powertrain loads. Conventional vehicles can glide when placed in neutral during downhill operation. Hybrids, however, must have the ability to rapidly connect the appropriate powertrain that best suits the driving situation while moving in and out of a gliding operating environment.

In summary, these seven concept fragments represent potential features of a hybrid vehicle. Some (such as power boost) are directly related to the driving experience, whereas others contribute to the overall performance without direct traceability to the operating experience of the driver. Having developed these concept fragments, we return to the question of the integrated concept. We will compose our integrated concepts from our two primary internal functions, energy storage and vehicle moving, and the seven concept fragments. How do these concepts interact? Do they conflict? Are they synergistic?

12.5 Evolve and Refine the Integrated Concepts

The third step in our concept development framework is evolving and refining the integrated concepts. One of the benefits of concept fragments is that they lend themselves to recombination and modularity. The intent of this third step is to search the space of concepts systematically, by enumerating integrated concepts from combinations of concept fragments. One way to analytically organize the search of the combinatorial space among the concept fragments is to structure the search according to the different possible mappings between functions and forms. Recall how in Chapter 7 we produced a variety of transport concepts using the morphological matrix. For example, having a propeller perform each of lifting, propelling, and guiding, we produced a helicopter. In Part 4 we showcase computational methods for reasoning exhaustively, but in this section we treat this topic analytically.

From the energy storage and powertrain concept fragments, we can begin by analyzing the spectrum of architectures within the "electric pathway" at various levels of electrification, including hybrid electric vehicles (HEVs), plug-in hybrid electric vehicles (PHEVs), and battery electric vehicles (BEVs).

Specifically, in the hybrid context, the relevant form-to-function mapping is whether the energy storage function is an input to the vehicle-moving function, implying that the form is shared between these two functions, or whether energy storage and vehicle-moving decompose to separate forms.

This mapping is referred to as parallel hybrids vs. series hybrids and is illustrated in Figure 12.8. *Parallel hybrid powertrain systems* are *additive* systems that combine both drives; by contrast, the *series hybrid powertrain system* works in a *sequential* manner, where the fuel-based powertrain provides power to the electric powertrain. The energy load consumption is the energy needed to propel the vehicle. Vehicles that can *mix* features of both parallel and series hybrid operating modes are referred to as *combined hybrid systems*. Combined hybrid vehicles may split the energy from the fuel converter into the series and parallel hybrid energy paths simultaneously, as in the case of power split hybrids. Or they may have a distinct switch that allows for only parallel operation, or only series operation, at any one time.

For brevity, we have excerpted the search process among the potential combinations of parallel and series hybrids, skipping directly to the integrated concepts. The conceptual typology of hybrid vehicle architectures can be described in a two-dimensional solution space of electrical driving range and degree of electrification, as depicted in Figure 12.9. The first dimension entails the electric distance the car can achieve with the electrical propulsion system

CHAPTER 12 • APPLYING CREATIVITY TO GENERATING A CONCEPT 275

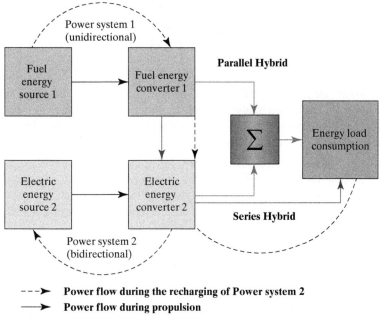

FIGURE 12.8 Conceptual illustration of a hybrid electric drive.

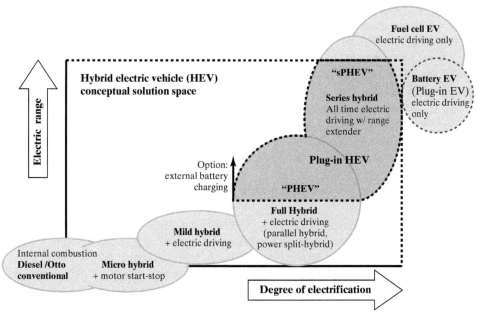

FIGURE 12.9 The hybrid electric vehicle conceptual solution space, showing the general vehicle concept areas based on electric range and degree of electrification. [13]   (Source: Carlos Gorbea, "Vehicle Architecture and Lifecycle Cost Analysis In a New Age of Architectural Competition," Dissertation, TU Munich, 2011)

alone, whereas the degree of electrification is determined by the ratio of cumulative peak electric motor power to maximum combined electric and engine power. A degree of electrification of zero describes a conventional internal combustion engine car with no electric system, and a degree of electrification of one describes a battery electric vehicle with no internal combustion engine installed.

There are seven integrated concepts that populate the car architecture solution space: conventional ICE, micro hybrid, mild hybrid, full hybrid, plug-in hybrid, battery electric vehicle, and fuel cell electric vehicle. These are obtained by combining concept fragments for energy storage and vehicle moving with the seven additional concept fragments describing internal functionality.

- **Conventional Internal Combustion Engine (ICE).** The conventional ICE car concept is the current dominant architecture. The most popular conventional cars feature a diesel or Otto cycle engine and are defined by a degree of electrification and electric driving range of 0 (zero), meaning no electrical propulsion system is installed.
- **Micro Hybrid Electric Vehicle (Micro HEV).** Micro hybrids achieve propulsion exclusively by an ICE but offer some functionality of hybrid vehicles, most importantly the motor start-stop function. The start-stop function of a micro hybrid can be achieved by a 12-V to 16-V electric battery system and requires a more robust starter-generator system. Some micro hybrids also exhibit limited regenerative braking. Micro HEVs are sometimes referred to as advanced ICE cars because they are essentially conventional cars with no electric driving capability and minimal electrification.
- **Mild Hybrid Electric Vehicle.** Mild hybrids differentiate themselves from micro hybrids in that they offer limited functionality in electric driving (some mild hybrids do not offer electric-only driving). Mild hybrids thus include the three key components of the electric drive system: a high-voltage battery, an electric starter-motor for propulsion, and a control system that determines when the electric and the combustion engine systems will work together. Mild hybrids are solely parallel systems that can offer additional functionality of motor assist and expanded regenerative braking.
- **Full Hybrid Electric Vehicle.** Full hybrids display larger degrees of electrification (10 to 30%) than mild hybrids and are characterized by short electric driving distances (for example, from 500 m to 3 km). The primary propulsion system still remains with the internal combustion engine, but the electric system can assist in providing power to the wheels. Most full hybrids exhibit parallel or combined configurations. Full hybrids exhibit all functions of mild hybrids and have a larger capacity for regenerative braking.
- **Plug-In Hybrid Electric Vehicle.** Plug-in HEVs are differentiated from other HEV types by the ability to charge the high-voltage battery externally through a battery charger and to plug into an external energy source. PHEVs come in a wide range of architectures, including parallel, series, and combined configurations, and they offer extended electric driving ranges (from 5 km to about 160 km). PHEVs offer a degree of electrification typically above 35%. The range extender (ICE plus generator) of series plug-in hybrids (sPHEV) can vary from large ICE power systems to small "limp home" emergency IC engines.
- **Battery Electric Vehicle (BEV).** The battery electric vehicle falls outside of the HEV solution space with a degree of electrification equal to 1. BEVs have no internal

combustion engine installed and are plug-in vehicles by definition: Electricity is their only energy source. The electric range of BEVs depends on the level of electrification installed.

- **Fuel Cell Electric Vehicle (FCEV).** Fuel cell cars exhibit an electrochemical cell that converts a source of fuel into an electrical current. FCEVs are primarily configured in a series architecture where the fuel cell power system provides energy to the electrical power system for propulsion.

In summary, these seven integrated concepts represent the decision facing us. These were generated by analytically combining functions and forms for the two primary functions and then combinatorially pairing the seven concept fragments. In this manner, we were able to generate coverage of the space of integrated concepts. Our focus here was on presenting the results; it is difficult to represent the creative process, particularly infeasible or conflicting options. In Part 4, we will examine methods for searching the space computationally, which provides the side benefit of tracking conflicting concept fragments and infeasible combinations.

12.6 Select a Few Integrated Concepts for Further Development

The last step in concept development is to select a few integrated concepts for further development. One of the primary aims of this discussion is to develop more than one concept for consideration. A decision with only one option is an imperative—there is nothing to decide! The role of the architect in the concept generation phase is often to *preserve* options, in the face of pressure to down-select early and focus resources on a single concept. The challenge of evaluating concepts is that they may not be detailed enough to understand the potential fit against goals—hence the pressure to down-select early.

Bazerman provides an excellent overview of biases in managerial decision making, [14] which are directly relevant to concept selection. For example, concepts that are easier to recall, concepts that confirm existing beliefs, concepts that contain more information or fidelity, and concepts we have anchored on or are familiar with are all subject to potential overrepresentation in our choices. For these reasons, we will attempt to be explicit in our judgments. We will illustrate a method for screening concepts for the purpose of our Hybrid Car, and then in Part 4 we will discuss a variety of other, more sophisticated methods.

We would like to apply two central screening criteria: a backwards-looking consideration of the prioritized goals and a forward-looking consideration of the potential for good architecture. Recalling that our list of goals was divided among critically important, important, and desirable goals, we could rate each of the concepts using a Pugh matrix (Table 12.1). A Pugh matrix [15] is a concept comparison tool, where each concept is rated according to a number of criteria. Here we use the goals developed out of the stakeholder analysis (Chapter 11) as the criteria. The intent of the Pugh matrix is to provide a simple scale, often showing just three choices: advantages, average, and disadvantages. Here we have chosen a five-level scale running from very advantageous (++) to having many disadvantages (− −).

Table 12.1 shows a qualitative comparison of the vehicle architecture types considered in this chapter, taking a full hybrid vehicle as a reference. Although the baseline Pugh matrix weights all

TABLE 12.1 | Qualitative comparison among various vehicle architecture concepts, taking a parallel full hybrid concept as the reference architecture

Evaluation Criteria	ICE	Micro Hybrid	Mild Hybrid	HEV (Reference)	PHEV	FCEV	BEV
Critically							
Must have an environmental satisfaction to the driver, and impact [as measured by EPA standards]	– –	– –	–	o	+	++	++
Must engage suppliers in long-term stable relationships with good revenue streams to them (mildly constraining) [as measured by supplier survey]	o	o	o	o	o	o	o
Must accommodate a driver (size) (constraining) [as measured by % American male and female]	o	o	o	o	o	o	o
Importantly							
Shall provide transport range [miles]	+	+	+	o	o		– –
Shall satisfy regulatory requirements (constraining) [certificate of compliance]	o	o	o	o	o	o	o
Shall carry passenger (size and number) [how many]	o	o	o	o	o	o	o
Shall provide stable and rewarding employment [turnover rate]	o	o	o	o	o	o	o
Shall have good fuel efficiency [MPG or MPGe]	– –	– –	–	o	+	++	++
Desirably							
Shall require modest investment from corporate, shall sell in volume, and shall provide good contributions and return (mildly constraining) [ROI, Revenue $]	++	++	+	o	– –	– –	– –
Shall have desirable handling characteristics [skidpad acceleration in g's]	+	+	o	o	–	– –	– –
Shall carry cargo (mass, dimension, and volume) [cubic feet]	+	+	o	o	–	– –	–
Shall be inexpensive [sales price not to exceed]	++	+	+	o	–	– –	– –

++ Very Advantageous; + Some Advantages; o Average; – Some Disadvantages; – – Many Disadvantages

criteria equally, we could also weight them according to their importance (Chapter 11). Overall, the benefits of electrification lie in reduced tank-to-wheel emissions, better fuel consumption, an enhanced ecological image, and support from government incentives that make these new electrified powertrain architectures more attractive. The disadvantages of electrified powertrains lie in increased weight, reduced driving range, higher manufacturing costs, and commercial risks related to the replacement or servicing of the high-voltage battery. These disadvantages are the technological risks that first movers in the new-vehicle market must be ready to improve on to gain and secure an advantage in the market. If the first down-selection was conducted qualitatively with Pugh analysis, we might also apply a quantitative comparison across the good concepts, as discussed in Part 4.

Notice that not all goals differentiate between concepts (a backwards consideration). For example, we would have to move to a level of much greater detail to determine which of the concepts were more likely to engage suppliers in stable, long-term relationships. Therefore, this goal appears to be related more to the development process than to the concept decision. The concept decision will have to be made according to the subset of goals that offer differentiation.

When considering the concept, one should explicitly consider whether it will lead to an elegant architecture (a forward consideration). This is arguably the function of the entire text, but some simple criteria should come easily to mind, such as subjective criteria for whether the solution is simple and pleasing. Further, we could examine whether the functions are mapped to form in a one-to-one mapping, as in Suh's axiomatic design, or the mapping is more complex. This may be difficult to determine until some architectural definition is done, so iteration is required. We could consider other dimensions of decomposition, such as whether form maps to development teams or manufacturing stations. A subjective down-selection, according to criteria like elegance, is often the final step in choosing two or three concepts for further development.

In summary, the exit from the concept development procedure should identify a small number of concepts for further development, based on a screening against important architecturally discriminating descriptive goals, and on a consideration of how the concepts lead to architectural elegance.

12.7 Summary

The purpose of this chapter is to apply creativity to generating a variety of concepts to choose from. Using both structured and unstructured creativity, we have worked to build out a number of concepts. This was based on two key ideas: More concepts will spur ideas for yet more concepts, and choosing among many concepts is more likely to yield a successful architecture than building out only one concept.

Figure 12.10 shows the number of concepts ballooning and then down-selected to a small number for more detailed trades. Specifically,

1 is the solution-neutral function.

N is the development of the concepts (Step 1).

(3 to 4)N is the expanding of each of the N concepts to reveal 3 to 4 internal functions, and the development of concept fragments for these (Step 2).

(3 to 4)N! is the combining to get all of the possible combinations of the fragment (Step 3).

10 to 20 is the more qualitative down-selection (part of Step 4).

5 to 7 is a quantitative down-selection (part of Step 4).

2 to 3 is the final outcome of the concept development stage (part of Step 4).

With a view to building a more complete picture of the architecture, we will now move from concept to architecture in the remainder of the text. The primary role of the architect in this last phase is to manage the investment in complexity, so we proceed in the next chapter to define complexity. As we've seen here, decomposition of the solution is central to the concept and the architecture, so in the next chapter we will use decomposition as a tool to manage complexity.

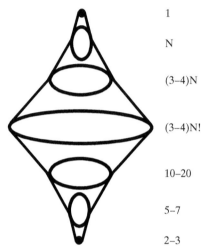

FIGURE 12.10 By expanding concepts and recombining concept fragments, we balloon the number of concepts under consideration. Subsequently, we use screening criteria to reduce these to a small number for further design.

Box 12.2 *Case Study: Architectural Competition*

Early Architectural Competition in the Automotive Industry

Architectural competition [16] refers to differentiating a product from others in the market based on product architecture. In the early years of the automotive industry, three different concepts (electric, steam, and internal combustion) competed to dominate the market. At this early stage, automakers (large and small) innovated around the basic structure of a car, but with significantly different concepts. The market exhibited a time of architectural

CHAPTER 12 • APPLYING CREATIVITY TO GENERATING A CONCEPT

FIGURE 12.11 Early vehicle advertising for an electric car [17]. (Source: Baker Electrics)

competition where a variety of powertrain elements were linked in various ways to enable the function of propelling the car.

Customers during the early 1900s had to decide which car architecture best met their mobility needs. For example, electric cars were marketed to female drivers for their ease of use and minimal maintenance, whereas internal combustion engine cars (ICEs), and steam cars were attractive to male drivers seeking power and speed. The advertisement in Figure 12.11 is evidence of architectural competition in the early automotive market.

The electric car advertisement on the left presents a car fit for aristocrats, claiming that electric range is no longer an issue and showing a female driver at the wheel in the countryside. The advertisement on the right, dating from 1904, is more technical in nature, boasting 100 miles on one filling of the tanks—that is, both the fuel tank and the water tank in the steam car.

The two vehicles support similar specific functionality at an aggregate system level—namely, driving. However, they exhibit very different vehicle architectures. The mapping of functions and components for the steam powertrain architecture had several decades of evolutionary development from steam engine locomotives, which then translated into a new product system in cars. The electric car featured some components similar to those of the steam car, along with new components configured differently in both form and function.

What factors triggered architectural competition and dominance in the past? During the early modern automotive time period (1885 to 1915), steam was clearly the dominant architecture, with its origins dating back to the late 1780s. The shift toward architectural dominance was triggered primarily by a series of technological breakthroughs that sought to improve one of the main weaknesses of steam power: its dependence on the availability of water. These technological breakthroughs included the internal combustion engine and the electric motor, both of which resolved the water availability issue of steam power. These two technologies were initially focused on solutions for the rail and electric power generation markets before they entered into the nascent automotive industry.

As competition grew among steam, electric, and internal combustion engine cars, price and quality advantages became more acute. Steam cars were more expensive, but they offered rapid acceleration, reaching 0 to 65 km/h (0 to 40 mi/h) in less than 10 seconds. The highest power and torque in a steam engine occur when the vehicle starts from rest. However, early steam models could not go more than 2 miles without replenishing their water tank, it and took more than 20 minutes for the boiler to build up enough pressure to drive.

The early electric car had a price comparable to that of a steam car and required an electric power source, something found mostly in major cities at the time. It was limited in range to less than 64 km (40 mi) per charge and limited in speed to 32 km/h (20 mi/h). This was because of the low energy density of early lead acid batteries, at roughly 15 to 20 Wh/kg (watt-hours per kilogram), compared to the 12,200 Wh/kg contained in gasoline fuel. However, the key advantage of the early electric car was its simplicity of operation, with no complicated shifting mechanisms, and essentially no maintenance or uncomfortable emissions.

Early performance of the ICE car was on par with electric cars but inferior to that of steam cars. Steam proponents would often refer to the "internal explosion engine" to imply that the ICE car was less safe than the proven steam car. Although an ICE car could be started in less than a couple of minutes, many motorists sustained injuries in starting the vehicle with the external hand crank—a genuine safety issue.

Dominance of ICE Architectures

Figure 12.12 shows architecture performance for automobiles between 1885 and 2008, divided into three periods: initial architectural competition, architectural dominance of the ICE, and renewed architectural competition. Figure 12.12 shows a shakeout in the market that allowed one architecture—the ICE car—to dominate all others. The principles and architecture of the internal combustion engine (ICE) design that has dominated the market since the mid 1930s have remained relatively unchanged. [18] Because the entire market adopted this dominant architecture, the basic risk of not knowing which architecture would prevail was completely eliminated. This allowed manufacturers to focus on innovation at the subsystem level, rather than the overall system architecture level.

The dominance of the internal combustion engine emerged only after two key events: first, the dramatic decrease in price achieved through assembly line production of ICE cars, and second, the development of the electric starter. These price and quality improvements

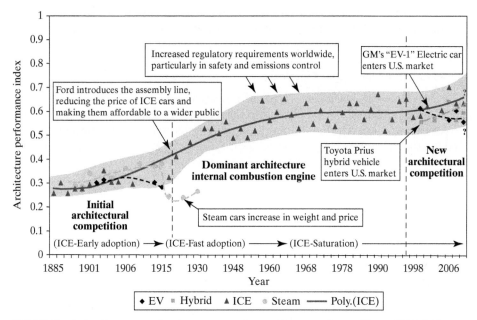

FIGURE 12.12 Performance of various automotive architectures from 1885 to 2008. [19] (Source: Gorbea, C.; Fricke, E.; Lindemann, U., "The Design of Future Cars in a New Age of Architectural Competition," ASME 2008 International Design Engineering Technical Conferences and Computers and Information in Engineering Conference IDETC/CIE. Brooklyn, New York, 3-6 August 2008. Reprinted with permission).

to the ICE car made it affordable to the masses and a better solution than all other architectures in the market by 1920.

Steam cars built from 1920 to 1930 remained a high-end market product. Significant improvements were introduced to compete with ICE cars, such as the integration of a condenser to recycle water, and a spark ignition starter that could provide the car with enough starting pressure within one minute. However, these systems added tremendous weight and increased the overall price of the already low production volumes. The steam car that had dominated since the 1780s eventually disappeared completely from the market by 1930.

During the architectural dominance of the ICE, incremental innovation flourished. Generations of cars exhibited improvements to major subsystems only, while sticking with the same basic architecture. During this time of architectural dominance, automakers paid little attention to alternative powertrains and focused primarily on generating core competencies that supported the optimization of the dominant architecture.

Future Architectural Competition

In Figure 12.12, the current time period (1998 to 2008) shows a renewed focus on vehicle architecture. The key historical events that marked the beginning of this new age were the reintroduction of electric vehicles into the market and the first mass-produced hybrid electric cars. Presently, new architectures are appearing in the automotive market, most notably a revival of electric and hybrid electric vehicles. Will the future of the ICE car mirror that of the steam car? At the moment, some auto manufacturers are trying to shift their focus from incremental innovation to architectural innovation. The shift has not come easily, because most organizations have been structured around the major subsystems within the automobile. Most auto manufacturers have invested in developing their core competencies in areas specific to the design of internal combustion engine cars.

The shift toward architectural competition is significant because it can place established firms in jeopardy if they are not able to adapt to the new competitive landscape that is developing. This was the case for most steam car manufacturers during the 1920s that failed to adapt to market changes. The electric car was a loser in 1910. Could it become a winner again in a new period of architectural competition?

References

[1] R.W. Weisberg, *Creativity: Understanding Innovation in Problem Solving, Science, Invention, and the Arts* (New York: John Wiley & Sons, 2006).

[2] A.F. Osborn, "Applied Imagination," 1953. http://psycnet.apa.org/psycinfo/1954-05646-000

[3] E. De Bono, *Six Thinking Hats* (Penguin, 1989).

[4] Agência AutoInforme: Siena Tetrafuel vai custar R$ 41,9 mil http://www.webmotors.com.br/wmpublicador/Noticias_Conteudo.vxlpub?hnid=36391 Accessed on 3 June 2009.

[5] M. Phenix, "BMW's Hybrid Vision: Gasoline and Steam: This Novel Concept Uses Your Car's Wasted Heat to Enhance Power and Fuel Economy," *Popular Science* 268, no. 3 (2006): 22.

[6] R. Harmon, "Alternative Vehicle-Propulsion Systems (Electric Systems, Gas Turbines and Fuel Cells)," *Mechanical Engineering-CIME* 114, no. 3 (1992): 58.

[7] J. King, *The King Review of Low Carbon Cars: Part I–The Potential for CO_2 Reduction* (London: UK HM Treasury, 2007). ISBN: 978-1-84532-335-6

[8] R. Lache, D. Galves, and P. Nolan (Eds.), *Electric Cars: Plugged In, Batteries Must Be Included* (Deutsche Bank Global Market Research, 2008).

[9] A.S. Brown, "Fuel Cells Down the Road?" *Mechanical Engineering-CIME* 129, no. 10 (2007): 36.

[10] D. Naunin, *Hybrid, Batterie- und Brennstoffzellenelektrofahrzeuge: Technik, Struckturen und Entwicklung* (Renningen: Expert Verlag, 1989). ISBN: 3816924336, 9783816924333

[11] M. Ehsani, A. Emadi, and Y. Gao, *Modern Electric, Hybrid Electric, and Fuel Cell Vehicles: Fundamentals, Theory, and Design* (CRC Press, 2009). ISBN: 1420053981.

[12] D.B. Sandalow, *Plug-In Electric Vehicles: What Role for Washington?* Washington, DC: Brookings Institution Press, 2009). ISBN: 0815703058, 9780815703051

[13] C. Gorbea, "Vehicle Architecture and Lifecycle Cost Analysis in a New Age of Architectural Competition." Dissertation, TU Munich, 2011.

[14] M. Bazerman and D.A. Moore, "Judgment in Managerial Decision Making," 2012. https://research.hks.harvard.edu/publications/citation.aspx?PubId=9028&type=FN&PersonId=268

[15] G. Pahl and W. Beitz, *Engineering Design: A Systematic Approach* (New York: Springer, 1995).

[16] R.M. Henderson and K.B. Clark, "Architectural Innovation: The Reconfiguration of Existing Product Technologies and the Failure of Established Firms," *Administrative Science Quarterly* 35, no. 1 (1990): 9–30.

[17] G. Farber, *American Automobiles* http://www.american-automobiles.com/Electric-Cars/Baker-Electric.html Accessed on 18 December 2009.

[18] J.M. Utterback, *Mastering the Dynamics of Innovation* (Cambridge, MA: Harvard Business School Press, 1996). ISBN: 0875847404, 9780875847405

[19] C. Gorbea, E. Fricke, and U. Lindemann, "The Design of Future Cars in a New Age of Architectural Competition," ASME 2008 International Design Engineering Technical Conferences and Computers and Information in Engineering Conference IDETC/CIE. Brooklyn, New York, 3–6 August 2008.

Chapter 13
Decomposition as a Tool for Managing Complexity

13.1 Introduction

Once the concept has been selected and the architect begins to define the full function-to-form mapping, the scope and information content of the development project can grow exponentially. Resources for each subsystem are added to the project during the design phase, marketing can begin shaping the message for the product, manufacturing begins to plan in earnest, and so on. In Section 3.2 we introduced complexity as an measure of a system, but now we focus on how to manage complexity.

We will argue that the most important task in this phase is the architect's management of complexity. Development programs that appear complicated are very difficult to conduct systems analysis on, whether that systems analysis is thermal modeling of a spacecraft (a classic systems analysis), evaluating the handling of a sports car, or optimizing the computation time for new software. It is the architect's responsibility to produce system descriptions that the relevant departments, functions, and stakeholders are able to comprehend and contribute to.

A central theme of this chapter is that complexity exists and is inherently neither good nor bad. We must consider it an investment to be made to reach system goals.

In Section 13.2, we introduce some of the attributes of complexity, to help the architect understand where complexity investments are merited. In the next section, we focus on decomposition as a key lever available to the architect in managing complexity. The chapter concludes with a case study on complexity and decomposition in the Saturn V and Space Station *Freedom* projects.

13.2 Understanding Complexity

Complexity

We defined complexity in Chapter 3 as the property of having *many interrelated, interconnected, or interwoven elements and interfaces.* Most of the systems that we deal with professionally are complex, hence the growth of system architecture in practice. Complexity is driven into systems by asking more of them: more function, more performance, more robustness, more flexibility. It is also driven into systems by asking them to work together and interconnect—monitoring power demand at a manufacturing site relative to the grid's capacity,* connecting customer order systems directly to manufacturing sequencing, instant communications among co-workers, and the like.

---

*As an example of grid solutions, EnerNOC enters into agreements with firms to reduce energy consumption on demand (in order to stay within grid capacity) in return for payment.

Underlying this definition of complexity is the idea that complexity scales with the number of entities in the system (a team of 30 people is more complex than a team of 10 people), as well as with their relative connectivity (10 people working together on one task is more complex than 10 people performing identical parallel tasks).

There are many potential definitions of complexity, but none of them has become operationally useful in a broad range of contexts. We present several of these subsequently for comparison, but it is our view that the simplest possible definition is perhaps the most conceptually useful, as shown in Box 13.1.

Box 13.1   Definition: *Complexity*

Let's look at several simple measures of complexity.

Number of things: N1
Number of types of things: N2
Number of interfaces: N3
Number of types of interfaces: N4

The most often quoted measure of complexity is the number of things (N1). For example, Microsoft has one million servers in its data centers. [1] Boeing has used the number of parts in a 747 (6 million, a large minority of which are rivets) in marketing and supplier literature to emphasize the complexity of the aircraft. [2]

Although "parts number counting," or counting the number of types of things (N2), seems like a relatively limited view of complexity, organizations stock parts by part number, sequence manufacturing by line-side part numbers, and create product designs by part number. For example, Home Depot stores 40,000 SKUs (stock keeping units) per store. [3]

Analogously, N3 and N4 capture the number of interfaces and the number of types of interfaces. The electrical system of a building is more complex if it has more electrical outlets—interfaces (N3). The complexity would rise again if there were various types of outlets (N4): some U.S. standard, some UK standard, some European, some Australian, and so on. In this case we would need multiple types at each location, or we could use appliances only at certain locations in the building where the proper outlet was located.

The simplest measure of complexity that captures all of these is $C = N1 + N2 + N3 + N4$. The variations on this theme are abundant. For example, Boothroyd and Dewhurst [4] suggest that formal complexity can be calculated according to $C = (N_1 \times N_2 \times N_3)^{\frac{1}{3}}$.

Our simple measure of complexity has the advantage that it formulates complexity as an absolute and quantifiable system property (once "measure" and "atomic level" are defined). This allows us to compare architectures: If one architecture has more things, more types of things, more connections, and more types of connections than the other, it is objectively more complex.

Many organizations use the word "complexity" but, when asked to define it, produce long, all-encompassing, qualitative, non-operational definitions. For example, the authors worked with a vehicle manufacturer that defined part sharing as a driver of complexity (essentially the *number*

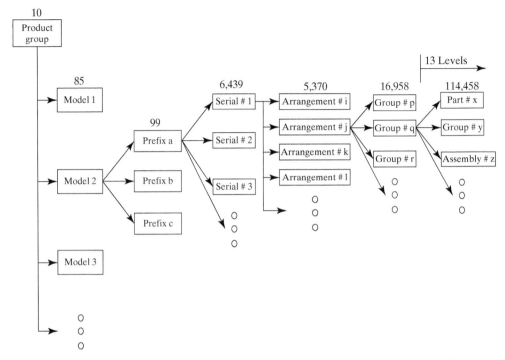

FIGURE 13.1 Product hierarchy created by a vehicle manufacturer to define a measure of complexity in terms of part sharing across products.

and types of entities). Part sharing was defined with respect to a product hierarchy with 13 levels, itself a nontrivial decomposition, shown in Figure 13.1. The firm then attempted to define the number of entities shared at each level of the hierarchy (part numbers shared across groups, across arrangements, across models, and so on). The net result of this analysis was not an incorrect definition of complexity, but it was so cumbersome (in part because the product decomposition was not universally agreed upon) that the measure went unused. In essence, the measure of complexity was too complex to be operational.

Complexity is an absolute and quantifiable system property, once a definition is agreed upon. Regardless of whether our simple definition or another definition is chosen, we believe that it is valuable to regard complexity as a measurable quantity. Complexity should not be viewed as an outcome of the design process, but rather as a quantity to be managed. Our experience suggests that when complexity is viewed as an intractable topic, it is more likely to be ignored, at the cost of increased complexity!

In addition to having a measure of complexity, in this chapter we discuss several ideas related to complexity: *apparent complexity* (also known as complicatedness), and *essential versus gratuitous complexity*.

Complex versus Complicated

We have asserted that one of the roles of the architect is to manage the evolution of complexity in the system. This is not to say that complexity is a negative property of the system. Complex systems today, such as regional transportation infrastructures and networked communication systems, deliver significantly more value than their simple counterparts, such as bicycles and letter mail. Complexity is an investment that the architect makes, where the payout is the performance of the system. This payout must be weighed against the challenges that complexity brings. Growing complexity is a fact of life.

The problem is that more complex systems are more difficult for humans to understand. In Chapter 3, we introduced the idea of *complicated* as a measure of our ability to perceive and comprehend complexity. Complicated things have high *apparent complexity (or complicatedness)*; our human processor is more easily confused, or is less able to make sense of things (Box 13.2 Principle of Apparent Complexity). For example, we might argue that the structure on the left in Figure 13.2 is less complicated than the structure on the right. We can confidently assert that the pyramids were structurally less complicated, but we leave comparing the emergent system aesthetics of the two to the reader's judgment and taste.

> Box 13.2   Principle of Apparent Complexity
>
> *"Less is more."*
>
> Ludwig Miles van der Rohe, New York Herald Tribune, June 28, 1959
>
> *"More is different."*
>
> Phillip Anderson, Science, 1972
>
> The complexity of modern systems exceeds the human ability to comprehend. Create decomposition, abstraction, and hierarchy to keep the apparent complexity within the range of human understanding.
>
> - Apparent complexity is akin to how "complicated" the system is; complicated systems are hard for humans to understand.
> - Complexity has the potential to scale combinatorially with the number of features of robust functionality and/or the number of objects. One cannot scale up the processes and techniques used in simple systems to address complicated systems because of human cognitive limitations.

a)   b)

FIGURE 13.2 Complicatedness in civil architecture. The pyramid is less complicated than the Taj Mahal, even though they perform the same function—entombing royalty. Clearly, in civil architecture there are aesthetic reasons for "complicatedness."   (Source: (a) Mrahmo/Fotolia (b) Omdim/Fotolia)

FIGURE 13.3 More complicated network. (Source: AP Photo/Bela Szandelszky)

It is important to separate complexity from complicatedness in architecture. Complicated architectures seem intractable or unmanageable. This is why our definition of complicatedness takes into consideration the ability of the human to perceive and understand complexity. Consider Figure 13.3 and Figure 13.4. The same function is delivered in both cases, and the inputs and outputs at an electrical layer are identical, but one architecture is less complicated and more amenable to analysis and modification.

FIGURE 13.4 Less complicated network. (Source: Andrew Twort/Alamy)

Complicated things have high *apparent complexity*. What causes apparent complexity? At first glance, complicatedness should scale with complexity; that is, more entities, more connections, and more types of entities and connections should all lead to more complicatedness. Most people would describe the *organization* of the cables in Figure 13.3 as the key attribute, and the system of Figure 13.4 is less complicated because it is organized. The two figures represent the same physical entities, but even if we scaled up the *organized* network and scaled down the *disorganized* network, we might well imagine that the now-larger, more capable organized network would still be less complicated than its disorganized smaller brother. In summary, complexity and complicatedness do not necessarily grow at the same rate in a system, an observation that allows us to hope we can increase complexity while limiting complicatedness.

The word "organized" is a very fuzzy term; for example, telling a junior network engineer to organize the backroom network does not lend itself to a step-by-step process. To unpack this backroom network, we'll begin by using the tools of Chapter 3, starting with decomposition and abstraction. The network can be decomposed into entities such as cables, routers, and racks. It is easier to see this decomposition in the "organized" figure (Figure 13.4), because the bundling of network cables tells the viewer that they belong to a similar group. Individual cables in the bundle can be treated with the same analysis: A cable can be treated as an instance of a class of entities that share similar functionality. If we introduce the abstraction of a "bus," represented by a bundle of cables tied together, the network appearance becomes simpler still, because we imagine that the same type of information is flowing along the bus, compared with the necessity of treating each cable separately in the complicated figure.

Is Figure 13.4 at maximum organization? It is hard to say without detailed knowledge of the function of the network. Could we develop a less complicated representation? Absolutely, depending on what information it is important to convey. We could draw a network diagram, but rather than showing point-to-point connectivity for cables, we could draw a layered diagram, with one of the layers being the physical connectivity layer. This representation might provide more information about the functional interaction among components or the higher-level uses of the network, but it abstracts out the detailed mapping of cables. In a sense, it "presumes" that the cables will be routed correctly, which releases the viewer from the burden of figuring it out. Whether this is a good or a bad attribute depends entirely on the use of the diagram or picture.

Consider low-level programming languages such as machine-level languages and assembly. They are hard to work with, even for experienced programmers. To reduce their difficulty, high-level languages like JAVA and MATLAB were developed. Systems become substantially more complex with the introduction of complex new technology such as compilers, interpreters, middleware, and graphical interfaces. These systems are more intuitive and less complicated to use. They present a system image that is vastly less complicated. Imagine inverting a matrix using assembly code; MATLAB does it in one instruction. For some users this investment in complexity is merited, but for others it is unmerited in light of performance compromises or scalability challenges.

Essential Complexity

Complexity is an absolute property of the system as we have defined it. Furthermore, the architect must invest in complexity in order to deliver more performance and functionality. This seems to beg the question: Is there a minimum complexity required to deliver a certain function and performance? Yes. We call this the *essential complexity* (Box 13.3).

> **Box 13.3   Principle of Essential Complexity**
>
> *"Complexity is inherently neither good nor bad."*
>
> Joel Moses
>
> *"Make everything as simple as possible. But not simpler."*
>
> Albert Einstein
>
> *"All things being equal, choose the simpler solution."*
>
> Occam's Razor
>
> *"It is the last lesson of modern science, that the highest simplicity of structure is produced, not by few elements, but the highest complexity."*
>
> Ralph Waldo Emerson
>
> Functionality drives essential complexity. Describe the required functionality carefully, and then choose a concept that produces low complexity.
>
> - Essential complexity is the minimum complexity that must be incorporated in the system to deliver the robust functionality within a concept.
> - Gratuitous complexity is complexity added over and above the essential complexity for any given robust functionality within any concept. Gratuitous complexity should be avoided.
> - "Follow the road of simplicity to the point where it crosses the road of full functionality, [and] you will then be at the intersection that is elegance."
>
> Daniel Kern, MIT 2001

For example, consider a powered liftgate, available on many hatchbacks and SUVs. At first it might seem that the minimum-complexity system would be a motorized liftgate with (1) a switch that interrupts power to the raising motor and (2) logic to reverse the direction of the motor. However, this simple switch would not capture the functionality of the system in Figure 13.5—namely, the function that prevents a user from opening the liftgate when the vehicle is moving.

Figure 13.5 illustrates the conditions that might be checked after the switch is depressed, before the liftgate is lifted or lowered. The circles illustrate processes, and the lines represent the objects of the system, the signals. The performance of the liftgate system in this SUV is partially measured by the physical liftgate (speed, smoothness of operation, and so on), but it is also partially measured by the safety of the system (for example, breadth of checks) conducted and accuracy of the checks. Although this representation appears complicated, it is hard to argue, upon closer inspection, that this diagram contains any gratuitous functions or forms. Does it represent only the essential complexity of the system, [5] as discussed in Box 13.3 Principle of Essential Complexity? Does it appear that this system has inherited previous architectures from legacy systems, and thus bears gratuitous complexity from history?

CHAPTER 13 • DECOMPOSITION AS A TOOL FOR MANAGING COMPLEXITY 293

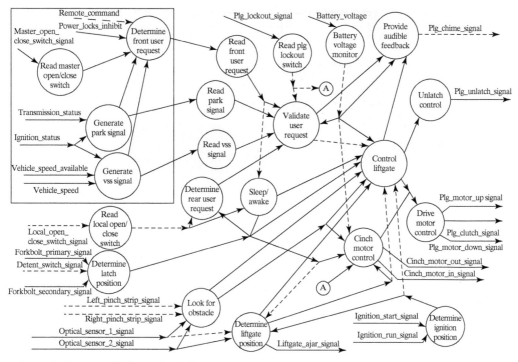

FIGURE 13.5 Powered liftgate logic diagram.

A system's level of essential or gratuitous complexity is independent of its apparent complexity. The *Concorde*'s flight deck, shown in Figure 13.6, is visually overwhelming and requires extensive training to operate. This architecture has high apparent complexity—the complexity of the system is apparent. However, given the success of the *Concorde*, it could be argued that the complexity was essential to the delivery of performance with the technology of the era.

FIGURE 13.6 Cockpit of the *Concorde*. Is the complexity essential for delivering the *Concorde*'s functionality?. (Source: Holmes Garden Photos/Alamy)

FIGURE 13.7 Smartphone with touchscreen, showing the low apparent complexity of the user interface masking the internal complexity of the underlying hardware. (Source: Yomka/Shutterstock)

By contrast, the Smartphone shown in Figure 13.7 has very low apparent complexity, a hallmark of good user interface design for consumer products. It requires little or no training to operate. There is substantial complexity underlying this interface, but that complexity is hidden away. Without further detailed analysis, it is difficult to assess whether the internal systems are at a level of essential complexity or have gone beyond that to include gratuitous complexity. However, no one would argue that the *Concorde*'s user interface should be condensed into a hierarchic menu, four buttons, and a scrolling function.

We've argued loosely that decomposition and abstraction can be used to reduce the apparent complexity of a system, but we've also argued that apparent complexity and actual complexity are on independent axes. In fact, apparent complexity and actual complexity can be related. Consider the system shown in Figure 13.8. On the left we have five entities and six connections. If we break the system into two groups of three and two entities, respectively, we can reduce the apparent complexity of the system, as shown on the right-hand side, where we illustrate each group as an entity, and their connectivity with one connection.

However, if we take a look at the middle of Figure 13.8, we can see that something fundamental has occurred. Our decomposition introduces an interface between the two groups: The system now has six entities and nine connections!

In our effort to reduce the apparent complexity of the system, we have increased the complexity of the system above the essential complexity in the original representation. This result has far-reaching implications. It implies that apparent complexity can be reduced, but sometimes at the cost of increasing actual complexity. The architect must decide whether to invest in actual

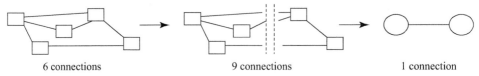

6 connections   9 connections   1 connection

FIGURE 13.8 Decomposition can increase the complexity of the system by creating additional interfaces. The investment in complexity must be weighed against the benefits of decomposition and potential modularity.

complexity (in order to bring down the apparent complexity), or not to invest, as discussed in Box 13.4 Principle of the 2nd Law.

> **Box 13.4   Principle of the 2nd Law**
>
> *"There are two ways of constructing a software design: one way is to make it so simple that there are obviously no deficiencies, and the other way is to make it so complicated that there are no obvious deficiencies."*
>
> C.A.R. Hoare
>
> *"Software is like entropy. It is difficult to grasp, weighs nothing, and obeys the second law of thermodynamics; i.e., it always increases."*
>
> Norman Augustine
>
> The actual complexity of the system always exceeds the essential complexity. Try to keep the actual complexity close to the essential complexity, and the apparent complexity within the range of human understanding.
>
> - The actual complexity is the real measured complexity of the system. It exceeds the essential complexity because of the extra complexity needed to create abstraction, decomposition, and hierarchy.
> - Actual complexity can also increase because of the creep of gratuitous complexity.

The history of interface design includes many examples where attempts to reduce the apparent complexity by simplifying interfaces failed. Witness the first generation of BMW's iDrive. The design intent was to reduce the apparent complexity of switches on the dashboard, consolidating the controls under a single button that could be twisted, angled, and depressed. The system was designed to operate on a menu screen. Users found that the menu was much deeper than desired, [6] forcing them to descend five levels to reach the heater controls! The press quipped, "iDrive? No, you drive, while I fiddle with this controller." [7] The decrease in apparent complexity of the switches caused an unacceptably complicated menu interface—a net increase in apparent complexity.

13.3 Managing Complexity

A number of tools for managing complexity have been presented in this text. Creating abstractions and defining entities and relationships were introduced in Chapter 2. Chapter 3 introduced decomposition, hierarchy, and certain logical relationships such as class/instance, type/specialization, and recursion. Combined, these tools allow the architect to represent different aspects of the architecture, while removing the irrelevant details to produce a minimum-complexity representation.

In the creation of architecture, we have to ask how these tools can be used to manage the evolution of complexity in the system. The architect is responsible for choosing a decomposition of the system, for determining which abstractions are relevant and useful, and for determining how the choice of hierarchy will impact the system design.

At this stage, the complexity of the system is expanding, and domain-specific vocabularies, methods, and tools are resurgent. It is not our intent to define a process so explicit that we can "turn the crank," or follow blindly. Rather, our intent is to illustrate the relevant principles. We will organize the discussion around what is perhaps the most powerful tool—and the most important decision—decomposition.

Choosing a Decomposition

Modern engineering is entwined with the use of the Bill of Materials, a decomposition of the design down to the purchased components (or sometimes below, for purchased sub-assemblies). Although this decomposition is very concrete, it does not necessarily lend itself to functional decomposition of the concept. The intent of a functional decomposition is to determine which subsystems or entities will accomplish which functions, to determine which functions are necessarily emergent and cannot be decomposed, and (as we will see) to carry the intent of the architecture down through the design. These objectives are not well served by a purely formal decomposition, which can fool the user into thinking that the management of the pieces will produce the desired emergence.

Decomposition is an explicit choice in system architecture [8] (Box 13.5 Principle of Decomposition), one of the highest-leverage decisions. Good decomposition will facilitate interface testing and highlight important subsystem couplings. The wrong decomposition will leave some subsystems (or functions) with enormous tasks and others with minor tasks. Worse still, the wrong decomposition will leave out relevant functions, systems, or operating conditions. For example, consider decomposing the operations of a system by nominal get ready, get set, go, get unset, get unready. This decomposition will focus attention on the typical operating case, while omitting fault conditions, contingencies, commissioning, and the like.

> ### Box 13.5 Principle of Decomposition
>
> *"Divide et impera"* [Divide and conquer]
>
> Roman saying
>
> Decomposition is an active choice made by the architect. The decomposition affects how performance is measured, how the organization should be set up, the potential for supplier value capture, and how the product can evolve, among many other things. Choose the plane of decomposition to align as many of these factors as possible, in order to minimize the apparent complexity of the system.
>
> - Decomposition involves choices not only of how many and where to make internal boundaries, but also (and most important) in what plane to consider internal boundaries.
> - The choice of the plane of decomposition (form, function, operation, supplier, and so on) in which to consider internal coupling is a key decision.

Many people assume that the core problem in large, complex systems is integration of the component parts. In fact, integration is often made unnecessarily difficult because the problem was poorly partitioned in the architecting phase.

Thus decomposition is important, but how do we judge the "goodness" of a decomposition? The fact is that the decomposition at Level 1 cannot be evaluated until we descend to Level 2. Consider decomposing a class of 30 students (Level 0) into two groups (Level 1): Student 1 and Students 2–30. We won't know whether this is useful until we create the decomposition at Level 2, such as decomposition by student height or by visual impairment. For student height, our decomposition is not useful; Student 1 may skew the distribution of heights. However, if Student 1 is visually impaired, it is a useful decomposition if we're looking to assess needs for legibility of presentation material. (This example is merely an analogy, and we've taken creative license by conflating the notion of decomposition with the statistical notion of sampling.)

We call this principle *2 Down, 1 Up* (Box 13.6 Principle of "2 Down, 1 Up"). It embodies the idea that in a hierarchic system, the real information about how to cluster or group the system one level down from the reference is present in the structure and interaction *two* levels down. As suggested by Figure 13.9, if we start at a reference Level 0, the wrong thing to do is to immediately try to decide how to cluster elements at Level 1. Instead, we should propose a trial decomposition at Level 1, work down to Level 2, identify the structure and interactions there, and place boundaries to cluster the Level 1 elements based on this insight.

Chapter 8 contains a worked example demonstrating this principle, the air transportation service. Starting from the Level 0 concept (flying a traveler with an airplane), we develop the

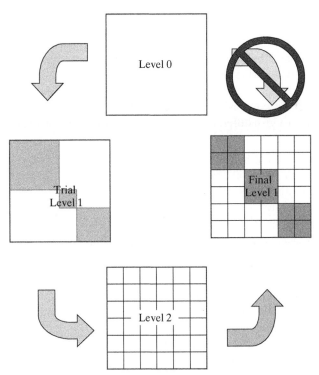

FIGURE 13.9 Illustration of "2 Down, 1 Up" in decomposition. The best way to test the goodness of a first-level decomposition is to evaluate it at the second-level down.

> **Box 13.6   Principle of "2 Down, 1 Up"**
>
> *"Whoever knows he is deep, strives for clarity; whoever would like to appear deep to the crowd, strives for obscurity. For the crowd considers anything deep if only it cannot see to the bottom: the crowd is so timid and afraid of going into the water."*
>
> Friedrich Nietzsche
>
> The goodness of a decomposition at Level 1 cannot be evaluated until the next level down has been populated and the relationships identified (Level 2). Select the modularization at Level 1 that best reflects the clustering of relationships among the elements at Level 2.
>
> - The modularization should consider maximizing the internal couplings within a cluster and minimizing the couplings among clusters. This is often, but not always, advantageous.
> - The real information about clustering at Level 1 is contained in the interrelationships at Level 2.
> - Therefore, if decomposing from Level 0 to Level 1, go all the way to Level 2 (2 Down), identify the appropriate clustering, and use this for the final Level 1 decomposition (1 Up).

Level 1 trial architecture (Figure 8.2). The plane of this organization is *sequence*—what happens first, second, and so on. These Level 1 processes are then decomposed to Level 2 and shown in Tables 8.2 and 8.3. Clustering at Level 2 is examined, and the alternative modularization of Level 1 (Table 8.4) is presented, which emphasizes functional interactions. Neither the original sequence-based Level 1 decomposition nor the final function-based decomposition is inherently right or wrong. They just give different views and a sense of the alignment of clustering on these two different planes.

Experience suggests that humans seem comfortable managing 7 +/− two elements at each level of decomposition (Chapter 2). More than nine elements inhibit the ability of the designer to design, and make cross-element analysis combinatorially more difficult, shifting workload up to the architect. Fewer than five elements at each level create deep decompositions, because more levels are needed to represent all components. The deeper the decomposition, the more difficult it will be to think vertically.

Modularity and Decomposition

The choice of decomposition is linked to the modularity of the resulting system. Modularity is often expressed as a desirable property of a system, in the sense that more modularity is better, but rarely is modularity clearly defined.

The idea of *modularity* refers to the interfaces of the system. A system is modular if its interface(s) allow old modules to be removed, and new modules to be inserted at the interface. It is most commonly brought up in the context of having an open number of variants, rather than a closed number of variants. Lego's interface enables an *open* number of variants (the number of

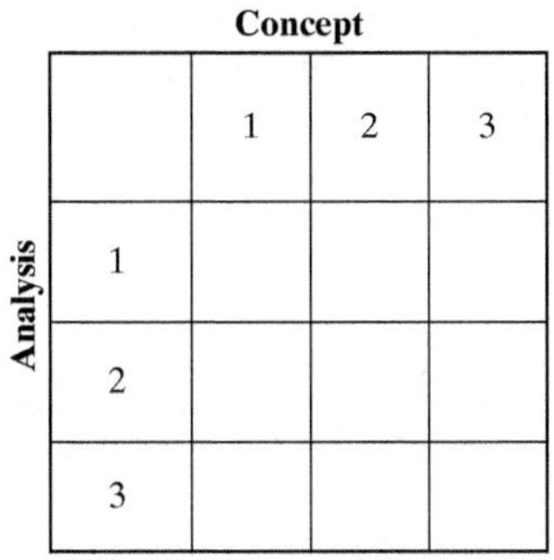

	Concept		
	1	2	3
Analysis 1			
2			
3			

FIGURE 13.10 Potential decomposition planes for a technical report.

shapes that can be built by combining many Lego blocks); although it is not infinite, the number of variants is large and unforecasted.

The related ideas of common components and platforms typically refer to a *closed* set of variants. For example, we might plan an alternator mounting bracket to accept the three alternator variants currently used in a product line. Modularity and commonality are not fully distinct ideas. A system with a lot of common components can be modular, but whereas commonality focuses on parts or code sharing for cost-saving reasons, modularity often focuses on operational and design flexibility.

In Part 2, we discussed the use of Design Structure Matrices (DSMs) to represent the internal connections of a system, and we explored their use in the modularization of a system (Chapter 8). DSMs are powerful tools. They allow the architect to analyze the system subject to a clear objective (minimizing the interconnections between subsystems). Here, we will expand on this analysis to ask an important question: Is minimizing interconnections the right principle for decomposing the system?

The two key ideas in decomposition and modularization are the *number of elements, chunks, or functions* (the easy part) and the *plane of decomposition* (the hard part).

Consider a technical report as an example of architectural decompositions, in which we report on three technical concepts, and three types of analysis were conducted on each concept. How should the report be organized? The chapters could be organized by concept, with all three analyses presented in each (see Figure 13.10). This can help the reader perform a broad evaluation across the three analyses, and it might even help the reader make judgments about aspects of the concept not covered by the analysis. In contrast, the chapters could be organized by analysis, where all three concepts are laid out for direct comparison against the metric of the analysis. This decomposition plane affords the reader an opportunity to understand the breadth of concepts for which the analysis might be relevant, and it may promote creativity about stretching the analysis to other concepts. Of course, hybrids are also possible; we might provide an overview of the analysis and then a further three chapters on the three concepts themselves.

The purpose of this example is to point out that the plane of decomposition, not the number of chunks, is the key decision. The choice of decomposition plane impacts the reader's or user's understanding. It creates a lens that enables creativity and emergence along one plane but represses them in other planes. The choice of the plan of decomposition is a strong lever on the elegance of the resulting architecture (Box 13.7 Principle of Elegance).

> **Box 13.7   Principle of Elegance**
>
> *"Pulchra sunt quae visa pacent."* [All that pleases the eye is beautiful.]
>
> Thomas Aquinas
>
> *"When I am working on a problem, I never think about beauty but when I have finished, if the solution is not beautiful, I know it is wrong."*
>
> R. Buckminster Fuller
>
> Elegance is appreciated internally by the architect when a system has a concept with low essential complexity and a decomposition that aligns many of the planes of decomposition simultaneously. Elegance is appreciated externally by a user when the system embodies a sense of aesthetics, quality, and low apparent complexity. Architect systems with these attributes that produce elegance.

Having spent a significant portion of this text on the distinction between form and function, it might be tempting to argue that the form of the system represents Option 1 for decomposition, and the function of the system represents Option 2 for decomposition. We will show that there are many more possible decompositions.

Let's first review decomposition by form and by function. Form decomposition allocates one or several elements of form to each Level 1 entity, with the end result that form is clearly divided (see Figure 13.11 left). For example, we could divide our Hybrid Car by the body, chassis, drivetrain, and interior. This form decomposition has the advantage of being more concrete. It also makes certain properties (such as mass) sum linearly, without emergence. Then we can determine the mass of the system by adding the masses of all the elements.

Function decomposition implies that we list the system functions at Level 1 as the entities (see Figure 13.11, right). For a Hybrid Car, we might list these functions as support and comfort passengers, store energy, produce power, isolate passengers from external vibrations, and so on. Function decomposition highlights how the decomposition decision can influence the eventual system performance: It will highlight certain types of emergence. For example, if Level 1 for our Hybrid Car includes the function "isolating passengers from external noise," that would force us to evaluate potential sources of vibration and potential causes of attenuation at a system level. The driver does care whether vibration comes from the powertrain, the road, the wheels, or the exterior trim. This is a classic systems problem, and the decomposition either enables or disables the architect's "emergence lens."

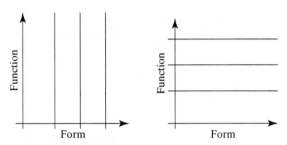

FIGURE 13.11 Form decomposition concept (left) and function decomposition concept (right).

There is no reason to believe that form and function are the only two decomposition planes possible for the system. The firm that builds our Hybrid Car may well operate from a stable form decomposition embodied in the Bill of Materials and a stable function decomposition embodied by the engineering organization (vibration, structures, and so on). But what happens if the competitive advantage of the firm relies on its ability to integrate the components of two dissimilar suppliers? This would suggest that the enterprise should explicitly define the interfaces between Supplier A and Supplier B components.

In 1968, computer scientist and programmer Melvin Conway postulated that "organizations which design systems are constrained to produce designs which are copies of the communication structures of these organizations." [9] The modularity of the system is inextricably intertwined with the organization that develops it. Conway's Law, as it is known, formulates the problem inversely from the way we have framed it (the product mirroring the organization versus the organization mirroring the product), but the best formulation is that the two are coupled. Conway's Law has been cited as a storyline for the Mars Climate Orbiter failure, where two teams used different units (one used metric, the other, imperial), resulting in the orbiter initiating maneuvers at the wrong altitude above the Martian surface. More recently, researchers have tested Conway's Law, finding strong evidence in support of a causal link between organization and product in software product modularity. [10]

It is easy to argue that modularity is desirable—that some modularity is good, and more modularity is better. However, modularity carries a cost. Figure 13.8 illustrated how additional interfaces increase the complexity of a system. Even where interfaces already exist, building modularity into an interface implies meeting a range of performance goals (different structural loads, additional pins for connectors that are sometimes used, and the like). Computer scientist Alan Perlis, known for pioneering work in programming languages, has said, "Wherever there is modularity there is the potential for misunderstanding: Hiding information implies a need to check communication." [11]

In Box 13.8 Potential Planes for Decomposition of Systems, we represent 12 potential decomposition planes, by no means an exhaustive list, along with some guidance on each. The eventual sustained competitive advantage of the system is greatly influenced by the decomposition and its resulting modularity: Decompose at each step with a principle or factor in mind, and with a holistic view of the system lifecycle. Alignment of the planes of decomposition can produce more elegant decompositions, where possible.

Box 13.8 Methods: Potential Planes for Decomposition of Systems

Delivered function and emergence	Be careful with decompositions that spread important externally delivered functions over many elements, because they require significant functional interface management and may complicate the emergence of function.
Form and structure	Cluster elements of form that have high connectivity or important spatial relationships. Don't place interfaces in areas of high connectivity, because that will cause actual complexity to rise.
Design latitude and change propagation	Components whose designs are tightly coupled should be grouped, in order to allow more design latitude within the module, minimize the constraints on the designer, and minimize propagation of design changes across module interfaces.
Changeability and evolution	Place interfaces so that modules can be combined to form platforms and support product evolution. The internal interfaces will become the stable features of the architecture.
Integration transparency	Create interfaces that allow easy testing and provide integration visibility and understanding of what occurs at the interface.
Suppliers	Create interfaces at points that will allow suppliers to work independently and creatively, driving new technology into the design. Suppliers can define modularity or defy modularity!
Openness	If module interfaces will be open to third parties, place interfaces in such as way as to balance potential advantages from innovation and network effects, against the potential drawbacks of information sharing.
Legacy components	Design that involves reuse of legacy components will often encounter constraints on decomposition and challenge interface design.
Clockspeed of technology change	Create interfaces that will allow technologies that evolve at different rates to be changed out asynchronously.
Marketing and sales	Enable differentiating features, as well as cosmetic or minor product refreshes, to be added without architectural change. Create modules that can be flexibly combined later in manufacturing to allow product customization.
Operations and interoperability	Decompose to allow operator touch points and wear parts to be easily delineated and accessed for training, maintenance, and repair.
Timing of investment	Modularize to phase the development spending. Allow the first investments to deliver some value, and subsequent ones to add additional value.
Organization	Match the modularization of the system and the organization (Conway's Law).

13.4 Summary

Managing complexity is an important task of the architect. The apparent complexity of the system influences how well the system can be designed, tested, operated, and reused. We distinguish between the apparent complexity of the system—how complicated it is—and the essential complexity of the system. The essential complexity is driven by the functionality of the system; the more we ask of the system, the more complexity we drive into it.

The architect must manage the actual complexity of the system by creating representations of the architecture that are understandable and useful to many different stakeholders. The architect can invest in abstractions, hierarchy, decomposition, and recursion, all of which can reduce the apparent complexity at the cost of potentially increasing the actual complexity.

The decomposition of the system is one of the architect's most powerful decisions. A poor decomposition can break important lines of communication between sub-teams or can hurt the future modularity of the system. The goodness of a decomposition can be judged by the Principle of 2 Down, 1 Up: The decomposition must be tested by decomposing a second level down and then examining whether the aggregation at the first level suits the system. Many different planes of decomposition are possible. The architect must judge which emphasis is important, choose a plane of decomposition, and then perform the 2 Down, 1 Up procedure.

Box 13.9 *Case Study: Decomposition of Saturn V and Space Station* Freedom

To illustrate the importance of alignment in various planes of decomposition, we present a comparison between two complex space systems: the Saturn V launch vehicle, the Apollo launcher of the 1960s and Space Station *Freedom*, the proposed space station design of the 1980s. We will see that the Saturn V provides for good alignment in different planes of decomposition, but the Space Station presents a more complex picture. We assert that the relative success of these two major NASA programs can be at least partially attributed to this difference in decomposition.

Saturn V Launch Vehicle

The Saturn V (see Figure 13.12) was the launch vehicle developed by NASA between 1962 and 1968. It was designed to launch the Apollo Command Service Module, the Lunar Module, and its crew from the surface of Earth into Earth orbit and then toward the Moon. Its solution-neutral function was therefore to accelerate a payload by a certain change in velocity (called Delta-v in the rocket business).

The Solution-Neutral Function array in Table 13.1 shows a very simple 3 by 3 Design Structure Matrix (DSM) of the vehicle, organized by elements of form (the stages). This DSM array is a way to visualize what decomposition this solution-neutral function plane would suggest. In this case, part of the solution-neutral function emerges from each element: Delta-v1 comes from Stage 1, and so on. When it is possible to cleanly decompose the solution-neutral function in this way, it strongly suggests a strategy for decomposition. However, as we will see in the next example, it is often not possible to decompose solution-neutral function.

FIGURE 13.12 Saturn V launch vehicle, showing (from bottom to top) the first, second, and third stages and the Apollo Command Service Module. The Lunar Module is concealed just below the Command Service Module. (Source: Courtesy of the National Aeronautics and Space Administration, NASA History Office and Kennedy Space Center)

In the Supplier array and the Operation array of Table 13.1, we examine two other planes of decomposition of form—by supplier and operation. The development of the stages was assigned to three different suppliers, as indicated by the Supplier array. Ground testing was also independently carried out on the three stages in parallel. The nominal operations are to fire the first stage, drop the first stage, fire the second stage, drop the second stage, and then fire the third stage. The third stage was dropped on the way to toward the Moon. This simple sequential set of operations is shown in the fourth array of Table 13.1. Note that while any one stage is operating, no other attached stage is active.

The Solution-Neutral Function array, Supplier array, and Operation array of Table 13.1 suggest a sparse uncoupled relationship among elements of form and a strong potential alignment of planes of decomposition. How should the internal functions be allocated in order to reinforce this sparse connectivity? The principal internal functions of the Saturn V were propelling (creating thrust), guiding the vehicle (creating guidance torques), carrying internal structural loads, and reacting external aerodynamic loads. The key architectural decision was to replicate each of these internal functions in every stage. For example, Stage 1 created thrust, created guidance torques, and carried internal and aerodynamic loads. Likewise for Stage 2 etc. The only important internal interface, shown by the L in the off-diagonal entry of the Internal Function array of Table 13.1, was the structural load interface, which transmitted force to accelerate the next stage up in the stack.

TABLE 13.1 | Decomposition of Saturn V rocket

Solution-Neutral Function

	Stage 1	Stage 2	Stage 3
Stage 1	Delta $\nu 1$		
Stage 2		Delta $\nu 2$	
Stage 3			Delta $\nu 3$

Internal Function

	Stage 1	Stage 2	Stage 3
Stage 1	TGLA		
Stage 2	L	TGLA	
Stage 3		L	TGLA

T - creating thrust
G - creating guidance torques
L - carrying structural load
A - reacting aerodynamic loads

Supplier

	Stage 1	Stage 2	Stage 3
Stage 1	Boeing		
Stage 2		North American	
Stage 3			McDonnell

Operation

	Stage 1	Stage 2	Stage 3
Stage 1	Fire 1		
Stage 2		Fire 2	
Stage 3			Fire 3

The replication of function across stages created some sub-optimality: for example, a single guidance system could have been used for the whole vehicle, but instead one was created for each stage. However, this replication carried enormous advantages in terms of alignment of planes of decomposition, especially from the perspective of suppliers. Each supplier could integrate and test its stage relatively independently, given the very simple structural interfaces. The three stages were assembled in the Vehicle Assembly Building at the launch complex at Kennedy Space Center. When the stages came together, the inter-stage interface consisted only of a bolt pattern and some control and sensor wires.

The result was that the Saturn V was the largest launch vehicle ever created and had a perfect launch record. There are certainly many reasons for this success, including the resources associated with the Moon race and the exceptional experience base of the space engineers of the 1960s. But a major contributor to this extraordinary performance was alignment of these planes of modularization, resulting in simple interfaces, clear mapping of form to function, decentralized testing at each contractor site, and simple integration.

Space Station Freedom

Space Station *Freedom* (SSF) was a concept for a semi-permanent Earth-orbiting station, announced by President Reagan in 1984. The SSF (see the NASA illustration shown in Figure 13.13) was never built. For over a decade it underwent a number of design changes, and eventually it evolved into the International Space Station in 1993.

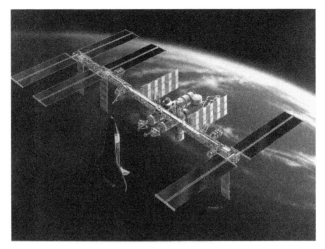

FIGURE 13.13 Space Station *Freedom*, showing (from left to center) the solar arrays, connecting truss, and habitation and laboratory modules. (Source: NASA)

Although SSF had many goals, it can be inferred that the solution-neutral function was to provide a laboratory for science, particularly micro-gravity science and Earth observation. Originally, there was also a plan to provide a base for space operations, including satellite servicing and assembly of missions beyond Earth orbit. Cementing international relationships with allies was another strong cold-war political function. The SSF incorporated elements from Europe, Canada, and Japan.

The form could be assigned to three types: the solar array, the truss that connected other main elements and housed many systems, and the modules that included habitation modules, laboratory modules, and connecting nodes. These three types of elements are visible in Figure 13.13.

The Solution-Neutral Function array of Table 13.2 shows that this is not a system in which the solution-neutral function can be allocated to elements of form: rather, each function emerges from all of the form. The solution-neutral function plane will therefore not give us much guidance in modularization.

The modularization of Space Station *Freedom* was incredibly complicated. Each piece of form was part of an element (e.g., truss, laboratory), part of a system (e.g,. thermal control, power), and part of a launch package (e.g., flight 1, flight 2).

In an attempt to capture this complexity, the Internal Function array of Table 13.2 shows the mapping of five important internal functions to the types of form. Power is generated by the arrays, but the power system was responsible for power control and distribution, so it extended to all elements. Likewise, all elements carried and exchanged structural load. Experiments were conducted externally on the truss, and internally in the modules. Only attitude control and crew support were largely localized to the truss and modules, respectively.

Four separate NASA center/supplier teams were responsible for both elements and systems. Thus team responsibility was spread across much of the station. For example,

TABLE 13.2 | Decomposition of Space Station *Freedom* as of 1987

Solution-Neutral Function

	Solar Array	Truss	Modules
Solar Array	ISM	ISM	ISM
Truss	ISM	ISM	ISM
Modules	ISM	ISM	ISM

I - international relationships
S - space operations
M - micro-gravity science

Internal Function

	Solar Array	Truss	Modules
Solar Array	PL	PL	P
Truss	PL	PLEA	PLE
Modules	P	PLE	PLEC

P - provide power
L - carry stuctural load
E - conduct experiments
A - attitude control
C - house crew

Supplier

	Solar Array	Truss	Modules
Solar Array	R	R	R
Truss	R	RM	RM
Modules	R	RM	RMB

R - Rocketdyne
M - McDonnell Douglas
B - Boeing

Operation

	Solar Array	Truss	Modules
Solar Array	1	1	
Truss	1	123	23
Modules		23	23

MB - 1
MB - 2   1987 manifest
MB - 3

Rocketdyne was the builder of the solar arrays and also of the distributed power system. This mixing of elements (form) and systems (function) caused many complications. For example, the Johnson Space Center/McDonnell Douglas team had control over "human factors" aspects of the man-rated systems in the Marshall Space Flight Center/Boeing habitat module. The development flow required parts to be shipped from the system builder to a NASA center (for system test), on to other NASA centers (for element tests), and to the launch site (for pre-launch tests). Despite the long development time of the program and this complex hardware flow, many mating components were not planned to meet until on-orbit. These three organizing dimensions for the Space Station *Freedom* (systems/function, elements/form, and operations/launch packages) meant that each component was influenced by groups with competing priorities and organizational responsibilities. The Supplier array of Table 13.2 represents an effort to capture this complexity.

Finally, the operational axis represented in the Operation array of Table 13.2 defied each of the other possible alignments. Elements were manifested for launch on certain Shuttle flights to maximize the payload utilization. A great deal of on-orbit assembly was planned. The Operation array of Table 13.2 shows the manifest of the first three assembly flights based on a 1987 manifest, which should give some sense of the non-alignment of assembly operations with other considerations.

The arrays of Table 13.2 show somewhat of a crisis in decomposition. Each of the factors separately (function, supplier, and operation) shows a highly coupled system, which is difficult to modularize. Combining the arrays, there is no obvious alignment of the planes of decomposition.

The result was that in ten years of Space Station *Freedom* development, roughly $10 billion was spent with only modest progress. There are several potential explanations for this failure, including inadequate funding and an overly complex management system, but the poor architectural decomposition was a major contributor.

In 1993 the program was dramatically restructured. A single prime contractor (Boeing) was overseen by a single NASA center (Johnson Space Center), with the rest of the suppliers subcontracting through Boeing. Critically, the redesign aligned launch packages with elements, eliminating one of three dimensions of heterogeneity. Functions were more closely aligned with elements, and the schedule was modified to enable more ground integration testing.

From these cases, we see that elegant architectures (Box 13.7 Principle of Elegance) have decompositional alignment that allows modularization of all of the factors in the same way, providing simple interfaces. Good architectures like that of the Saturn V accommodate some irregularities. Poor architectures have no alignment, and they require difficult choices of what modularization to use and what to leave to a more complex set of interfaces.

References

[1] http://www.microsoft.com/en-us/news/speeches/2013/07-08wpcballmer.aspx

[2] http://www.boeing.com/787-media-resource/docs/Managing-supplier-quality.pdf

[3] https://corporate.homedepot.com/OurCompany/StoreProdServices/Pages/default.aspx

[4] G. Boothroyd and P. Dewhurst, *Product Design for Assembly* (Wakefield, RI: Boothroyd and Dewhurst, 1987).

[5] F.P. Brooks, *The Mythical Man-Month* (Reading, MA: Addison-Wesley, 1975. Enlarged and republished in 1995). http://www.inf.ed.ac.uk/teaching/courses/rtse/Lectures/mmmWikipedia090302.pdf

[6] J.G. Cobb, "Menus Behaving Badly," *New York Times,* May 12, 2002. Retrieved January 18, 2008. http://www.nytimes.com/2002/05/12/automobiles/menus-behaving-badly.html

[7] A. Bornhop. "iDrive? No You Drive, While I Fiddle with This Controller," *Road and Track,* June 1 2002. http://www.roadandtrack.com/car-reviews/page-2—2002-bmw-745i

[8] D.L. Parnas, "On the Criteria to Be Used in Decomposing Systems into Modules," *Communications of the ACM* 15, no. 12 (1972): 1053–1058.

[9] Conway, Melvin E. (April, 1968), "How Do Committees Invent?" *Datamation* 14, no. 5: 28–31.

[10] A. MacCormack, C. Baldwin, and J. Rusnak, "Exploring the Duality between Product and Organizational Architectures: A Test of the 'Mirroring' Hypothesis," *Research Policy* (Elsevier, 2012).

[11] Perlis, A. J. (September 1982). "Epigrams on Programming." ACM SIGPLAN Notices (New York, NY, USA: Association for Computing Machinery) 17 (9): 7–13. doi:10.1145/947955.1083808

Part 4
Architecture as Decisions

Parts 2 and 3 presented principles and methods for analyzing and synthesizing architecture. Because of the complexity inherent in architecture, many of the steps involved in analysis and synthesis are difficult to conduct exhaustively. For example, we considered only seven concepts for the Hybrid Car in Chapter 12, and we did not exhaustively search the space of concept fragment combinations. Although some aspects of these tasks can be reduced to computation, great care must be taken when distilling the outputs of computational models to yield recommendations. With these thoughts in mind, in Part 4 we embark on architectural decision support, where we will investigate how methods and tools can support, but not replace, the role of the architect.

Part 4 will introduce computational methods and tools that can be useful to the system architect. These are taken from the fields of decision analysis, global optimization, and data mining.

Our intentions are twofold. On the one hand, methods and tools can augment the architect's analysis, helping to reduce ambiguity, employ structured creativity, and manage complexity. However, the second purpose of Part 4 is to create mental models for the reader (such as the tradespace mental model). It has been our experience that some of the constructs and modes of reasoning presented in this part are useful for supporting architectural decisions, even without actually building analytic models.

Part 4 is organized as follows. In Chapter 14, we illustrate the system-architecting process as a decision-making process. We provide an overview of decision support, with an eye to restricting the field of view to the correct subset of decisions. In theory, every stroke of the painter's brush is a decision, but not all decisions are *architectural decisions*. We illustrate these ideas with a simple case study of the Apollo mission-mode decision, to demonstrate the power of these methods and tools.

In Chapter 15 we examine how to synthesize information from an architecture tradespace—that is, the result of evaluating a set of architectures. We discuss how to combine tools such as Pareto analysis, design of experiments, and sensitivity analysis to gain a better understanding of the main architectural tradeoffs, sensitivities, and couplings between decisions. In particular, we explore how to structure a complex architectural tradespace around a hierarchy of decisions. We will argue that developing the hierarchy of decisions is a key lever on managing complexity, and that knowing which decisions have the greatest impact is at least as important as finding the model's prediction of the best architecture.

In Chapter 16, we show how to encode architectural decisions in a model that can be used by a computer to automatically generate and explore a tradespace. We start by introducing the idea of system architecting problems and the canonical types of decisions that the architect makes. Based on this study, we propose six patterns of architectural decisions, and we introduce their corresponding formulation as optimization problems. We look at the mathematical structure of these problems and discuss similarities to, and differences from, classical optimization problems (knapsack problem, traveling salesman problem). A single class of tools—heuristic optimization algorithms—is introduced as an approach to solving architectural problems computationally.

The appendices describe additional tools that can be used to support the main functions of decision support (representing, simulating, structuring, and viewing). In particular, we discuss in the appendices how to automatically enumerate and evaluate architectures using rule-based systems, and how to structure the tradespace using clustering algorithms.

Chapter 14
System Architecture as a Decision-Making Process

The authors would like to acknowledge the substantial contributions of Dr. Willard Simmons to this chapter.

14.1 Introduction

The job of a system architect is to transform a set of needs and goals into a system architecture. For complex systems, the task of architecting is challenging, because the complex relationships between design parameters and their alternatives introduce a massive search space that challenges both humans' and computers' abilities to exhaustively process the space. This chapter argues that a system architecture can be effectively represented as a set of *interconnected decisions*. It should be no surprise that we consider these decisions as a system—they have entities (the decisions) and relationships (the connections between the decisions). These decisions are an intermediate system, between the system of needs and the final architecture. Most important, these decisions can be used both cognitively and computationally to reduce the perceived complexity of the architecting task.

Let's start with an example. Consider the process of architecting the Apollo project of the 1960s. In his famous May 1961 speech, President Kennedy stated that the United States would send a man to the Moon and return him safely to Earth by the end of the decade. [1] This and other needs and goals were then transformed into an architecture for the system. System architects achieved this transformation by identifying, and making decisions to reduce, the candidate space of architectures.

NASA reduced the perceived complexity of the tradespace by identifying key decisions that would define the mission-mode, which described how and where the various elements would meet in space and how the crew would shift between elements. Historical evidence [2] shows that progress in the Apollo program was limited until June 1962, when the decision was made to choose Lunar Orbit Rendezvous as the Apollo mission-mode, setting the program on a path to the successful Moon landing in 1969. [3]

There are two major decisions in this architecture: Will there be rendezvous and docking operations in Earth orbit (called Earth orbit rendezvous, or EOR)? And will there be rendezvous and docking operations in lunar orbit (called lunar orbit rendezvous, or LOR)? If there is neither EOR nor LOR, it is called the direct mission-mode, and a single huge spacecraft is launched into Earth orbit, travels to Moon orbit, descends on the Moon's surface, ascends into Moon orbit, and returns to Earth. This option minimizes the number of vehicles developed. In the LOR mode, two spacecraft are injected into lunar orbit, and only one descends to the Moon. After completing the mission on the Moon's surface, it ascends and is assembled with the second vehicle that carries the propellant required to return to Earth. This option is the most fuel efficient. Finally, in the EOR mission-mode, two spacecraft are assembled in Earth orbit and then travel to the surface of the Moon. This option is preferable from a risk standpoint, because the riskiest operation (the assembly of spacecraft) occurs near Earth, which greatly facilitates the return of the astronauts in case of failure. A fourth mission-mode includes both EOR and LOR.

The mission-mode involving lunar orbit rendezvous (LOR) was chosen for the Apollo mission. By making this decision, the system architects provided design engineers with a stable category of acceptable designs that were eventually refined into a single detailed design.

In this chapter, we present the notion of *system architecting as a decision-making process*—a perspective that can dramatically change how early design activities are conducted. In the introduction to this text, we highlighted the idea that the first ten decisions taken in an architecture determine a majority of the performance and cost. Even with only two choices per decision, there are $2^{10} = 1024$ possible architectures in this design space. Much of what we discussed in Parts 1 through 3 of this text helps system architects apply experience and heuristics to winnow this space. In this chapter we will show how typical decision support tools, such as decision trees, can make it easier to comb through large combinatorial spaces and help the architect to make informed choices. It goes without saying that this is not intended to replace the judgment of the architect, but to augment it. We seek to develop a set of methods and tools that are primarily aimed at reducing complexity but can also help in resolving ambiguity and thinking creatively. In other words, we are trying to optimize the functional allocation between humans and computers in the system architecting task.

14.2 Formulating the Apollo Architecture Decision Problem

Heuristics for Decisions

Formulating the Apollo project as a decision-making problem requires selecting a set of decisions (with their corresponding alternatives) and a set of metrics. This section describes three heuristics for formulating decisions and shows how they are applied to the Apollo architecture problem.

The first heuristic for formulating the decision problem is to *carefully set the boundaries of the architectural space under consideration* (see Principle of the System Problem Statement in Chapter 11). What range of architectures should be included in the Apollo analysis? The highest-level specification of the Apollo system can be derived from President Kennedy's 1961 speech: "landing a man on the Moon and returning him safely to the Earth." This statement sets as the minimum scope that Apollo should land at least one man on the Moon. But architectures that extend to a much larger scope of issues (Mars exploration, space stations, and so on) may not have been appropriate, considering the limited schedule NASA was given.

One decision that NASA faced was the size of the crew. In the early 1960s, landing even one man on the Moon was an extremely ambitious goal. [4] However, because a lunar mission is a relatively long and complex space mission, it may have been too risky to send a lone astronaut on the voyage. Again considering schedule constraints, it was probably inappropriate to consider missions with large teams of astronauts, such as von Braun's proposed Conquest of the Moon. [5] In this retrospective analysis, we will bound the number of crew members to at least one and no more than three.

Another decision was that of the mission-mode. The feasibility and reliability of in-space rendezvous and docking was a heavily debated topic at the beginning of the Apollo project. Earth orbit rendezvous was considered of lower technical risk but also of lower benefit. John C. Houbolt, a NASA engineer, showed that lunar orbit rendezvous was challenging but technically feasible. He argued that missions including rendezvous and docking in lunar orbit should be considered, because they provided opportunities for saving mass and launch cost. [6] Therefore, in this retrospective analysis we will consider both EOR and LOR.

The second heuristic states that *the decisions should significantly influence the metrics by which the architecture is evaluated*. This seems obvious, but when architecture decision models are created, some of the decisions are often found to have low impact on the metrics (that is, the metrics are relatively insensitive to these decisions), implying that these decisions could be dropped. Two metrics that are considered important in space missions are the total mass of the mission elements and the probability of mission success. Both strongly depend on the mission-mode and the crew size, as well as on the fuel types to be used for spacecraft maneuvers. [7]

The third heuristic states that *the decision model should include only architectural decisions*. For Apollo, decisions related to the mission-mode directly drive the function-to-form mapping. For example, if the mission-mode includes lunar orbit rendezvous, the concept for the mission includes two vehicles: one crew vehicle that has a heat shield so that it can re-enter Earth's atmosphere, and a lunar lander vehicle that is specialized for descent to the surface of the Moon. Sometimes decisions indirectly influence architecture. For example, the fuel types influence the kind of engine used and the requirements for propellant tanks. The application of this third heuristic led to the removal of some decisions from our model; these include the decision on the location of the launch site, which does not substantially change the architecture.

Apollo Decisions

After considering the three heuristics, we selected a set of nine decisions for the Apollo study (see Figure 14.1). The process of creating a decision model is iterative in nature. For example, if a model shows that architectures that differ only in Decision X produce the same metric scores, then either Decision X can be eliminated from the model, or a new metric should be included that reflects the effect of this decision on metrics.

The set of nine decisions shown in Figure 14.1 includes decisions related to the mission-mode, the crew size, and the rocket propellant type used for Apollo. The figure also shows the range of allowed alternatives for each decision. Such tables are called morphological matrices, and in Section 14.5 we will discuss them as a tool to organize architectural decisions.

The first five decisions listed and illustrated in Figure 14.1 are related to the mission-mode: *EOR* (Will there be rendezvous and docking in Earth orbit?); *earthLaunch* (Will the vehicles go into orbit around the Earth or launch directly to the Moon?); *LOR* (Will there be rendezvous and docking in lunar orbit?); *moonArrival* (Upon arrival, will the vehicles go into orbit around the Moon or descend directly to the Moon's surface?); and *moonDeparture*

shortID	Decision	units	alt A	alt B	alt C	alt D
EOR	Earth Orbit Rendezvous	none	no	yes		
earthLaunch	Earth Launch Type	none	orbit	direct		
LOR	Lunar Orbit Rendezvous	none	no	yes		
moonArrival	Arrival At Moon	none	orbit	direct		
moonDeparture	Departure From Moon	none	orbit	direct		
cmCrew	Command Module Crew	people	2	3		
lmCrew	Lunar Module Crew	people	0	1	2	3
smFuel	Service Module Fuel	none	cryogenic	storable		
lmFuel	Lunar Module Fuel	none	NA	cryogenic	storable	

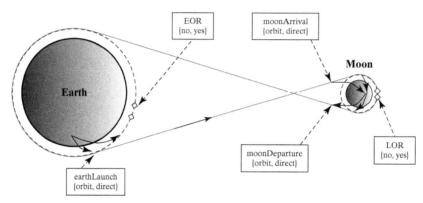

FIGURE 14.1 Mapping of historical Apollo mission-modes to the nine decisions. Note that the combinations of decision assignments listed must also satisfy the logical constraints shown in Figure 14.2.

(Upon ascent, will the vehicles go into lunar orbit or proceed directly toward the Earth?). All five of these decisions indicate alternative maneuvers at different points of the mission. By combining one alternative from each of the five decisions, a mission-mode can be defined.

The four remaining decisions are related to crew size and fuel type. The choice from *cmCrew* sets the size of the Command Module crew, and *lmCrew* includes the choices for the size of the Lunar Module crew (which of course is zero if there is no lunar module). The choices for *smFuel* and *lmFuel* set the fuel of the Service Module and the Lunar Module (NA if it does not exist). "Cryogenic" indicates a higher-energy LOX/LH2 propellant, and "storable" represents a lower-energy but higher-reliability hypergolic propellant.

Constraints and Metrics

In addition to defining the decisions, a complete description of the architecture model contains constraints and metrics. The constraints capture available knowledge about the system and the relationships between the decisions. *Logical constraints* are those that identify combinations of decisions that are not possible. Table 14.1 shows the logical constraints in the Apollo example. For instance, constraint d says that the Lunar Module crew must be smaller than or equal to the Command Module crew—you cannot create astronauts in the vicinity of the Moon!

TABLE 14.1 | Constraints in the Apollo example

Id	Name	Scope	Equation
a	EORconstraint	EOR, earthLaunch	(EOR == yes && earthLaunch == orbit) || (EOR == no)
b	LORconstraint	LOR< moonArrival	(LOR == yes && moonArrival == orbit) || (LOR == no)
c	moonLeaving	LOR, moonDeparture	(LOR == yes && moonDeparture == orbit) || (LOR == no)
d	lmcmcrew	cmCrew, lmCrew	(cmCrew ≥ lmCrew)
e	lmexists	LOR, lmCrew	(LOR == no && lmCrew == 0) || (LOR == yes && lmCrew > 0)
f	lmFuelConstraint	LOR, lmFuel	(LOR == no && lmFuel == NA) || (LOR == yes && lmFuel ! = NA)

Table 14.1 shows that the key decision is LOR. If it is yes, there cannot be direct descent at lunar arrival or direct lunar departure (constraints b and c), and there must be a Lunar Module with a crew of at least 1 and a propellant (constraints e and f).

There are also weaker forms of "reasonableness constraints" that encode things you would *probably* not do together. For example, if you were to construct an international effort to go to the Moon, you would probably not have the United States build a lander, and a second nation also build one. That would be wasteful of resources and hence unreasonable. But there is nothing *logically* incorrect about it.

In assessing the effectiveness of architectures, we usually find that there are metrics that quantify some measures of performance, some measures of cost, and some measures of developmental and operational risk. Our Apollo case is iso-performance, which means that all architectures provide for landing at least one crew on the lunar surface, satisfying Kennedy's goal. Therefore, the two metrics that we use in assessing the potential success of the Apollo project are (1) operational risk and (2) initial mass to low Earth orbit (IMLEO), a proxy for cost. The IMLEO is calculated for any architecture using the rocket equation [8] and the parameters taken from Houbolt's original documents. [9]

The risk metric was based on Table 14.2, which directly links decisions with the probability of successful operation. The overall probability of success is obtained by multiplying the individual probabilities together for the operations represented in any one architecture. The risks metric contains four categories of risk: high (0.9 probability of success), medium (0.95), low (0.98), and very low (0.99). The risk factor for each operation is assessed on the basis of documents written in the early 1960s and interviews with key decision makers.

Metrics are another way in which decisions can be linked to each other. For example, the probability of mission success is computed by multiplying all individual probabilities. Thus each decision is linked through this metric to all other decisions.

Computed Apollo Architecture

Figure 14.2 shows the results of possible architectures for the Apollo program, which were obtained by exhaustively calculating the outcomes (IMLEO and probability of mission success) for all combinations of decisions that are not logically constrained. Among the best solutions closest

TABLE 14.2 | Table used to compute the risk metric in the Apollo example with the probability shown in brackets below each alternative.

shortID	Decision	alt A	alt B	alt C	alt D
EOR	Earth Orbit Rendezvous	no	yes		
risk		(0.98)	(0.95)		
earthLaunch	Earth Launch Type	orbit	direct		
risk		(0.99)	(0.9)		
LOR	Lunar Orbit Rendezvous	no	yes		
risk		(1)	(0.95)		
moonArrival	Arrival at Moon	orbit	direct		
risk		(0.99)	(0.95)		
moonDeparture	Departure from Moon	orbit	direct		
risk		(0.9)	(0.9)		
cmCrew	Command Module Crew	2	3		
risk		(1)	(1)		
lmCrew	Lunar Module Crew	0	1	2	3
risk		(1)	(0.9)	1	1
smFuel	Service Module Fuel	cryogenic	storable		
risk		(0.95)^(burns)	(1)		
lmFuel	Lunar Module Fuel	NA	cryogenic	storable	
risk		(1)	(0.9025)	1	

to the Utopia point (low mass and high probability of success, identified with the "U" symbol in Figure 14.2), we've highlighted eight architectures. What is remarkable is that these eight mirror the three main proposals considered at the time: von Braun's Direct mission, Houbolt's LOR concept, and the Soviet mission design.

Point designs 1 and 2 are "direct" missions with three and two crew members, respectively. A direct mission-mode implies that the mission has neither lunar orbit rendezvous nor Earth orbit rendezvous, and no lunar module. These types of missions were among the ones initially proposed by von Braun. [10] They have high mission reliability, at the cost of very high IMLEO.

Point designs 3 to 8 are architectures that include lunar orbit rendezvous maneuvers. Point design 3 matches the actual configuration of Apollo: It has three crew members in the

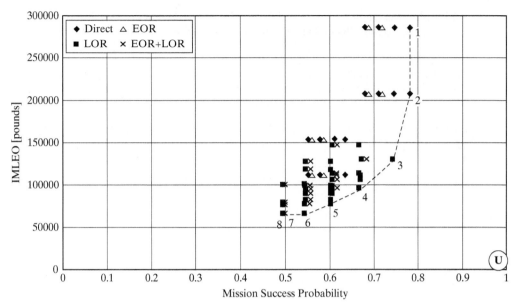

FIGURE 14.2 Apollo tradespace plot comparing IMLEO to probability of mission success. Each point in the plot indicates a logically feasible combination of decision assignments. The dashed line indicates the best architectures. The "U" symbol indicates the Utopia point, an imaginary point with perfect scores for all metrics.

command module, has two crew members in the lunar module, and uses storable propellants for both the service module and the lunar module. [11] It represents a reasonable compromise between mass and risk. Point 8, which is the minimum-mass configuration, uses two crew members in a command module and one crew member in a lander with cryogenic propellants. Point design 8 is the point that most closely models the proposed Soviet lunar mission's architecture. [12] Searching computationally through the possible architectures, we surface the three primary choices considered during the mission-mode decision process in the 1960s, and we come to understand the essential tradeoff between mass and risk. We will see in the next chapter how to mine this type of chart for useful information that can help us structure the system architecture process.

14.3 Decisions and Decision Support

According to R. Hoffman, the word "decide" comes from the Sanskrit word *khid 'ati*, meaning "to tear," the Latin word *cædare*, meaning "to kill" or "cut down," and also the Latin word *decædare*, which means "to cut through thoroughly." [13] In contemporary English, a *decision* is "the passing of judgment on an issue under consideration" [14] or a purposeful selection from mutually exclusive alternatives. *Decision making* is "goal-directed behavior in the presence of options" [15] that culminates in one or more decisions. The key ideas in decision making are that there is a situation with multiple alternatives; a selection is made that separates the solution space; and there is some expected benefit that will be achieved by making this decision.

Decision support is about assisting decision makers in making a decision. Many decision support processes can be described by Herbert Simon's four-phase process, [16] which includes:

1. **Intelligence Activity:** "Searching the environment for conditions calling for a decision."
2. **Design Activity:** "Inventing, developing, and analyzing possible courses of action."
3. **Choice Activity:** "Selecting a particular course of action from those available."
4. **Review Activity:** "Assessing past choices."

According to Simon, decision makers tend to spend a large fraction of their resources in the Intelligence Activity phase, an even greater fraction of their resources in the Design Activity phase, and small fractions of their resources in the Choice Activity phase and the Review Activity phase.

Simon's research makes the complementary observation that there are two "polar" types of decisions: programmed decisions and non-programmed decisions (these have sometimes been called "structured" and "unstructured" decision problems by subsequent authors [17]). Examples and characteristics of these two types of decisions are shown in Figure 14.3.

Programmed decisions are "repetitive and routine decisions" where a procedure for making decisions for this type of problem has been worked out *a priori*. Examples of programmed decisions range from simple to very complex. For instance, deciding how much to tip a waiter, deciding the optimal gains of a control system, and deciding the routing of all aircraft flying over the United States can all be considered programmed decisions. In each case there is a known

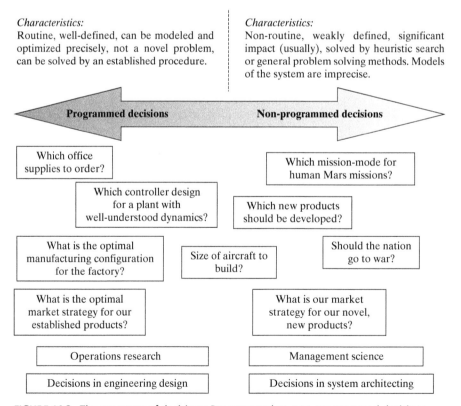

FIGURE 14.3 The spectrum of decisions: Programmed vs. non-programmed decisions.

and defined approach that a decision maker can follow to arrive at a satisfactory choice. Models exist for the behavior of the problem, and the objectives are clearly definable. Note that classifying a decision as programmed does not imply that it is an "easy" decision, but only means that a method to solve it is known and available. Furthermore, the classification "programmed" does not quantify the amount of resources necessary to use the prepared methodology. For many engineering problems, this predetermined routine will be difficult to implement or expensive to compute. [18] Simon considers programmed decisions to lie in the domain of Operations Research.

On the other end of the spectrum are non-programmed decisions. These decisions are novel, ill-structured, and often consequential. An example is the decision of the mission-mode for Apollo. Simon maintains that non-programmed decisions are generally solved by creativity, judgment, rules of thumb, and general problem-solving methods such as heuristics. The examples given in Figure 14.3 include deciding whether a nation should go to war, deciding what market strategy to adopt for new, unproven products, and deciding on the mission-mode for human Mars missions. Simon considers non-programmed decisions to lie in the domain of management science.

In many cases, decisions that have been thought to be non-programmed become programmed once someone is clever enough to invent a programmed method to solve that problem. Perhaps a better name for non-programmed decisions is "not-yet-programmed decisions." An example of people becoming able to systematically program a previously non-programmed problem occurred when Christopher Alexander introduced the concept of "pattern language" as a systematic way to develop civil architectures. The pattern language catalogs the elements of an architecture as reusable triples made up of the context in which they are relevant, the problem they are intended to solve, and the solution they provide. [19] We will discuss the notion of patterns further in Chapter 16.

It is evident that there is sometimes an opportunity to "program" decisions that were previously thought to be "non-programmed." The goal of the remainder of this chapter and the next is to develop a decision support system to comprehensively and efficiently examine a solution space using rigorous analysis as suggested by Simon, rather than using heuristics.

14.4 Four Main Tasks of Decision Support Systems

Decision support consists of assisting decision makers in making a decision. Today, many commercial software programs (such as Decision Lens®, TreePlan®, and Logical Decisions®) provide decision support systems by implementing some form of automation to support the decision-making process.* The goal of these systems is to enhance the efficiency of decision makers by providing tools to quantitatively and qualitatively explore a space of alternatives for single or multiple decisions.

In constructing a decision support system (DSS) for architecture, it is useful to characterize the tasks. We assert that the task of Simon's Design Activity applied to architecture can be described by four "layers": representing, structuring, simulating, and viewing [20]:

- The **representing** layer includes methods and tools for representing the problem for the human decision maker and encoding the problem for computation. The matrix

---
*The Institute for Operations Research and the Management Sciences (INFORMS) maintains a list of decision analysis software at its website. http://www.orms-today.org/surveys/das/das.html

in Figure 14.1, showing the choices for each Apollo decision, is an example of representing. This morphological matrix is useful because it provides a simple representation of the decisions and choices to be considered. However, it does not provide any information related to constraints (in the structuring layer) or preferences (in the simulating layer). Other ways of representing architectural spaces include trees, graphs, OPM, and SysML.

- The **structuring** layer involves reasoning about the structure of the decision problem itself. This includes determining the order of decisions and the degree of connectivity between decisions. For example, in the Apollo example, one cannot choose to have an Earth Orbit Rendezvous and simultaneously choose to launch directly to the Moon (see Table 14.1). This type of logical constraint, and other types of couplings between decisions, can be represented using Design Structure Matrices (DSMs) containing bilateral interactions between decisions. DSMs are described in more detail in the next section.
- The **simulating** layer is used to determine which combinations of decisions will satisfy logical constraints and calculate the metrics. The simulating layer is thus about evaluating the ability of a system architecture to satisfy the needs of the stakeholders. For example, in the Apollo example, the simulating layer computes the two metrics shown in Figure 14.2. A variety of tools are used for system simulation, ranging in complexity from simple equations to discrete-event simulation. [21]
- The **viewing** layer presents, in a human-understandable format, decision support information derived from the structuring and simulating layers. For example, the chart shown in Figure 14.2 provides some viewing support by graphically representing the evaluation results from all the architectures in the tradespace. We will discuss tradespaces and how to mine information from them in detail in Chapter 15.

Note that we include in Simon's Design Activity the methods and tools that involve representing, structuring, simulating, and viewing. The actual architecture selection process is the goal of Simon's Choice Activity, and the steps leading to this Choice Activity are discussed in Chapter 16.

14.5 Basic Decision Support Tools

We have already seen, in Parts 2 and 3, examples of tools that are useful in some of the four layers of decision support. For example, we used Object Process Methodology (OPM) to represent not only system architectures but also specific decision-making problems, such as function or form specialization (see, for example, Figure 7.2). We introduced morphological matrices as a means to represent simple decision-making problems, such as concept selection (see, for example, Table 7.10). We also used Design Structure Matrices in Part 2, mostly to represent the interfaces between entities of a system. In Part 4, we focus on the other aspects of decision support: structuring, simulating, and viewing.

We start this section by revisiting morphological matrices and DSMs, showing that these tools provide limited support to structuring, and no support to simulation and viewing. Then we introduce decision trees as a widely used decision support tool for decision making under uncertainty, providing some support to the structuring and simulating layers. (Other widely used decision support tools, such as Markov Decision Processes, are not described in this section because they are seldom used for system architecture purposes.)

Morphological Matrix

The morphological matrix was introduced in Chapter 7 as a way to represent and organize decisions in a tabular format. The morphological matrix was first defined by Zwicky as a part of a method for studying the "total space of configurations" (morphologies) of a system. [22] Since then, the use of morphological matrices as a decision support tool has grown. [23] Figure 14.1 includes the morphological table for the Apollo example.

A morphological matrix lists the decisions and associated alternatives, as shown in Figure 14.1. An architecture of the system is chosen by selecting one alternative (labeled "alt") from the row of alternatives listed to the right of each decision. Note that alternatives for different decisions do not have to come from the same column. For example, a configuration of the Apollo system could be: number of EOR = no (alt A), LOR = yes (alt B), command module crew = 2 (alt A), lunar module crew = 2 (alt C). An example of an expanded or more explicit form of the morphological matrix was given in Table 7.10.

In terms of decision support, the morphological matrix is a useful, straightforward method for representing decisions and alternatives. It is easy to construct and simple to understand. However, a morphological matrix does not represent metrics or constraints between decisions. Thus it does not provide tools for structuring a decision problem, simulating the outcome of decisions, or viewing the results.

Design Structure Matrix

As introduced in Chapter 4, a Design Structure Matrix is actually a form of decision support. The term Design Structure Matrix (DSM) was introduced in 1981 by Steward. [24] DSMs are now widely used in system architecture, product design, organizational design, and project management.

A DSM is a square matrix that represents the entities in a set and their bilateral relationships (see Table 4.4). These entities can be the parts of a product, the main functions of a system, or the people in a team, as demonstrated in the different examples throughout Part 2.

When a DSM is used to study the interconnections between decisions, each row and column corresponds to one of the N decisions, and an entry in the matrix indicates the connections, if any, that exist between the two decisions. The connections could be logical constraints or "reasonableness" constraints, or they could be connections through metrics.

For example, in Table 14.3, the letter a at the intersection between "EOR" and "earthLaunch" indicates that there is a connection between these two decisions as the result of the constraint labeled "a" in Table 14.1. Likewise, the letters b through f indicate the other five constraints. A blank entry in the intersection indicates that there is no direct connection between these decisions imposed by the constraints.

A connection could mean that a metric depends on those two variables. For example, Table 14.4 shows the connection between the decisions given by the IMLEO metric. The entries in the matrix may have different letters, numbers, symbols, or colors to indicate different types of connections.

TABLE 14.3 | DSM representing the interconnection of decisions by logical constraints for the Apollo case

		1	2	3	4	5	6	7	8	9
EOR	1		a							
earthLaunch	2	a								
LOR	3				b	c		e		f
moonArrival	4			b						
moonDeparture	5			c						
cmCrew	6							d		
lmCrew	7			e			d			
smFuel	8									
imFuel	9			f						

TABLE 14.4 | DSM representing the interconnection of decisions by the IMLEO metric in the Apollo case

		1	2	3	4	5	6	7	8	9
EOR	1									
earthLaunch	2									
LOR	3						I	I	I	I
moonArrival	4									
moonDeparture	5									
cmCrew	6				I			I	I	I
lmCrew	7				I				I	I
smFuel	8				I			I		I
imFuel	9				I			I	I	

In Table 14.5, the DSM of Table 14.3 has been partitioned and sorted so as to minimize interactions across blocks of decisions when possible. These blocks represent sets of decisions that should be made approximately simultaneously, because they have couplings with each other (for example, LOR with Lunar Module crew, Service Module crew, Service Module fuel, Lunar Module fuel). A partitioning procedure and example are given in Steward's papers. More generally, clustering algorithms can be used for partitioning (see Appendix B).

A DSM provides information in the representing layer and the structuring layer of decision support. It represents the decisions (but not their alternatives) and their interconnections. In combination with partitioning or clustering algorithms, a DSM can be sorted to show which sets of decisions are tightly coupled and which sets are less tightly coupled or not coupled. This information informs both system decomposition and the timing of architectural decisions, if the order of the decisions in the matrix is taken to represent the sequence in which they are made. [25]

TABLE 14.5 | Sorted DSM of the interconnections of decisions by logical constraints for the Apollo case

		1	2	3	4	5	6	7	9	8
EOR	1		a							
earthLaunch	2	a								
LOR	3				b	c		e	f	
moonArrival	4			b						
moonDeparture	5			c						
cmCrew	6							d		
lmCrew	7			e			d			
imFuel	9			f						
smFuel	8									

Decision Trees

A *decision tree* is a well-known way to represent sequential, connected decisions. A decision tree can have three types of nodes: decision nodes, chance nodes, and leaf nodes. Decision nodes represent decisions, which are controllable by the decision maker and have a finite number of possible assignments represented by branches in the tree from that node. Chance nodes represent chance variables, which are not controllable by the decision maker and also have a finite number of possible assignments, which are also represented by branches. The endpoints, or "leaf" nodes, in the decision tree represent a complete assignment of all chance variables and decisions.

When these decisions are architectural decisions, a path through the decision tree essentially defines an architecture. Thus, decision trees without chance nodes can be used to represent different architectures. Figure 14.4 shows a decision tree for the Apollo example concerning the five decisions related to the mission-mode. All nodes in this chart are decision nodes except for the final node in each branch, which is a leaf node. The first three of the constraints of Table 14.1 have also been implicitly included in the tree; for example, if earthLaunch = yes, there is no option of EOR = yes. This branch has been "pruned" from the tree by applying the logical constraint.

Note that this decision tree representation explicitly enumerates all the combinations of options (that is, all architectures), whereas the morphological matrix enumerated architectures only implicitly, by showing the alternatives for each decision. A limitation of decision trees is immediately visible in Figure 14.4: The size of a decision tree grows rapidly with the number of decisions and options, which results in huge trees, even for modest numbers of decisions.

In addition to representing architectures, decision trees can be used for evaluating architectures and selecting the best ones. This requires computing the metrics for each architecture. Sometimes, it is possible to compute a metric incrementally while following a path in the tree. For example, the probability-of-success metric in the Apollo example can be computed in this way, because the contribution of each decision to the overall probability of success is given by Table 14.2, and the metric is constructed by multiplying the individual contributions together.

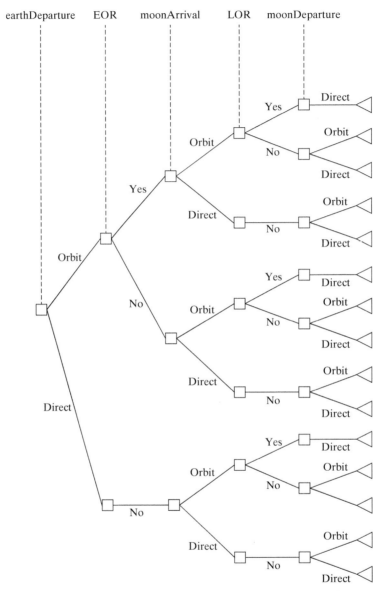

FIGURE 14.4 Simple decision tree with only decision nodes and leaf nodes (no chance nodes) for the Apollo mission-mode decisions.

In most cases however, there is coupling between decisions and metrics that is not additive or multiplicative, such as in the case of the IMLEO metric, a highly nonlinear function of the decision alternatives. In these cases, the metric for every leaf node (such as IMLEO) is usually computed after all the decision alternatives have been chosen. This approach may be impractical in cases where the number of architectures is very large, as we will see in Chapter 16.

The leaves or architectures represented on decision trees are usually evaluated using a single metric. In such cases, it is customary to combine all relevant metrics (such as IMLEO and probability of success p) into a single metric representing the utility of an architecture for stakeholders [such as $u = \alpha u(IMLEO) + (1 - \alpha) u(p)$]. The weight α and the individual utility functions $u(IMLEO)$ and $u(p)$ can be determined by means of multi-attribute utility theory. [26]

In decision trees that don't have chance nodes, choosing the best architecture is straightforward. It is more complicated in the presence of chance nodes, because leaf nodes no longer represent architectures but, rather, combinations of architectures and scenarios. It is thus necessary to choose the architecture that has the highest expected utility by working backwards from the leaf nodes. Let's illustrate that with an example.

Assume that, instead of having a single value for the risk factor of each decision in Table 14.2, we had two values: an optimistic one and a pessimistic one. What is the best architecture given this uncertainty? Figure 14.5 shows the addition of the chance nodes corresponding to the last decision in the tree for the Apollo case (only a very small part of the tree is actually shown). Note that the actual tree has effectively doubled in size just through the addition of chance nodes corresponding to the *moonDeparture* decision. We compute the risk metric for all the new leaf nodes by traversing the tree from start to finish and multiplying all the risk factors together (the results are the R_i values in Figure 14.5). At this point, we can't simply choose the leaf node with the lowest risk, because it does not represent an architecture; it represents an architecture for a given scenario—that is, a value of the risk factors. Instead, we need to assign probabilities to the branches of the chance nodes (such as 50%-50% everywhere), and compute the expected value of the risk metric at each chance node. Then, it makes sense to choose the Moon departure option with the lowest risk at each decision node. If we add chance nodes to all the other decisions, we can follow the same two steps. First, we compute the expected risk factor at each chance node, and then we pick the option with the best risk metric. We perform these steps for the *LOR* decision, then the *moonArrival* decision, and so forth, until all decisions have been made. This process yields the architecture that has best performance on average across all scenarios. This architecture is given by the combination of all the optimal decisions.

In short, the best architecture in a decision tree can be found by applying these two steps (expectation for chance nodes and maximization/minimization for decision nodes), starting from the leaf nodes and going backwards through the tree. [27]

Even though they are commonly used as general decision support tools, decision trees have several limitations that make them impractical to use in large system architecture problems. First, they require pre-computation of a *payoff matrix* containing the utilities of all the different options for each decision for all possible scenarios, which may be impractical for many problems. More generally, their size grows exponentially with the number of nodes in the problem.

Second, the method assumes that the payoffs and probabilities for a given decision are independent from the rest of the decisions, which is often unrealistic. For example, consider that instead of modeling the uncertainty in the risk factors, we wish to model the uncertainty in the ratio between propellant mass and structure mass for any of the vehicles used in the architecture. This is a very important parameter of the model, because it can drive the IMLEO metric. However, this parameter does not directly affect any of the decisions the way the risk factor did. It affects the IMLEO metric through a complex mathematical relationship based on the mission-mode

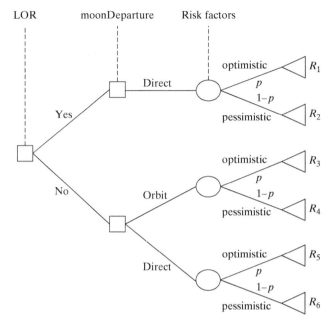

FIGURE 14.5 Fragment of the Apollo decision tree, including a chance node for the risk factors. In this tree, decision nodes are squares, chance nodes are circles, and leaf nodes are triangles. R_i are the values of the risk metrics for each architecture fragment, and p is the probability of being in an optimistic scenario for each individual decision. (Note that these probabilities could in general be different for different chance nodes.)

decisions. This ratio is thus a "hidden" uncontrollable variable that affects multiple decisions. Modeling this in a decision tree is very difficult, or else it requires collapsing all the mission-mode decisions into a single decision with 15 options (the number of leaf nodes in Figure 14.4).

In summary, decision trees provide a representation of the decision and can represent some kinds of structure, but they are of limited use in simulation.*

14.6 Decision Support for System Architecture

In the previous section, we discussed how three standard decision support tools can be applied to system architecture. At this stage, the reader may wonder whether there is anything special about system architecture decisions at all. How are architecture decisions different from other decisions?

Box 14.1 summarizes the characteristics of architectural decisions and metrics. The combination of these characteristics makes existing decision support tools very hard to apply to

---

*Decision networks are a more general version of decision trees where the tree structure condition is relaxed to allow arbitrary topologies between decision nodes, chance nodes, and leaf nodes.

Box 14.1 Insight: Properties of Architectural Decisions and Metrics

The following properties characterize architectural decisions and metrics and differentiate them from other types of decisions.

- **Modeling breadth versus depth.** Architecture decision support focuses on modeling breadth—that is, analyzing a large space of very different architectures at relatively low fidelity—whereas design decision support focuses on modeling depth—that is, analyzing a smaller number of designs with higher fidelity. Note that there is an inherent tradeoff between modeling breadth and modeling depth.
- **Ambiguity.** Ambiguity is present in both design and architecture problems. However, architecture problems suffer from larger and more varied sources of uncertainty (unknown outcomes due to random events) and ambiguity (inaccuracies or fuzziness in statements), simply because they occur early in the development process. This often makes the use of typical probabilistic techniques such as Monte Carlo simulation impractical or inadequate for system architecture. [28]
- **Type of variable.** In decision analysis and optimization, decisions are classified into three types: continuous, discrete (those that can take only integer values), and categorical (those that can take any value from a discrete set of symbols representing abstractions). [29] Architectural decisions are most often categorical variables, are sometimes discrete variables, and are rarely continuous variables. The reason is that architectural decisions often consist of choosing among different entities of form or function, or among different mappings between function and form, which are inherently categorical in nature. We will have a deeper discussion about classes of architecture decisions and how to model them in Chapter 15. For example, the Apollo mission-mode decision is really about choosing which vehicles (form) perform which maneuvers (function). Conversely, design decisions tend to be continuous, are sometimes discrete, and are less likely to be categorical.
- **Subjectivity.** Architecture problems often deal with some subjective metrics, which reflect the ability of the system to provide value to the stakeholders. Dealing with subjective metrics leads to the use of techniques such as multi-attribute utility theory or fuzzy sets. [30] Furthermore, subjectivity highlights the need for traceability. Architecture decision support should provide an explanation that justifies the assessments and can trace it back to expert knowledge or judgment if necessary.
- **Type of objective functions.** Architecture problems sometimes use relatively simple equations in their objective functions, but often they also tend to resort to simple look-up tables and if-then structures, such as the Apollo risk metric. This is because of (1) the breadth of architectures considered, which results in the need to apply different strategies to evaluate different types of architectures; (2) the low modeling fidelity, which leads to replacing complex computations with heuristics; and (3) the subjectivity of some metrics, which can be captured by rules that "set" some metric of interest for an architecture when a number of conditions are met.
- **Coupling and emergence.** Architecture decision problems typically have a relatively low number of decisions compared to design problems. However, the dimensionality of the corresponding architecture space is often extremely large due to combinatorial explosion. What this means is that architecture variables are often extremely coupled. This is important because it precludes the utilization of some very powerful methods that exploit decouplings in the structure of decision problems (for example, the Markov property in dynamic programming techniques).

architecture decision support, and this has motivated the development of decision support tools for system architecture. [31]

In response to the characteristics listed in Box 14.1, architecture decision support systems are more interactive and less automatic, and they often employ tools from the fields of knowledge reasoning and engineering (such as knowledge-based systems for incorporation of expert knowledge and explanation).

14.7 Summary

We began Part 4 by introducing a key idea: System architecting is a decision-making process, so we can benefit from decision support tools. We started off with an example based on the Apollo program to illustrate how we formulate a system architecting problem as a decision-making problem. We argued that the goal of these tools is to support—not replace—the system architect, given that system architecting requires creativity, holism, and heuristic approaches for which humans are much better suited than machines.

We described four fundamental aspects of decision support systems (representing, structuring, simulating, and viewing), and we showed how Parts 2 and 3 have used decision support tools mostly for representing, but not for structuring, simulating, or viewing. We discussed three basic decision support tools that support representing and that provide some limited structuring, simulating, and viewing capabilities. These are: morphological matrices, design structure matrices, and decision trees. We discussed in particular some of the limitations of these tools for system architecture.

This led us to ask about the differences between system architecture and other decision-making processes. We concluded that although some of the standard decision support tools remain applicable, there is a need for tools that can deal with subjectivity, ambiguity, expert knowledge, and explanation. The next chapters will discuss some aspects of the tools we use for the four aspects of decision support in the context of system architecture. We will first take a deeper look at architectural tradespaces like the one in Figure 14.2. We will be working mostly in the structuring layer of the Design Activity, and we will see how to obtain useful knowledge about the system architecture by applying simple data processing techniques to tradespaces.

References

[1] J.F. Kennedy, "Special Message to the Congress on Urgent National Needs." Speech delivered in person before a joint session of Congress, May 25, 1961.

[2] See I.D. Ertel, M.L. Morse, J.K. Bays, C.G. Brooks, and R.W. Newkirk, *The Apollo Spacecraft: A Chronology,* Vol IV, NASA SP-4009 (Washington, DC: NASA, 1978) or R.C. Seamans Jr., *Aiming at Targets: The Autobiography of Robert C. Seamans Jr.* (University Press of the Pacific, 2004).

[3] T. Hill, "Decision Point," *The Space Review,* November 2004.

[4] J.R. Hansen, *Spaceflight Revolution: NASA Langley Research Center from Sputnik to Apollo,* NASA SP-4308 (Washington, DC: NASA, 1995).

[5] W. von Braun, F.L. Whipple, and W. Ley, *Conquest of the Moon* (Viking Press, 1953).

[6] J.C. Houbolt, "Problems and Potentialities of Space Rendezvous," Space Flight and Re-Entry Trajectories. International Symposium Organized by the International Academy of Astronautics of the IAF Louveciennes, June 19–21, 1961, *Proceedings,* 1961, pp. 406–429.

[7] W.J. Larson and J.R. Wertz, eds., *Space Mission Analysis and Design* (Microcosm, 1999).

[8] V.A. Chobotov, ed., *Orbital Mechanics* (AIAA, 2002).

[9] J.C. Houbolt, *Manned Lunar-landing through Use of Lunar-Orbit Rendezvous,* NASA TM-74736, (Washington, DC: NASA, 1961).

[10] J.R. Hansen, *Enchanted Rendezvous,* Monographs in Aerospace History Series 4 (Washington, DC: NASA, 1999).

[11] C. Murray and C.B. Cox, *Apollo* (South Mountain Books, 2004).

[12] J.N. Wilford, "Russians Finally Admit They Lost Race to Moon," *New York Times,* December 1989.

[13] Hoffman, R. R., "Decision Making: Human-Centered Computing," *IEEE Intelligent Systems,* vol. 20, 2005, pp. 76–83.

[14] "Decision," *American Heritage Dictionary of the English Language,* 2004.

[15] S.O. Hansson, *Decision Theory: A Brief Introduction* (KTH Stockholm, 1994).

[16] H.A. Simon, *The New Science of Management Decision.* The Ford Distinguished Lectures, Vol. 3, (New York: Harper & Brothers, 1960).

[17] E. Turban and J.E. Aronson, *Decision Support Systems and Intelligent Systems* (Prentice Hall, 2000).

[18] C. Barnhart, F. Lu, and R. Shenoi, R., "Integrated Airline Schedule Planning," *Operations Research in the Airline Industry,* Vol. 9, Springer US (1998), pp. 384–403.

[19] C. Alexander, *A Pattern Language: Towns, Buildings, Construction* (Oxford University Press, 1977).

[20] D. Power, *Decision Support Systems: Concepts and Resources for Managers* (Greenwood Publishing Group, 2002), pp. 1–251; R.H. Bonczek, C.W. Holsapple, and A.B. Whinston *Foundations of Decision Support Systems* (Academic Press, 1981); and E. Turban, J.E. Aronson, and T.-P. Liang, *Decision Support Systems and Intelligent Systems* (Upper Saddle River, NJ: Prentice Hall, 2005).

[21] J. Banks, J.S. Carson II, B.L. Nelson, and D.M. Nicol, *Discrete-Event System Simulation* (Prentice Hall, 2009), pp. 1–640 and B.P. Zeigler, H. Praehofer, and T.G. Kim, *Theory of Modeling and Simulation* (Academic Press, 2000), pp. 1–510.

[22] F. Zwicky, *Discovery, Invention, Research through the Morphological Approach* (Macmillan, 1969).

[23] G. Pahl and W. Beitz, *Engineering Design: A Systematic Approach* (Springer, 1995), pp. 1–580; D.M. Buede, *The Engineering Design of Systems: Models and Methods* (Wiley, 2009), pp. 1–536; and C. Dickerson and D.N. Mavris, *Architecture and Principles of Systems Engineering* (Auerbach Publications, 2009) pp. 1–496. See also T. Ritchey, "Problem Structuring Using Computer-aided Morphological Analysis," *Journal of the Operational Research Society* 57 (2006): 792–801.

[24] D.V. Steward, "The Design Structure System: A Method for Managing the Design of Complex Systems," *IEEE Transactions on Engineering Management* 28 (1981): 71–74.

[25] S.D. Eppinger and T.R. Browning, *Design Structure Matrix Methods and Applications* cambridge. MA: The MIT Press, 2012), pp. 1–352.

[26] Multi-attribute utility theory was developed by economists Keeney and Raifa in the seventies. See: R.L. Keeney and H. Raiffa, *Decisions with Multiple Objectives: Preferences and Value Trade-Offs* (New York: Wiley, 1976), p. 592. Since then, it has been widely applied in systems engineering. See also: A.M. Ross, D.E. Hastings, J.M. Warmkessel, and N.P. Diller, "Multi-Attribute Tradespace Exploration as Front End for Effective Space System Design," *Journal of Spacecraft and Rockets* 41, no. 1 (2004): 20–28.

[27] An example of the application of this algorithm can be found in C.W. Kirkwood, "An Algebraic Approach to Formulating and Solving Large Models for Sequential Decisions under Uncertainty," *Management Science* 39, no. 7 (1993): 900–913.

[28] Alternative tools for dealing with ambiguity include again fuzzy sets, or simple interval analysis. See, for example, J. Fortin, D. Dubois, and H. Fargier, "Gradual Numbers and Their Application to Fuzzy Interval Analysis," *IEEE Transactions on Fuzzy Systems* 16, no. 2 (2008): 388–402. An application of interval analysis to system architecture appears in D. Selva and E. Crawley, "VASSAR: Value Assessment of System Architectures Using Rules," in Aerospace Conference, 2013 at Big Sky.

[29] Note that categorical variables are often implemented as integer variables in optimization problems, but that does not make them discrete variables. One cannot define a gradient-like concept in categorical variables unless some meaningful distance metric is defined within the discrete set of abstractions, which is often not trivial.

[30] The foundational paper on multi-attribute utility theory is R.L. Keeney and H. Raiffa, *Decisions with Multiple Objectives: Preferences and Value Trade-Offs* (New York: Wiley, 1976), p. 592, and an example of a recent application to system architecture can be found in A.M. Ross, D.E. Hastings, J.M. Warmkessel, and N.P. Diller, "Multi-Attribute Tradespace Exploration as Front End for Effective Space System Design," *Journal of Spacecraft and Rockets* 41, no. 1 (2004): 20–28. Concerning fuzzy sets, the foundational paper is L.A. Zadeh, "Fuzzy Sets," *Information and Control* 8, no. 3 (1965): 338–353. An application to conceptual design can be found in J. Wang, "Ranking Engineering Design Concepts Using a Fuzzy Outranking Preference Model," *Fuzzy Sets and Systems* 119, no. 1 (2001): 161–170.

[31] See, for example, Koo's development of the Object Process Network meta-language for systems architecture, focusing on the simulating layer: B.H.Y. Koo, W.L. Simmons, and E.F. Crawley, "Algebra of Systems: A Metalanguage for Model Synthesis and Evaluation," *IEEE Transactions on Systems, Man, and Cybernetics–Part A: Systems and Humans* 39, no. 3 (2009): 501–513. Simmons's Architecture Decision Graph is a related tool that focuses on the structuring layer: W.L. Simmons, "A Framework for Decision Support in Systems Architecting" (PhD dissertation, Massachusetts Institute of Technology). Selva's VASSAR methodology is a third example, focusing on the structuring and simulating layers: D. Selva and E. Crawley, "VASSAR: Value Assessment of System Architectures using Rules," in Aerospace Conference, 2013 at Big Sky.

Chapter 15
Reasoning about Architectural Tradespaces

The authors would like to acknowledge the contributions of Dr. Willard Simmons to this chapter.

15.1 Introduction

Architectural wisdom is often an understanding of the tradeoffs between decisions. Chapter 14 presented basic decision support tools for representing and simulating architectures. Chapter 16 and the appendices introduce more advanced tools for these purposes. In this chapter, we focus on the tasks of structuring and viewing in support of decisions.

Making architectural decisions is rarely as simple as encoding an architecture as a decision-making problem and then choosing the best architecture according to a single metric. Rather, architectural decisions often involve understanding the main tradeoffs among decisions, evaluating the weighting between metrics in view of the solutions they lead to, and understanding the coupling among decisions. This chapter is devoted to analyzing and mining information and insight from an architectural tradespace such as the Apollo tradespace explicitly represented in Figure 14.2. Indeed, this tradespace can be a powerful tool, but more important, it is a gateway to a number of valuable mental models.

The chapter is organized as follows. We start with a simple example of a tradespace of car engines in Section 15.2: We explain what a tradespace chart is and how it is different from other tools to represent a space of system architectures. Section 15.3 focuses on the idea of dominance and the Pareto frontier: how to compute it, and what it means for an architecture to be non-dominated. In Section 15.4, we move beyond the concept of Pareto frontier and discuss how the structure of the tradespace (for example, the presence of stratification or of clusters of architectures) can provide information to system architects. Once we have our results, we ask ourselves how sensitive those results are to the assumptions that we have made in the model, which leads to the introduction of sensitivity analysis in Section 15.5. That leads to a more general discussion about the "importance" of architectural decisions in Section 15.6, which presents a framework for organizing architectural decisions: How do we identify the subset of key architectural decisions? Is there a sequence in which these decisions should be made?

15.2 Tradespace Basics

In general, a tradespace is a representation of a set of architectures in a space defined by two or more metrics. For example, see Figure 15.1, which illustrates a tradespace for different automotive engines. It shows two axes—performance (in horsepower) and cost (in dollars)—and a cloud of points between them, where each point represents one engine. What can we learn from this representation?

First, we can see that there is a range of both performance and cost, with many choices within each of them. Notice that we do not have any engines represented below 250 hp or above 540 hp; either we do not have the technology to work outside these bounds, or there are other constraints not illustrated. If we were looking for an engine that produced 330 to 350 hp, we could narrow the list of options significantly.

Second, we have condensed everything about these engines down to two metrics: performance and cost. There is simplicity in this representation; it is comprehensible. Decision makers can make reasonable tradeoffs between these two axes, in a way that they cannot among the 12 metrics we identified for the Hybrid Car in Chapter 11.

We have alternately argued in this text for detail and simplicity in metrics. In Chapter 16, we will argue for simplicity, maintaining that there should be an objective function that aggregates the needs of stakeholders into one or two metrics—for example, using a weighting function. However, we argued in Chapter 12 for a detailed list of goals and their metrics. How do we reconcile these competing views? Here are the suggestions we offer for presenting information on tradespaces:

- Ensure that all metrics are transparent (stakeholder satisfaction according to a weighted metric is less transparent; miles per gallon is more transparent).
- Avoid, if possible, single aggregate metrics that are likely to hide the weightings among various metrics at the next level down.
- Identify the two or three metrics that represent the important tradeoffs in the architecture, and thus should ideally be presented to the decision maker as options, providing the decision maker with more visibility and authority.

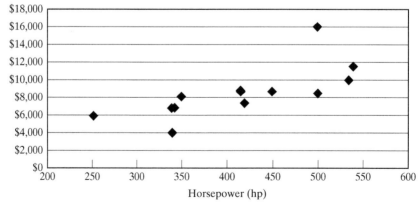

FIGURE 15.1 Engines tradespace, using performance (hp) and cost as metrics. [1] (Source: Drawn from http://www.fordracingparts.com/crateengine/Main.asp)

- Avoid presenting a long list of undifferentiated metrics. Allow the decision maker to see the key tradeoffs between high-priority metrics based on important needs and constraints in the goals, and then use goals of lower importance to differentiate among remaining options.

A tradespace representation is fundamentally different from a comparison of point designs. *Tradespaces* include numerous architectures, represented at lower fidelity and evaluated with a few simple key metrics. Choosing among them requires identifying a few preferred architectures that are likely to resolve key tensions and lead to acceptable designs. The remainder of this chapter will present approaches to reasoning through the metrics associated with tradespaces. *Point designs* represent significant detailed design effort for a small number of solutions. Point designs are characterized by many detailed metrics, like the ones we derived for the Hybrid Car (Table 15.1). Choosing between them implies scrutinizing them individually, but also by pairwise comparison.

TABLE 15.1 | Point design comparison for two Hybrid Cars

	Point Design 1	Point Design 2
Critically		
Must have an environmental satisfaction to the driver, and impact [as measured by EPA standards]	**128 g CO_2/km**	200 g CO_2/km
Must engage suppliers in long-term stable relationships with good revenue streams to them (mildly constraining) [as measured by supplier survey]	Average 4/5 survey score	Average 4/5 survey score
Must accommodate a driver (size) (constraining) [as measured by % American male and female]	94% of males 98% of females	**95% of males 99% of females**
Importantly		
Shall provide transport range [miles]	**200 km**	140 km
Shall satisfy regulatory requirements (constraining) [certificate of compliance]	Satisfies	Satisfies
Shall carry passenger (size and number) [how many]	5	5
Shall provide stable and rewarding employment [turnover rate]	2%/year	2%/year
Shall have good fuel efficiency [L/100km]	**4.2 L/100 km**	5.2 L/100 km
Desirably		
Shall require modest investment from corporate, shall sell in volume, and shall provide good contributions and return (mildly constraining) [ROI, Revenue $]	**24% ROI**	19% ROI
Shall have desirable handling characteristics [skidpad acceleration in g's]	**0.95 g**	0.82 g
Shall carry cargo (mass, dimension, and volume) [cubic feet]	1.1 m^3	**1.5 m^3**
Shall be inexpensive [sales price not to exceed]	$32,000	**$28,000**

In the point design representation given in Table 15.1, the entry with the more favorable design value is shown in boldface for each metric. Note that not all metrics are architecturally distinguishing; some produce equivalent metric outputs.

People use several heuristics to sort through a decision between two point designs:

- Set aside the metrics that are identical or nearly identical in value: supplier scores, regulatory compliance, number of passengers, and so on.
- Attempt to set aside metrics that are effectively equivalent in value: 99% versus 98% of females, 0.82 g versus 0.95 g on the skidpad (this means that the car will handle well enough as a sedan. But it is not a high-performance sports car).
- Focus on the metrics associated with the *critical* needs: CO_2, mileage. This is why we prioritize goals—to guide tradeoffs among the abilities of architectures to satisfy goals.
- Try to identify new constraints that would eliminate a choice; for example, anything less than 160 km of range is unacceptable.
- Finally, if all else fails, we often make decisions on the basis of a single metric. For example, all else being equal, we would prefer a cheaper car, so we choose Point Design 2.

Comparing two point designs is different from employing a tradespace. Even though the comparison of Table 15.1 enabled us to go deeper into the metrics, it encouraged us to make the decision on the basis of a single driving metric. Table 15.1 disabled our ability to reason in two or three dimensions.

Our third observation about tradespaces is that they present a large number of choices. This is a trivial statement, but it counters a common mental model that there are usually few design options available. If we had begun with the objective of designing a 335-hp engine and listing requirements, we might have quickly backed ourselves into one or two options. Thinking about a tradespace means thinking in a paradigm where choices are enumerated, structured, assessed, and down-selected. Just like solution-neutral function (Chapter 7), tradespaces help steer us away from focusing on a single design. Tradespaces are not easy to construct—witness the complexity we invested in Chapter 14—but we believe that they are feasible and transformational for complex systems that can bear this investment.

The rest of this chapter delves deeper into how we can mine a tradespace for valuable information.

15.3 The Pareto Frontier

The Pareto Frontier and Dominance

The Pareto frontier [2] is a set of architectures that form the "edge of the envelope," colloquially speaking. Because we have two or more metrics represented in a tradespace, it is unlikely that any single architecture is uniquely "the best." Rather, the Pareto frontier, also known as the Pareto front, showcases the architectures that are "good" and represent good tradeoffs between the metrics.

Consider, for example, the simple tradespace given in Figure 15.2, which contains three engines from Figure 15.1 (A, B, and C). Note that performance and cost are in tension: Low cost and high performance are good, but increasing performance generally requires increasing cost, as Figure 15.1 confirms.

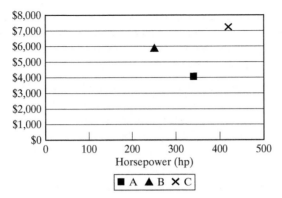

FIGURE 15.2 A simple tradespace with a few engines, using performance (hp) and cost as metrics.

The key concept for the Pareto front is *dominance*: The set of non-dominated architectures is called the Pareto frontier. An architecture A_1 *strongly dominates* another A_2 if A_1 is better than A_2 in *all* metrics. If A_1 is at least as good as A_2 in all metrics, and better than A_2 in at least one metric, we say that A_1 *weakly dominates* A_2. If no architecture dominates an architecture A_1, then A_1 is said to be *non-dominated*.

In the simple example of Figure 15.2, engine B is objectively worse than A: Engine A can provide more power than B at lower cost. Engine B is dominated by engine A. Engines A and C are not dominated by any other engine; that is, they are non-dominated and on the Pareto frontier. Engine B is not on the Pareto frontier. There is no objective way of assessing, on the basis of this information, which of A and C is better; we need to include other information in order to refine our preferences.

The Pareto Frontier of a GNC System Example

To see what a Pareto front looks like in a more complex case, we will now introduce an example based on a guidance, navigation, and control (GNC) system. GNC systems are an essential component of autonomous vehicles—cars, aircraft, robots, and spacecraft. They measure the current state (position and attitude of the vehicle), compute the desired state, and create inputs to actuators to achieve the desired state. A key feature of such systems is extremely high reliability, which is often achieved by having multiple (and sometimes interconnected) paths from sensors to actuators.

We return to our observation from Chapter 14 that, in general, the screening metrics for a system are some combination of performance, cost, and developmental or operational risk. In the GNC case, we assume that all architectures give adequate performance, and we focus on mass and reliability. Mass is important for all vehicles because they need to carry their own weight, and mass has a significant impact on cost. Reliability is a key concern in system-critical autonomous systems, especially those where remote maintenance is very costly or impossible.

We further assume that we have three types of sensors available (A, B, C), with increasing mass and reliability (more reliable components are generally heavier); three types of computers (A, B, C), also with increasing order of mass and reliability; and three types of actuators (A, B, C). We have one generic type of connection, which we initially assume is massless and of perfect reliability. For this example, we can define the decisions as the number and type of sensors, of computers, and of actuators, each up to three. Initially there is no penalty for choosing different types of, say, actuators. Another architectural decision is the number and pattern of the connections, with the constraint that every component must be connected to at least one other component.

To be more concrete, an autonomous car could have three different sensors (sonar sensors to determine range to cars in front, a GPS receiver, and a cornering accelerometer), three computers, one of a different type (one FPGA and two microprocessors), and three similar actuators (an electric motor for the front wheels, a larger electric motor for the rear wheels, and a gas engine that can directly drive the front wheels).

Given the symmetry of the problem in the connections between sensors and computers, and between computers and actuators, we will look only at the sensor–computer part of the problem. Figure 15.3 shows some valid architectures.

For each architecture, boxes represent sensors (S) and computers (C) of different types (A, B, C), and lines represent interconnections between them. Figure 15.4 shows the tradespace of mass versus reliability for the 20,509 valid architectures, with the Pareto frontier highlighted with a solid line.

Let's assume that we want to be in the region of the space with at least 0.99999999 (colloquially known as "8 nines") of reliability. We can zoom in on that region of the tradespace, and see what some architectures on this region of the Pareto front look like. This is shown in Figure 15.5.

Four architectures that lie on the Pareto frontier are represented on the righthand side of Figure 15.5. For example, the top-right architecture consists of three sensors of type C and three computers of type C, with all possible connections between them. This architecture achieves the highest reliability at the highest mass, as expected.

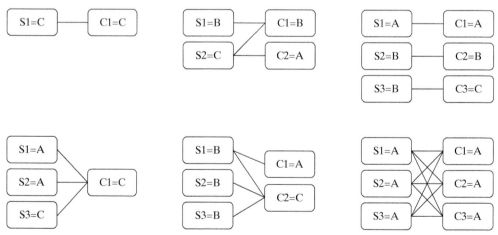

FIGURE 15.3 Pictorial representation of some valid architectures for the GNC problem. Architectures consist of a number of sensors of type A, B, or C, connected in a topology to a set of computers of type A, B, or C. Note the high level of abstraction at which we work, allows us to formulate a decision-making problem that captures the main trades of interest—in this case, mass versus reliability.

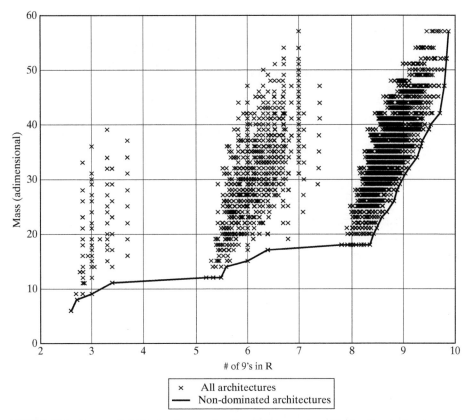

FIGURE 15.4 Mass–reliability tradespace for the GNC example, with the Pareto front highlighted. The number of nines (# of 9's) in the reliability is shown instead of the reliability value (for example, 3 nines is equivalent to R = 0.999, and R = 0.993 is equivalent to 2.15 nines).

When analyzing a Pareto plot, our first instinct must be to look for the *Utopia point,* an imaginary point on the tradespace that would have the best possible scores in all metrics. In Figure 15.4 and Figure 15.5, the Utopia point is at the bottom right of the plot, ideally with maximum reliability and minimum mass. Non-dominated architectures typically lie on a convex curve off the Utopia point. Visually, we can quickly find the approximate Pareto frontier of a tradespace plot by moving from the Utopia point toward the opposite corner of the tradespace plot; the first points that we find—those that are closest to the Utopia point (engines A and C in Figure 15.2)—usually form the Pareto frontier. Note, however, that the numerical distance from an architecture to the Utopia point cannot be used to determine whether one architecture dominates another.*

Despite the density of architectures in the tradespace enumerated in Figure 15.5, the Pareto front is not smooth. Unless there is a simple underlying function governing the tradeoff

---

*In fact, any measure of distance is just another way of expressing preferences among conflicting objectives. For example, Euclidean distance gives equal weights to the squares of the distances of the metrics to the Utopia values, which in practice gives disproportionally greater weight to higher values of distances.

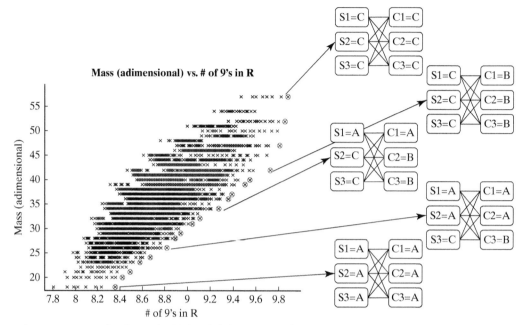

FIGURE 15.5 Zooming in on the high-reliability region of the mass–reliability tradespace. Some architectures on the Pareto frontier are pictorially illustrated.

(for instance, antenna diameter is a square root function of data rate in a communication system), it is rare for the Pareto front to be smooth. Irregularities are caused by discrete and categorical decisions and shifts between technologies. However, there is a shape to the front; it is roughly running up and right but "curving left," indicating that gains in reliability require successively greater mass investment. This type of "knee in the curve" around 9.7 R's is frequently interpreted as a balanced tradeoff in decisions.

The Fuzzy Pareto Frontier and Its Benefit

Dominated architectures are not necessarily poor choices. Some slightly dominated architectures may actually outperform non-dominated ones, judging on the basis of other metrics or constraints not shown in the plot, and this makes them excellent candidates for further study.

The concept of *fuzzy Pareto frontier* was introduced to capture a narrow swath of choices near the frontier. The fuzzy frontier can be an explicit acknowledgment of performance uncertainty, ambiguity in measurement, errors in the model, or the like. Moreover, architectures that are non-dominated in a few metrics (such as performance and cost) are sometimes the least robust in responding to variations in model assumptions. [3] When we are dealing with families of architectures, we sometimes find that one key member of the family is slightly off the frontier. The main idea is to relax the Pareto filter condition so that slightly dominated architectures that may subsequently be important are not discarded. A method for computing the fuzzy frontier is given in the next section. The fuzzy Pareto frontier represents our first cut at reducing the tradespace down to a manageable decision.

Mining Data on the Fuzzy Pareto Frontier

In Figure 15.5, all architectures on the frontier appear to be fully cross-strapped; that is, all sensors were connected to all computers. This suggests that fully cross-strapped architectures may be dominating the Pareto frontier. How can we perform this analysis in a systematic way?

Simply plotting the architectures with the same decision choice using the same color or marker is a simple way to view the results. For example, Figure 15.6 is a view of the same tradespace, where non-dominated cross-strapped architectures are highlighted using squares. The Pareto frontier is still shown by the solid line. It is apparent from Figure 15.6 that all architectures on the Pareto front are fully cross-strapped.

When graphical methods are not conclusive, a simple quantitative method is to compute the relative frequency with which a certain architectural decision D is set to a certain alternative $D1$ inside and outside the (fuzzy) Pareto frontier. If the percentage of architectures with

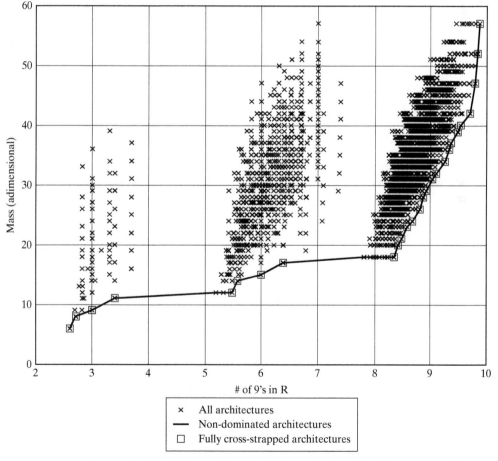

FIGURE 15.6 Tradespace for the GNC example, where all non-dominated architectures are shown to be fully cross-strapped.

TABLE 15.2 | Percentage of architectures on the GNC Pareto front that share a common architectural feature (that is, a particular decision choice or combination of decision choices) and other, related statistics

Architectural feature	% architectures with this feature in entire trade space	% architectures of the Pareto front that have this feature	% architectures with this feature that are on the Pareto front
Fully cross-strapped	2%	100%	7%
Homogeneous (one type of sensors and computers)	7%	33%	1%
Uses one or more sensors of type C	59%	56%	0.1%

alternative $D1$ is significantly higher inside the Pareto frontier than outside the Pareto frontier, this would suggest—though not prove—that there is a correlation between choice $D1$ and the non-dominated architectures.

Table 15.2 shows statistics on the Pareto frontier of Figure 15.5. Three features are analyzed:

- Fully cross-strapped versus not fully cross-strapped.
- Homogeneous (using a single type of sensor and a single type of computer) versus non-homogeneous.
- Architectures that have one or more sensors of type A versus those that don't.

Table 15.2 confirms that "fully cross-strapped" is a dominant architectural feature: Only 2% of the architectures enumerated in the tradespace are fully cross-strapped, and yet all non-dominated architectures are fully cross-strapped. Being fully cross-strapped appears to be a necessary condition for an architecture to lie on the Pareto front.

The case of homogeneous architectures is similar in that only 7% of architectures in the tradespace are homogeneous, but all of them lie on the Pareto front. However, this feature is not as dominant as the "fully but a disproportionately high percentage of non-dominated architectures are homogeneous" condition: Only 33% of the architectures on the Pareto front are homogeneous. In other words, homogeneity is not a necessary condition to lie on the Pareto front; there are ways to achieve non-dominance even for non-homogeneous architectures.

Conversely, 59% of the architectures in the tradespace use one or more sensors of type C, but about the same percentage (56%) of non-dominated architectures have this feature. This clearly suggests that having one or more sensors of type C is not a dominant feature.

Eliminating architectures that are not on the Pareto front is a means for down-selecting architectures, but it is not the sole method. Pareto dominance may not be useful for selecting architectures when the number of metrics increases. Indeed, in large-dimensional objective spaces, the number of non-dominated architectures is simply too high to be useful. Therefore, Pareto dominance needs to be seen as one filter used to discard poor (highly dominated) architectures, rather than as an algorithm sufficient by itself to obtain a manageable set of preferred architectures.

Pareto Frontier Mechanics

How do we compute the Pareto frontier in a general tradespace with thousands or millions of architectures and several metrics? If we generalize what we did in our trivial example of the engines in Figure 15.2, an algorithm for computing a Pareto front requires comparing each architecture to all other architectures in all metrics. Therefore, in a tradespace with N architectures and M metrics, the algorithm takes, in the worst case, $\binom{N}{2}M$ comparisons to find the non-dominated set.

Sometimes, it can be interesting to compute Pareto fronts in a recursive fashion in order to sort architectures according to their *Pareto ranking*. This process is called *non-dominated sorting*. An architecture that is non-dominated has a Pareto ranking of 1. To find the architectures with Pareto ranking 2, we simply discard architectures with Pareto ranking 1 and compute the new Pareto frontier by following the same procedure. The architectures on the new Pareto front have a Pareto ranking of 2. We repeat this process recursively until all architectures have been assigned a Pareto ranking. This algorithm takes on the order of N times $\binom{N}{2}M$. Faster algorithms exist for non-dominated sorting that keep track of the dominance relationships, at the expense of memory usage. [4]

Pareto ranking can be used as a metric for architecture selection. For example, given the Pareto rankings of a set of architectures, it might be a good idea to discard architectures that have very bad Pareto rankings (in other words, those that are heavily dominated).

A straightforward implementation of a fuzzy Pareto filter is based on non-dominated sorting. For instance, a fuzzy Pareto frontier may be formed with architectures that have a Pareto rank less than or equal to 3. This implementation of fuzzy Pareto fronts is mathematically elegant, but it does not take into account the distance in the objective space between architectures with different Pareto rankings. As a result, one might discard an architecture that is in a region of very high density of architectures, and is very close to the first Pareto frontier, on the basis of a high (that is, poor) Pareto rank. To address this problem, an alternative implementation of fuzzy Pareto fronts computes the true Pareto front and then discards all points that have a distance to the Pareto front greater than some threshold (such as 10% Euclidean distance in a fully normalized objective space).

15.4 Structure of the Tradespace

Pareto analysis is an important part of tradespace analysis, but it is by no means sufficient. There is a great deal to be learned from looking at the structure and features of the tradespace as a whole, rather than focusing only on the fuzzy Pareto frontier. Sometimes a trade space just looks like a cloud of points. However, most tradespaces have some structure; they have a shape. They have features such as holes, subgroups, and fronts. These are due to factors such as discrete metrics, different dynamic ranges of metrics, and physical laws limiting certain metrics.

A common feature of tradespaces is the presence of *clusters*—that is, the accumulation of architectures in relatively small regions in the objective space, leaving relatively large open areas, as shown in Figure 15.7 for the GNC example. In Figure 15.7, we observe three very large clusters in the mass–reliability space. Clusters are important because they suggest the presence of families of architectures that achieve similar performance in one or more metrics. Note, however, that distance in the objective space is not necessarily linked to distance in the architectural space. In other words, architectures in the same cluster can actually be very different from each other.

342 PART 4 • ARCHITECTURE AS DECISIONS

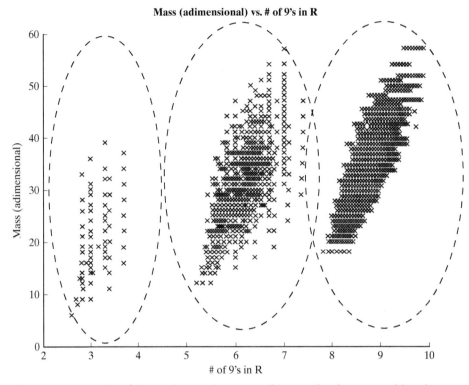

FIGURE 15.7 Example of clusters in a tradespace. In this case, the clusters are driven by an emergent property of the architecture, namely the minimum number from among the number of sensors and the number of computers.

It is useful to view clusters with similar architectural variables, which can be simply done by highlighting the points that share the same decision choice (for example, by using the same color or marker type). If a cluster in the objective space corresponds to a cluster in the architectural space, we have found a family of architectures with similar architectural decisions and similar metrics. This idea is expanded upon in the concept of set-based S-Pareto frontiers. [5]

Clusters appear when one architectural feature (which can be a decision or a property that emerges from a combination of decisions) dominates the objective space. For instance, in the GNC architecture example of Figure 15.7, the clusters are driven by the minimum number from among the number of sensors (*NS*) and the number of computers (*NC*), which drives the reliability of the system. This is clearly seen in Figure 15.8, where architectures use different markers according to their $min(NS, NC)$. The leftmost cluster consists exclusively of architectures with $min(NS, NC) = 1$ (circles on Figure 15.8), which means that the architectures there have either one sensor or one computer. The rightmost cluster consists exclusively of architectures with $min(NS, NC) = 3$ (dots), and all architectures with $min(NS, NC) = 2$ (crosses) are in the center cluster. Note that some dots appear in the center cluster. (Note that by changing the reliabilities of the components—the sensors and computers—we can completely separate the three clusters or make them much closer together, but the underlying structure will still be a tradespace with three clusters.)

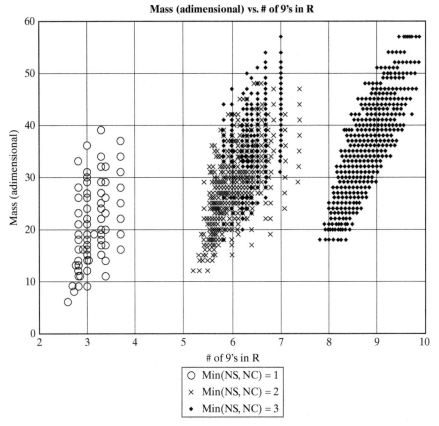

FIGURE 15.8 Same tradespace as in Figure 15.4, but here the architectures have different markers according to the minimum number from among their number of sensors (NS) and their number of computers (NC).

Although the brain is an excellent cluster identifier, we can also use clustering algorithms to divide the space into clusters. Clustering algorithms and their application to system architecture are described in Appendix B.

One of the most salient features of the general structure of many tradespaces is the presence of *stratification*. Strata are groups of points for which one of the metrics is constant while the other varies. In a two-dimensional tradespace, strata appear when points line up in a number of vertical or horizontal lines, with gaps in between them. Stratification appears, for example, when one of the metrics is discrete (number of development projects, for instance). More generally, even in the case of continuous metrics, this can occur when combinations of architectural options produce only a few distinct values of the metric. For example, in the GNC case, our mass metric is restricted to all the different sums of the masses of sensors and computers of types A, B, and C (there are 49 different values).

When a tradespace is stratified, most of the architectures are dominated by a handful of N non-dominated architectures, where N is typically equal to the number of strata (for example, the number of possible values of a discrete metric). This is because all the architectures in a stratum are dominated by one architecture at the edge of the stratum. For instance, in Figure 15.5, all architectures with a mass of 57 are dominated by the architecture highlighted in the top right corner. Furthermore, whole strata can be dominated by other strata. This is apparent in Figure 15.4, where several strata from the first cluster are dominated by strata from the second cluster.

Just as we can ask ourselves what features are shared between architectures on the Pareto front, we can also ask ourselves what features are shared between architectures within a stratum. Figure 15.9 is the same zoom on the high-reliability region of the tradespace that we showed in Figure 15.5, but now we highlight with squares the architectures that have a fixed choice of sensors (namely, three sensors of type A) and a fixed choice of computers (namely, three computers of type A). We see that these AAA-AAA architectures, which differ only in the way these elements are interconnected, form the bottom right stratum of the tradespace. This is easy to understand. Because, in this particular simulation, interconnections were assumed massless, one can obtain increasingly reliable architectures by adding connections to the same set of sensors and computers. All these architectures are dominated by the fully cross-strapped architecture.

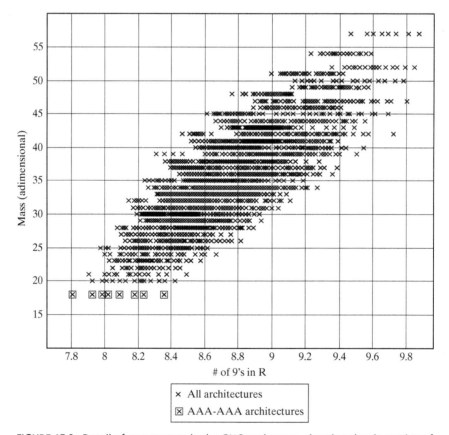

FIGURE 15.9 Detail of one stratum in the GNC tradespace, showing that it consists of different ways of connecting three sensors of type A with three computers of type A.

15.5 Sensitivity Analysis

At the conclusion of any analysis—here, our tradespace analysis—it is important to ask how sensitive the results are to modeling and human input assumptions. This would be near the end of what Simon called the "Design Activity." Answering this question in a systematic fashion is often more resource-consuming than solving the problem in the first place, but it is an essential part of the process.

A posteriori sensitivity analyses typically involve re-running a model under many different *scenarios,* where each scenario contains a different set of modeling assumptions. For instance, all the GNC tradespaces that we showed in this chapter were generated assuming that the mass of connections between elements is zero. If we make the connection mass non-zero, we obtain the tradespace shown in Figure 15.10 (compare it with Figure 15.5, which assumes connections with no mass). Note that the shape of the tradespace has changed considerably, because the horizontal strata formed by different ways of connecting the same components have disappeared. Moreover, architectures on the Pareto frontier are no longer fully cross-strapped. We see channelized architectures (one connection between sensors and computers), as well as architectures that have two connections per component.

Similarly, all the previous tradespaces assumed no penalty for architectures that used more than one type of component. The justification for such a penalty could be simply a desire to minimize the number of types of components of the architecture. Conversely, one might try to justify a negative value for the "dissimilar component" penalty by arguing that diversity reduces the risk of common-cause failures.

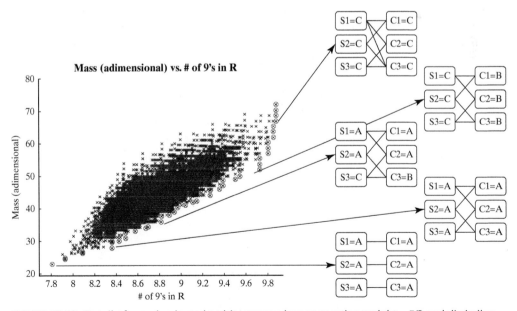

FIGURE 15.10 Detail of non-dominated architectures when connection weight = 5/3 and dissimilar component penalty = 0.

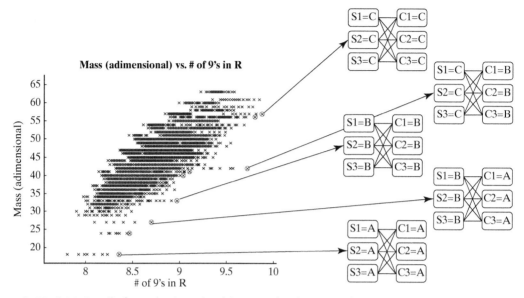

FIGURE 15.11 Detail of non-dominated architectures for the case with connection weight = 0 and dissimilar component penalty = 9. Homogeneous architectures now dominate the Pareto frontier.

Figure 15.11 shows what happens if we add a mass penalty of +9 (a large penalty) to architectures with dissimilar components. Remember that mass is assumed to be normalized with respect to the minimum component mass m_{ref}. Unlike the case of Figure 15.5, which had no penalty on dissimilar components, the Pareto frontier of Figure 15.11 is completely dominated by homogeneous architectures that use only one type of component.

Of course, these two tradespaces of Figure 15.10 and Figure 15.11 were obtained for two specific scenarios. In general, we will need to perform a more exhaustive and systematic scenario analysis to understand the robustness of the results; however, this could be extremely resource-consuming. If we wanted to analyze scenarios with values of connection weight between 0 and 2 in increments of 0.1, values of dissimilar component penalty between −2 and 10 in increments of 1, and three different values of the masses and reliabilities of each type of component, we would need to run the tool about 600,000 times. Fortunately, some techniques from design of experiments (such as Latin hypercubes and orthogonal arrays) can be used to identify a smaller subset of scenarios to run. These techniques are briefly introduced in Chapter 16.

Once the scenarios have been run, it is interesting to view the *robustness* of the results. Figure 15.12 shows results from the simulation for different values of normalized connection weight ranging from 0.0 to 2.0. The plot shows the percentage of architectures on the Pareto front that are fully cross-strapped as a function of the connection weight. For massless connections, we confirm our previous result that 100% of the architectures on the Pareto front are fully cross-strapped (the point on the top left corner of the chart). Interestingly, the percentage of fully cross-strapped architectures falls very quickly with connection weight. For a connection

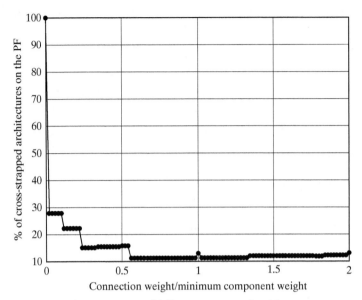

FIGURE 15.12 Percentage of fully cross-strapped architectures on the Pareto front as a function of connection weight.

weight of only 0.02, the percentage is already below 30%. This indicates that our result that fully cross-strapped architectures dominate the Pareto frontier is not a robust result, because when a parameter of the model changes even modestly, that result no longer holds. In fact, the converse is a more robust result: Fully cross-strapped architectures are not a dominating option unless the weight of connections is negligible compared to the weight of components.

Sometimes the result of the sensitivity analysis will indicate that the results are dependent on the value of a parameter. In these cases, it is necessary to study this part of the problem more deeply and consider refining the model. Suppose, for example, that we try to see the sensitivity to the dissimilar component penalty parameter. Figure 15.13 plots the percentage of architectures on the Pareto front that are homogeneous as a function of the normalized dissimilar component penalty (normalized by the maximum component mass), which varied between −2.0 and +2.0. In this range, the percentage of homogeneous architectures increases from almost 4% to 100% as the dissimilar component penalty (DCP) goes from −0.5 to 1.1. This allows us to define three regions of sensitivity. For DCP < −0.5, virtually no architectures on the Pareto front are homogeneous, and the result is insensitive to the penalty. For DCP > 1.1, 100% of architectures on the Pareto front are homogeneous, and the result is insensitive to the penalty. In between, the results are sensitive to the value of the DCP. Note that our default value of DCP = 0 is in the sensitive region, which suggests that this result (33% of architectures are homogeneous) is not robust.

Robustness of results can be used to assess the robustness of architectures to real variations in the technical and programmatic environment. For instance, if an architecture appears on the fuzzy Pareto front in most of the scenarios, it suggests that the result that this is a good

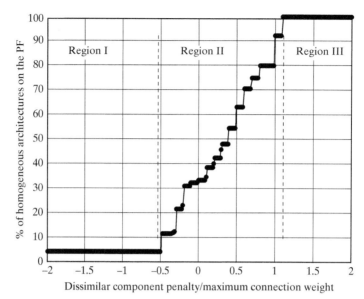

FIGURE 15.13 Percentage of architectures on the Pareto front that are homogeneous as a function of the dissimilar component penalty. Three regions can clearly be defined with different ranges for the parameter.

architecture is a robust result—that it holds for a range of different scenarios. Conversely, if an architecture that was non-dominated in the nominal scenario appears to be highly dominated in many scenarios, then it is clear that this is not a robust result. This situation is discussed in Box 15.1 Principle of Robustness of Architectures.

More systematically, we can compute average metrics of an architecture across all scenarios, or the percentage of scenarios in which an architecture attains a "good" score on a certain metric, and use that information to assess the robustness of an architecture in responding to changes in the modeling assumptions.

More sophisticated approaches to sensitivity analysis are possible. Probabilities can be assigned to the values of the model input parameters—and therefore to each scenario. For instance, in the GNC example, we may not know what the right value for the connection weight is, but we may have an idea of its probability distribution (such as pessimistic value = 0.5 with probability 20%, best guess = 0.2 with probability 60%, and optimistic values = 0.05 with probability 20%). These probabilities could be used to weigh the values of the different scenarios, rather than doing simple averages. More generally, probability distributions could be assigned to all model parameters, and then the probability distribution of the results could be obtained by Monte Carlo analysis.

For example, demand, the price of some primary materials (such as graphene), and the performance of key enabling technologies (such as an optical communications transceiver) are important uncertain parameters in designing many complex systems. One can model the uncertainty of these parameters using probability distributions. For example, based on current

> **Box 15.1 Principle of Robustness of Architectures**
>
> *The last 10 percent of performance generates one-third of the cost and two-thirds of the problems.*
>
> Norm Augustine's Law Number XV
>
> *Ninety percent of the time things will turn out worse than you expect. The other 10 percent of the time you had no right to expect so much.*
>
> Norm Augustine's Law Number XXXVII
>
> Good architectures need to respond to all manner of variations. They can respond to these variations by being robust (capable of dealing with variations in the environment) or by being adaptable (able to adapt to changes in the environment). Optimal architectures, in the Pareto sense, are often the least robust. Consider optimality, robustness, and adaptability in the choice of an architecture.
>
> - Variations can occur in the input assumptions to the architecting process, in the downstream execution of the design, and/or in the operating environment.
> - Architectures on the Pareto frontier are sometimes the least robust in responding to these variations. For example, optimal architectures may rely on a game-changing technology that has low technology maturity. What happens if that technology is not ready when the system is developed? Is the architecture flexible enough to accommodate a different technology? Is there a strong tradeoff between optimality and robustness in your industry? The architect must decide whether a narrow peak in performance is achievable.
> - Robustness can be assessed by analyzing the architectures under a variety of technical and market scenarios and then picking the architectures that consistently perform well under a broad range of circumstances. Using the fuzzy Pareto frontier instead of the Pareto frontier in the context of architecture optimization, as explained in this chapter, can also help identify the more robust architectures.

data, [6] one can assume that in five years the price of graphene will be at $1\$/cm^2$ with probability 0.6, at $10\$/cm^2$ with probability 0.2, and at $0.1\$/cm^2$ with probability 0.2. Demand in power distribution systems is also estimated using probability distributions to simulate the performance of the grid. [7]

However, this approach is generally very resource-consuming, because a large number of scenarios are likely to be required to accurately approximate the probability distribution of the results. Furthermore, defining probability distributions for all modeling parameters may be a challenging task in itself.

In practice, the Monte Carlo approach may be infeasible for many problems, especially when heavy computational power is not available. Instead, system architects typically have to conduct sensitivity analysis based on a reduced number of scenarios designed to cover most significant combinations of values for the relevant parameters. The techniques mentioned earlier for the design of experiments (orthogonal arrays and Latin hypercubes) can provide guidance in choosing those scenarios.

15.6 Organizing Architectural Decisions

Impact on Other Decisions

The information garnered from the simulation and visualization of the tradespace can help architects understand tradeoffs, but we argued earlier that this information needs to be condensed and organized to be most useful. The goal of this section is to structure the set of architectural decisions—that is, to identify decisions that need to be made with priority in mind and to study how these decisions are connected to each other.

The first step in organizing architectural decisions is identifying the "high-impact decisions." These are the decisions that are likely to have a strong impact either on metrics or on other decisions. In other words, there are two main reasons why an architectural decision may be "high-impact": a high sensitivity of the metrics, or a high degree of connectivity (or coupling) with other decisions. [8]

Architectural decisions are coupled. The *degree of connectivity* of a decision is a measure of how coupled this decision is to the other architectural decisions. For example, the choice of drive wheels for a car is highly connected to other decisions, such as front steering geometry, engine layout, and rear wheel suspension. This decision might be so important that we refer to "rear wheel drive architectures." Similarly, the choice of microprocessor in a computer drives many design decisions, including speed and amount of memory. A change in a decision with a high degree of connectivity will potentially affect a larger set of decisions than a change in a decision with a lower degree of connectivity. Connectivity also allows the impact of decisions to propagate through a system. [9]

How do we quantify this degree of connectivity? In artificial intelligence, in the context of constraint satisfaction problems, the degree of connectivity of variables is often computed as the total number of constraints that use that variable. [10] Simmons adapted this definition and counted the number of logical constraints that use a specific decision. [11] For instance, in the Apollo example, the constraints of Table 14.1 couple the LOR decisions through a number of logical constraints. (For example, if the *LOR* decision is set to "yes," then the *moonArrival* decision needs to be set to "Orbit"). Each logical constraint adds one to the connectivity of the decisions it contains. Counting this way, the LOR decision would have a connectivity of four.

A slightly different version of the same concept is based on the use of Design Structure Matrices (DSMs) shown in Table 14.3. We can construct a DSM by considering decisions as the entities and the constraints as relationships. The relationships are the number of logical constraints that combine two decisions. Table 14.3 shows four constraints for the LOR decision, but does not tell us what they are, as Table 14.1 does.

The reader may notice that this definition of connectivity depends on the formulation of the decisions in the architectural model. In the Apollo example, we could have decided to have a single decision for mission-mode with only the feasible options, as opposed to five binary decisions with three logical constraints to establish the mission-mode. This formulation would change the connectivity of decisions. This would essentially be a different choice of decomposition for the architectural decisions—one that combines a set of coupled individual decisions (*EOR, earthLaunch,* and so on) into a single *missionMode* decision. In essence, this is not different from the discussion of decomposition in Chapter 13, but is merely applied to decisions. As general guidance, if a set of decisions are highly coupled, such as the five binary mission-mode decisions, it is usually preferable to replace them with a single decision. This makes the model simpler by reducing the number of decisions and removing unnecessary constraints. Furthermore, it guarantees that we are not artificially infusing connectivity into some decisions by our choice of problem formulation. Indeed, decomposing the mission-mode decision in five binary decisions requires adding three constraints (a, b, and c in Table 14.1) to eliminate illogical combinations. These three constraints disappear when we use a formulation with a single mission-mode decision; therefore, the connectivity of the single mission-mode decision will be lower. Even though doing this would not substantially change our conclusions in the Apollo example, it is good practice to assess how much of the connectivity of the decision comes from the problem formulation. This topic will be discussed more in Chapter 16.

This definition of connectivity can be augmented by also considering "reasonableness constraints" and couplings due to metrics, as discussed in Section 14.2. In the Apollo example, one could argue that the mission-mode decisions are coupled to the choice of propellants through the IMLEO metric, as shown in Table 14.4. This kind of coupling is driven by interactions between decisions, and it is clearly not captured by the logical connectivity metric. Instead, it needs to be analyzed by looking at the sensitivity of the metrics to specific groupings of decisions and alternatives. We will come back to this point later in this section, and again in Chapter 16. [12]

Impact on Metrics

The *sensitivity* of a decision to a metric measures how much that decision affects that particular metric. One way of quantifying this sensitivity would be to compute the main effect, as defined in the field of design of experiments. [13] The *main effect* is a measure of the average change in systemwide properties produced by changing one binary variable in a decision problem. For example, given a group of architectures $x \in X$, a decision i that can take values $\{0, 1\}$, and a metric M, the main effect is given by

$$\text{Main effect (Decision } i, \text{Metric } M) \equiv \frac{1}{N_1} \sum_{\{x | x_i = 1\}} M(x) - \frac{1}{N_0} \sum_{\{x | x_i = 0\}} M(x)$$

where N_1 is the number of architectures for which decision i is set to 1, and N_0 is the number of architectures for which decision i is set to 0. For example, in the Apollo case, if we switched LOR from yes to no, we could use a model and the main effect analysis to identify how the architectures with LOR = yes perform compared to those with LOR = no.

This formulation based on main effects deals only with binary variables. We can extend this formulation to allow for decisions with more than two alternatives in the following way. [14] Given a decision i that can take values $k \in K$, we compute its sensitivity to a metric as the sum, for all possible values of the decision, of the difference in absolute value between the mean value of the metric across all architectures for which the decision takes that particular value, and the mean value of the metric across all other architectures. More precisely,

$$\text{Sensitivity (Decision } i, \text{Metric } M) \equiv \frac{1}{|K|} \sum_{k \in K} \left| \frac{1}{N_{1,k}} \sum_{\{x | x_i = k\}} M(x) - \frac{1}{N_{0,k}} \sum_{\{x | x_i \neq k\}} M(x) \right|$$

Note that this is nothing more than adding the main effects (in absolute value) of a set of $|K|$ "hidden" binary variables that tell us whether or not decision i is taking value k. This alternative formulation is a measure of the average magnitude of change in a metric that occurs when the assignment of a particular decision alternative is changed. (Note that in the special case of $k = 2$, the alternative formulation yields twice the main effect of the decision.)

The sensitivity of a decision to a metric is calculated for a given group of architectures. It could be the entire tradespace, the fuzzy Pareto frontier, or any other interesting region in the architecture tradespace. In practice, these options can yield different results, and they have their own advantages and disadvantages. In particular, choosing only the fuzzy Pareto front has the advantage of excluding very poor architectures from the analysis, but it will introduce a bias in the calculation of main effects for certain architectural decisions that are not well represented in the Pareto front. Sometimes, we may not have access to the entire tradespace, because our search algorithm may provide us only with the Pareto front or the fuzzy Pareto front.

If the entire tradespace is available, it might make sense to use it for the sensitivity analysis in the initial exploratory phase, where the main goal is to identify the driving variables from the original tradespace to reduce the size of the model. Conversely, it might make sense to restrict the analysis to the (fuzzy) Pareto front in later phases of the system architecture process, when we are more interested in the actual values of certain sensitivities important in the final selections.

Decision Space View

We now start to consider how information about the connectivity and sensitivity of decisions can be viewed, and how it can influence the decision process. Decisions can be plotted in a two-dimensional space representing the degree of connectivity versus sensitivity, forming what Simmons called the *decision space view*. The decision space view is a way to visualize both measures of a decision's impact in a single diagram. On the vertical axis is a measure of sensitivity of each decision to a given metric. On the horizontal axis is the degree of connectivity. A separate chart must be made for each metric. The decision space view for IMLEO in the Apollo program is given in Figure 15.14.

The interpretation of the decision space view is shown in Figure 15.15. The plot is divided into four quadrants in which a decision maker can prioritize decisions. The four quadrants in Figure 15.15 are a qualitative measure of the different types of decisions in an architecture.

The upper right quadrant (I) contains decisions that are both sensitive and strongly connected. These decisions should be made first. They influence the outcome of downstream decisions, and they have the greatest impact on the system properties—that is, they determine the performance envelope, system cost, and so on.

The upper left quadrant (II) contains decisions that are very sensitive but are weakly connected to other decisions. A change in these decisions will influence metrics strongly but will not influence many other decisions. The decisions that lie in this quadrant can be analyzed quite independently, apart from the other decisions. Decisions in this category should be given the second-highest priority.

The lower right quadrant (III) contains decisions that do not strongly affect metrics but are strongly connected to other decisions. The decision maker can wait to resolve these decisions until the highly sensitive ones (those in quadrant I or quadrant II) are resolved, in order to have a complementary set of decisions in the architecture. These decisions are classic detailed design decisions. They help specify how the product or system will deliver functionality, and they are

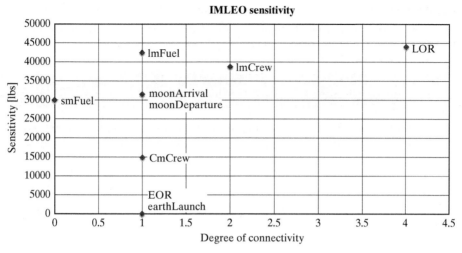

FIGURE 15.14 Decision space view for IMLEO in the Apollo program. Each dot in the chart represents a decision, and the position of the dot indicates how sensitive and how connected that decision is.

wrapped up in system-level issues, but they do not have a strong influence on the value proposition (benefit at cost). These decisions can be made late in the design phase.

The lower left quadrant (IV) contains decisions that are neither impactful nor connected to other decisions. The decisions in this quadrant can be processed in parallel or left until the end of the decision-making process, since they are largely independent of other decisions. They can, for example, be used to satisfy requirements of lower priority than those used to screen the decisions for sensitivity.

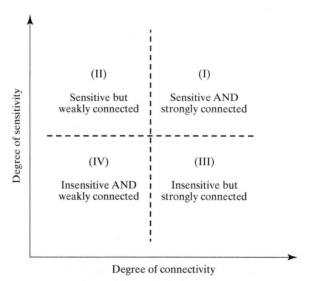

FIGURE 15.15 Organization of architectural decisions according to their sensitivity and degree of connectivity.

Sequencing Decisions

Building on this analysis, we can construct a hierarchy of architectural decisions, as discussed in Box 15.2 Principle of Coupling and Organization of Architectural Decisions.

> ### Box 15.2 Principle of Coupling and Organization of Architectural Decisions
>
> *"We are drowning in information, while starving for wisdom. The world henceforth will be run by synthesizers, people able to put together the right information at the right time, think critically about it, and make important choices wisely."*
>
> Edward O. Wilson
>
> Architecture can be thought of as the outcome of a relatively small number of important decisions. These decisions are highly coupled, and the sequence in which they are made is important. The sequence of architectural decisions can be chosen by considering the sensitivity of the metrics to the decisions and the degree of connectivity of decisions.
>
> - Architectural decisions are generally coupled. Couplings can be due to both constraints and metrics. The strongest couplings are due to hard constraints, when choosing a value for one decision precludes our choosing certain values for another decision.
> - When a decision is strongly coupled with many other decisions through constraints, we say that this decision has a high degree of connectivity.
> - Metrics have varying degrees of sensitivity to architectural decisions.
> - Decisions that fall in the "first quadrant" (high sensitivity and high degrees of connectivity) should be made first, because they have the greatest impact on the remaining options.

In addition to the sequence suggested above, which is based on connectivity and sensitivity, there are two other factors to consider in sequencing decisions. First, weakly coupled decisions should be treated in parallel, because there is a high cost associated with waiting for one piece of analysis to begin before pursuing another. Second, decisions that are highly coupled should be combined in a single trade study, so that the relevant interaction effects are considered and a thorough tradeoff can be performed. Thus, the interdependency within groups of highly coupled decisions should be maximized, and the dependency between groups of decisions that are lightly coupled should be minimized.

In our Apollo example, we can organize the nine decisions of Figure 15.14 as shown in Figure 15.16. The root node of the hierarchy is clearly the LOR decision, since it is the most coupled and the most sensitive decision. One level below, we find three clusters of highly coupled decisions that can be made in parallel: the decisions concerning the lunar module (*lmCrew, lmFuel*), those related to the mission mode on the Moon (*moonArrival, moonDeparture*), and those related to the design of the service and command modules (*smFuel, cmCrew*). These three different trade studies should be made in parallel, since they are coupled through the IMLEO mass matric. The EOR decision can be made last, because it really depends on which launch

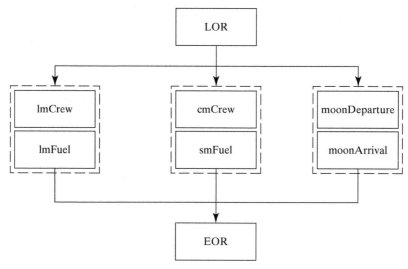

FIGURE 15.16 Organization of architectural decisions for the Apollo example.

vehicle is available at the time and is only loosely coupled to the rest of decisions. [15] For more complex problems, clustering algorithms can be used to suggest optimal sequencing, as discussed in Appendix B.

This example of organizing Apollo decisions illustrates that we can gain insight and process efficiency by structuring the architectural decisions as a hierarchic tree. This is a much more organic and tractable framing than simply using a large tradespace, which in essence asks the architect to make several decisions at once. Because these decisions take place at different times and potentially on the basis of different criteria, separating them mirrors the organic process more closely.

Our belief is that the outputs of this analysis—the ordered decision set and the identification of connected and sensitive decisions—make the analysis a valuable mental model in and of itself. If the analysis is not available, it may still be worth attempting to construct these two diagrams as a representation of knowledge about the architecture.

The main reason is that this representation has the advantage of helping to structure the architecting process. For example, we have identified which decisions should be made in tandem with other decisions, so a trade study for each set of coupled decisions would be desirable. This representation also speaks to the team structure, in that teams that influence the same decision will require more cross-team coordination effort, and by the same token, tightly coupled decisions should be matched with tightly coupled teams.

Concluding Remarks about Decisions and Sequence

The results of an architecture study are undoubtedly dependent on the architectural decisions and their formulation. Indeed, system architecting is an iterative process, and it is beneficial to refine the architectural model as we gain insight about which decisions are truly significant. For example, if we observe that a decision is mostly unconnected and has low sensitivity, we may want to get rid of this decision in the next iteration's model. Conversely, if we observe that a certain architectural decision has a very high impact, we can zero in on that decision and add finer options, or decompose it into multiple decisions.

A similar argument can be made for increasing or decreasing the fidelity of the metrics. If we observe that a certain metric at relatively high fidelity is not architecturally distinguishing, we can decide to get rid of that metric. Conversely, if a relatively low-fidelity metric is driving some of the results, it may be necessary to refine the metric.

We have described methods available for encoding some of these decisions in models, but the mainstay of a system architect's work remains judgment-oriented. Experience would suggest that the principles and methods that we have developed in support of computational system architecture are useful even in the absence of a model, to represent knowledge, to find patterns by structuring the representations, and to communicate the decisions, especially to decision makers who are comfortable with the decision framework.

Architectural competition (Chapter 12) exists in many industries today: mobile phones, UAV aircraft, and hybrid vehicles, to name just a few. A study of technology strategy suggests that industries consolidate under a dominant architecture after a period of architectural competition. Next they innovate on processes and components to earn incremental gains, but then they are disrupted as a new architecture emerges that removes constraints on performance gains that were imposed by the previous architecture. In industries with a dominant architecture, systems thinking is paramount in optimizing components to work within the whole. In industries without a dominant architecture, the architect has to work hard to resolve ambiguity in the absence of a presumed architecture. In both cases, we believe that reasoning about the architecture using the mental models presented in this chapter helps architects meet stakeholder needs, deliver value, and outperform the competition.

15.7 Summary

The main message of this chapter is that constructing an enumerable architecture model and exploring the corresponding tradespace of architectures can help system architects to organize architectural decisions and to understand the coupling between them and their sensitivity for the metrics of interest.

We provided a procedure to analyze an architectural tradespace. The first step is to look at the fuzzy Pareto frontier—that is, the architectures that represent the best tradeoffs between the metrics of interest. Looking at common features among architectures on the fuzzy Pareto front can immediately reveal dominant decisions. Moreover, the Pareto front is a good place to start looking for good architectures to study in more detail, even though the condition of non-dominance cannot by itself serve as a basis for down-selecting to a small set of architectures.

Further analysis of the structure of the tradespace can reveal additional information. For example, the presence of clusters or stratification in the tradespace may indicate that the tradespace is dominated by a small set of driving decisions. Finding the set of driving decisions is a byproduct of the tradespace mental model that can sometimes be even more important than finding a set of good architectures.

It is important to conduct sensitivity analysis on the results generated by an architectural model, especially when the uncertainty in the model parameters is large. In particular, sensitivity analysis can reveal architectures that are the most robust in responding to variations in the environment. Robust architectures are preferable because they are more likely to sustainably ensure value delivery to stakeholders over the system's lifecycle.

Finally, the system architect can produce a hierarchy of architectural decisions that suggests the sequence in which decisions should be made, based on the four-quadrants method. Decisions in the first quadrant are highly sensitive and highly coupled, so they should be made first. Decisions in the fourth quadrant are relatively insensitive and uncoupled, so they can be relegated to the bottom of the tree. The mid-levels of the tree are filled out by grouping highly coupled decisions in a single trade study and having trade studies that are weakly coupled run in parallel.

References

[1] Drawn from http://www.fordracingparts.com/crateengine/Main.asp

[2] The concept of dominance and the non-dominated front was introduced by Wilfredo Pareto in V. Pareto, *Cours d'economie politique* (Geneva: Librairie Droz, 1896).

[3] The concept of fuzzy Pareto frontiers was introduced in R. Smaling, and O. de Weck, "Fuzzy Pareto Frontiers in Multidisciplinary System Architecture Analysis," *AIAA Paper*, 2004, pp. 1–18. The Epsilon-dominance concept is similar; it was introduced in the domain of mean-variance financial portfolio optimization in D.J. White, "Epsilon-dominating Solutions in Mean-Variance Portfolio Analysis," *European Journal of Operational Research* 105 (March 1998), pp. 457–466. Then it was adapted for engineering (see, for example, C. Horoba and F. Neumann, "Benefits and Drawbacks for the Use of Epsilon-Dominance in Evolutionary Multi-objective Optimization," *Proceedings of the 10th Annual Conference on Genetic and Evolutionary Computation*, 2008, pp. 641–648).

[4] See K. Deb, A. Pratap, S. Agarwal, and T. Meyarivan, "A Fast and Elitist Multiobjective Genetic Algorithm: NSGA-II," *IEEE Transactions on Evolutionary Computation* 6 (April 2002), pp. 182–197.

[5] See Mattson, C. A., & Messac, A. (2003). Concept Selection Using s-Pareto Frontiers. AIAA Journal, 41(6), 1190–1198. doi:10.2514/2.2063 and Salomon, S., Dom, C., Avigad, G., Freitas, A., Goldvard, A., & Sch, O. (2014). PSA Based Multi Objective Evolutionary Algorithams, In O. Schuetze, C. A. Coello Coello, A.-A. Tantar, E. Tantar, P. Bouvry, P. Del Moral, & P. Legrand (Eds.), EVOLVE - A Bridge between Probability, Set Oriented Numerics, and Evolutionary Computation III (pp. 233–255). Heidelberg: Springer International Publishing. doi: 10.1007/978-3-319-01460-9.

[6] These numbers were taken from www.graphenea.com, last accessed on 4/8/2014.

[7] See for example Charytoniuk, W., Chen, M. S., Kotas, P., & Van Olinda, P. (1999). Demand forecasting in power distribution systems using nonparametric probability density estimation. IEEE Transactions on Power Systems, 14(4), 1200–1206.

[8] Simulation, Fifth Edition by Sheldon M Ross. Academic Press, San Diego, CA, 2013.

[9] There has been much emphasis recently in systems engineering on studying how changes propagate in a system. See, for example, M. Giffin, O. de Weck, G. Bounova, R. Keller, C. Eckert, and P.J. Clarkson, "Change Propagation Analysis in Complex Technical Systems," *Journal of Mechanical Design* 131, 2009/081001.

[10] This is called the degree heuristic in constraint satisfaction. See Chapter 6 of S. Russell and P. Norvig, *Artificial Intelligence: A Modern Approach*, 3rd ed. (Edinburgh, Scotland: Pearson Education Limited, 2009).

[11] See W.L. Simmons, "A Framework for Decision Support in Systems Architecting," PhD dissertation, Massachusetts Institute of Technology, pp. 66–71.

[12] The general problem of assessing the coupling between variables in a system can be tackled using statistical methods such as augmented linear regression including interaction effects, or clustering algorithms. See, for example, T. Hastie, R. Tibshirani, and J. Friedman, *The Elements of Statistical Learning: Data Mining, Inference, and Prediction* (New York: Springer, 2011).

[13] The field of design of experiments is concerned with the rigorous design and analysis of experiments of any kind. The concept of main effects and other concepts of the design of experiments, such as statistical validity, are described in detail in introductory texts such as D.C. Montgomery, *Design and Analysis of Experiments* (New York: Wiley, 2010).

[14] See J.A. Battat, B.G. Cameron, A. Rudat, and E.F. Crawley, "Technology Decisions under Architectural Uncertainty: Informing Investment Decisions through Tradespace Exploration," *Journal of Spacecraft and Rockets*, 2014 (Ahead of print).

[15] The analysis and figures of this subsection are adapted from J.A. Battat, B.G. Cameron, A. Rudat, and E.F. Crawley, "Technology Decisions under Architectural Uncertainty: Informing Investment Decisions through Tradespace Exploration," *Journal of Spacecraft and Rockets*, 2014 (Ahead of print).

Chapter 16
Formulating and Solving System Architecture Optimization Problems

16.1 Introduction

Chapter 14 introduced the mental model of the "system architecture as a set of decisions." We argued that one can build a model of a system architecture by encoding architectural decisions as variables with a set of allowed alternatives. We saw some examples of architectural models in both Chapter 14 (Apollo example) and Chapter 15 (guidance, navigation, and control systems) but did not discuss in detail how to build an architectural model. In this chapter, we study the formulation of these models.

Recall the definition of the term "system architecture":

System architecture is the embodiment of concept, the allocation of function to elements of form, and the definition of relationships among the elements and with the surrounding context.

Several decision-making problems are explicitly or implicitly mentioned in this formulation, such as defining the main elements of form and function, their mapping or allocation, and defining the relationship between elements. We begin this chapter by identifying a subset of recurring architecting tasks that can be formulated as programmed decisions or *architecture optimization problems*, shown in Table 16.1.

By studying the features of the architecture optimization problems resulting from these six tasks, we observe that they share some characteristics. We introduce six Patterns, or classes, of programmed architectural decisions: the DECISION-OPTION Pattern, the ASSIGNING Pattern, the PARTITIONING Pattern, the DOWN-SELECTING Pattern, the PERMUTING Pattern, and the CONNECTING Pattern. We will discuss the mapping of the six tasks to the six Patterns as we introduce them. Note that the fact that there are the same number of tasks as of Patterns should not lead the reader to think that there is a one-to-one correspondence between them. The mapping between tasks and Patterns is described in more detail later in the chapter.

Our intent is that these six Patterns will help the system architect identify opportunities to treat programmed decisions analytically. However, an important secondary intent is to identify mental models (like we made mental models for architectural decisions and the architectural tradespace) that help system architects reason about architectural decisions, regardless of whether the problem is ever encoded in a computational tool.

TABLE 16.1 | Tasks of the system architect that are amenable to the use of automatic decision support tools

Task	Consists of	Described in	In the Context of
1. Decomposing Form and Function	Choosing a system decomposition—that is, clustering elements of form or function	Chapter 3 Chapter 4 Chapter 5 Chapter 6 Chapter 8 Chapter 13	Managing complexity Form decomposition Function decomposition Architecture analysis Modularization Choosing a decomposition
2. Mapping Function to Form	Defining concept by assigning elements of function to elements of form	Chapter 6 Chapter 7 Chapter 8 Chapter 12	Architecture analysis Concept fragments Developing architecture from concept Refining integrated concept
3. Specializing Form and Function	Going from solution-neutral to solution-specific by choosing one among several alternatives for a certain element of form or function	Chapter 4 Chapter 7	Specialization of form Solution-neutral versus solution-specific
4. Characterizing Form and Function	Choosing one among several alternatives for the attributes of an element of form or function	Chapter 7 Chapter 13 Chapter 14	Characterization of function (data network example) Characterization of form (Saturn V example) Characterization of form (Apollo example)
5. Connecting Form and Function	Defining system topology and interfaces	Chapter 4 Chapter 5 Chapter 7	Connectivity in form Functional flow Sequence of operations
6. Selecting Goals	Defining scope by choosing among a set of candidate goals	Chapter 11	Prioritizing goals

We begin Chapter 16 by outlining, in Section 16.2, a generic formulation for system architecture optimization problems that will provide a common basis for expressing the Patterns. Section 16.3 introduces the example of NEOSS, an Earth observation satellite system, that is used throughout the rest of the chapter to illustrate several concepts. Section 16.4 describes the six Patterns individually, including their formulation and typical examples, and ends with a summary of the mapping among the six Patterns and the tasks presented earlier in this chapter. Section 16.5 provides some guidance in applying the Patterns to a real system architecture problem; it also discusses the issues of decomposition and overlap between Patterns. Finally, Section 16.6 focuses on solving system architecture optimization problems using full-factorial enumeration or heuristic algorithms.

16.2 Formulating a System Architecture Optimization Problem

Let's start with the Apollo example from Chapter 14. Recall that in Chapter 14 we defined nine decisions and a number of allowed values for each decision (such as EOR yes or no, command module crew of 2 or 3, service module fuel cryogenic or storable). We also had a way of measuring how good an Apollo architecture is by means of two metrics: total launched mass (IMLEO) and probability of mission success. We used the rocket equation to link mass to the decisions, and a risk table gave us the probability of mission success for each combination of decision variables.

This example can be formulated in such a way that we have a set of architectural decisions $\{d_i\} = \{d_1, d_2, \ldots, d_N\}$, each with a set of choices or allowed values $\{\{d_{ij}\}\} = \{\{d_{11}, d_{12}, \ldots, d_{1m_1}\}, \{d_{21}, d_{22}, \ldots, d_{2m_2}\}, \ldots, \{d_{N1}, d_{N2}, \ldots, d_{Nm_N}\}\}$. For example, d_1 corresponds to EOR and can take two values: $d_1(EOR) = \{d_{11}, = yes, d_{12} = no\}$. An architecture is then defined by choosing one value of each decision; for example, architecture $A = \{d_i \leftarrow d_{ij}\} = [d_1 \leftarrow d_{12}; d_2 \leftarrow d_{24}; \ldots; d_N \leftarrow d_{Nm}]$. For instance, the true Apollo architecture with LOR can be represented by the following array of values:

A = {EOR ← yes; earthOrbit ← orbit; LOR ← yes; moonArrival ← orbit; moonDeparture ← orbit; cmCrew ← 3; lmCrew ← 2; smFuel ← storable; lmFuel ← storable,

or, more compactly, if we omit the names of the decisions,

$$A = \{yes;\ orbit;\ yes;\ orbit;\ orbit;\ 3;\ 2;\ storable;\ storable\}$$

The next step in formulating a system architecture optimization problem is to express how well an architecture satisfies the needs of the stakeholders. We do this with a set of metrics $M = [M_1, \ldots, M_P]$ and a function $V(\cdot)$ that translates architectures (represented by arrays of symbols) into metric values. In the Apollo example, the function $V(\cdot)$ would contain the rocket equation and the risk look-up table, and the two outputs would be the values for IMLEO and probability of mission success. This is already an enormous assumption: Can a few metrics summarize all the important features of the architecture? Problems that can be computationally treated often require measures like this, and we'll want to evaluate the biases and assumptions of our metrics.

At this point, we can proceed to generating an architectural tradespace (such as those presented in Chapter 14 and Chapter 15) by enumerating all possible architectures and evaluating them according to the metrics defined. Alternatively, we can focus on finding the Pareto frontier by solving the following optimization problem:

$$A^* = \underset{A}{\operatorname{argmax}}\ M = V\big(A = \{d_i \leftarrow d_{ij}\}\big)$$

In other words, this optimization problem attempts to find a set of non-dominated architectures A^* in the space defined by the metrics M, computed from the architecture using the function $V(\cdot)$. Because there is more than one metric (such as IMLEO and probability of mission success), the non-dominated architectures A^* optimize the tradeoffs between the metrics M. Let's look at the components of the previous equation in more detail.

Architectural decisions are the variables that we optimize upon; we select different choices and measure how well the architecture performs as a result. As we discussed in Chapter 14, in most cases, architectural decisions will be represented using discrete and categorical variables (such as propellant type = {LH$_2$, CH$_4$, RP-1}), rather than continuous variables. This has important implications; working in a space with categorical variables precludes the use of gradient-based optimization algorithms. Furthermore, combinatorial optimization problems are usually harder to solve than continuous optimization problems. In fact, most combinatorial optimization problems are what computer scientists call NP-hard. Informally, this means that the time it takes to solve them grows exponentially with the size of the problem (the number of decisions, for example). In practice, these problems become impossible to solve exactly (that is, to find the real global optimal architectures) for relatively small numbers of decisions (say, 15 to 20). Moreover, there is a threshold for the number of decisions and options that we can solve; it is not far from the magic rule of 7+/– 2 decisions and options per decision. [1] Consequently, in formulating an architecture optimization problem, it is very important to choose the architectural decisions and their range of values carefully.

The value function $V(\cdot)$ links architectural decisions to metrics. It can be seen as the "transfer function" of the evaluation model, as illustrated in Figure 16.1.

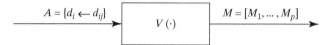

FIGURE 16.1 The value function takes an architecture as an input and provides its figures of merit (value) as the output.

The value function is a very concentrated summary of the stakeholder analysis of the system: one that takes a compressed version of the architecture as an input (only the values of all the decisions involved) and provides one or a few outputs. The challenge is to choose, from the set of metrics developed in the stakeholder analysis, a subset that is feasible and useful to model. Let's discuss how to do that.

Just as decisions need to be architecturally distinguishing (Chapter 14), metrics need to have sensitivity to decisions. We expect the metric to show a variation across different architectures. Indeed, if the metric gives the same output for all architectures, it isn't useful. For instance, if none of our Hybrid Car architectures are inherently safer than others, safety should not be used as a metric, because it is not architecturally distinguishing.

Just as there were limits to the number of decisions and options, here too we have practical limits in the number of metrics. When there are too many metrics, most architectures become non-dominated. If we have 100 metrics that all come from stakeholder analysis, every architecture that gets the maximum score in one single metric will be non-dominated. In practice, a number of metrics between 2 and 5 is usually adequate, with 2 or 3 metrics being preferable in most cases.

Value functions often have a subjective component that arises from fuzzy and ambiguous stakeholder needs that are hard to quantify, such as "community engagement." Even "scientific value" has a strong subjective component because of the uncertainty about whether a given scientific discovery will lead to subsequent scientific discoveries. Subjectivity in the evaluation of architectures is not inherently bad, but it can certainly hinder the decision-making process with consistency and bias issues. When we use subjective judgments in our value functions, it is important to maintain the traceability of our assessments to provide the rationale behind the scores. [2] Appendix C briefly discusses how knowledge-based systems, a technology from artificial intelligence, can help achieve this goal.

In addition to objective functions, most architecture optimization problems have constraints. In the Apollo example, we used constraints to eliminate nonsensical combinations of decisions (for example, LOR = yes and MoonArrival = Direct do not make sense together, because Lunar Orbit Rendezvous requires stopping at the Moon to assemble spacecraft before descent and therefore precludes direct arrival).

Constraints can also be used to eliminate architectures or families of architectures that are very likely to be dominated. For example, hybrid cars with a range of less than 5 km are unlikely to find a substantial market. Therefore, if at any point in the evaluation process this threshold is not achieved, the architecture can be eliminated.

In general, constraints can be used to express more or less stringent goals. Recall that in Chapter 11, we classified goals according to how constraining they were: absolutely constraining, constraining, or unconstrained. Here we formalize this idea. Computer scientists often distinguish between hard constraints and soft constraints, depending on their effect on the architectures that violate them. An architecture that does not satisfy a hard constraint is *eliminated* from the tradespace. An architecture that violates a soft constraint is somehow *penalized* (such as with a cost penalty) but is not eliminated from the tradespace.

A given constraint in a real architecting problem can be formulated as a hard or a soft constraint in a computational formulation. In some cases, hard constraints are preferable—for example, when a soft constraint might result in the evaluation of nonsensical architectures. This is the case in constraints that eliminate invalid combinations of decisions and options, such as the "LOR/Moon arrival" example discussed earlier. On the other hand, the constraint for the hybrid car having a range greater than 5 km could be implemented as a soft constraint. Soft constraints are generally implemented as penalties; the further past the soft constraint, the higher the penalty. Their goal is to speed up the search by driving the algorithm away from unpromising regions quickly.*

In summary, one can formulate a system architecture optimization problem by encoding the architecture as a set of decisions, defining one or more metrics that encapsulate the needs of the stakeholders, creating a value function mapping architectural decisions to metrics, and adding a set of hard or soft constraints as needed.

16.3 NEOSS Example: An Earth Observing Satellite System for NASA

We'll introduce an example of a complex system for use throughout the remainder of the chapter to formulate different architecture optimization problems. We call this system NEOSS, for NASA Earth Observing Satellite System (Figure 16.2). As its name indicates, NEOSS is a collection of satellites that carry a set of remote sensing instruments that provide observations of Earth's land, oceans, and atmosphere in the form of data products. [3]

The main stakeholders of NEOSS are the many scientists who will be the primary users of the data products produced by NEOSS. Most of the goals that flow logically from scientists' needs concern the attributes (spatial resolution, temporal resolution, accuracy) of different data products

---

*Although the distinction between hard constraints and soft constraints may be fairly clear from a stakeholder perspective or in cases where it eliminates a nonsensical architecture, this boundary is often blurred by optimization algorithms that transform constrained optimization problems into unconstrained optimization problems by using Lagrange multipliers.

FIGURE 16.2 ESA's Envisat (top) and NASA's A Train (bottom), two examples of potential NEOSS architectures. (Source: (a) Esa/epa/Corbis (b) Ed Hanka/NASA)

(soil moisture, atmospheric temperature, sea level height). Scientists are also very concerned about data continuity: They have lengthy, consistent records of data that they want to extend for as long as possible. A good value function for NEOSS should consider satisfaction of data product requirements from scientific communities, data continuity, and some notion of cost, schedule, and risk. [4]

What is the most important decision for NEOSS? The answer may seem intuitive. The most important decision is the selection, from the set of all candidate instruments, of remote sensing instruments that will be included on a spacecraft. For example, in order to measure the land's topography, we may choose to develop a laser altimeter, a synthetic aperture radar, or both. Note that we are effectively mapping a solution-neutral function (measuring topography) to a form (laser, radar, or both) and therefore to the corresponding solution-specific functions. In this example we have eight candidate instruments: a radar altimeter, a conical microwave radiometer, a laser altimeter, a synthetic aperture radar, a GPS receiver, an infrared spectrometer, a millimeter-wave sounder, and a high-resolution optical imager.

Although selecting instruments is the most important decision, the satellites themselves may have a strong impact on the stakeholder needs. For example, higher orbits lead to coarser spatial resolution but better coverage. Power limitations and interactions with other instruments aboard the satellite (such as electromagnetic interference) can impact value. Thus the assignment of instruments to satellites and orbits will also have a major impact on value. For this example we will allow constellations with 1 to 3 planes and 1 to 4 satellites per plane in true polar, sun-synchronous, and tropical orbits at 400, 600, and 800 km.

Finally, the launch dates of these satellites are also important, because they drive the ability of the system to meet data continuity needs. If an existing satellite taking an important measurement is going to be de-orbited soon, the value of a replacement is higher if its launch date avoids a gap in the data series.

Hence, if we separate the selection of instruments from their assignment to satellites, a first model for the architecture of NEOSS is given by three major groups of decisions, which we called instrument selection, instrument packaging, and satellite scheduling.

- **Instrument selection.** Given the set of eight candidate instruments, which instruments or sensors (sometimes called payloads in aerospace terminology) do we choose to perform the required measurements?
- **Instrument packaging.** Given the set of candidate orbits, and given the instrument set chosen in the instrument selection problem, which satellites and orbits do we choose to fly these payloads on? More precisely, do we operate one or many satellites (for example, single satellite, tandem, or constellation)? What altitude, inclination, and other orbital parameters do we choose? In the case of a constellation, how many orbital planes are the satellites in, and how many satellites do we have in each plane?
- **Satellite scheduling.** Given the satellite missions defined by the previous decisions, how do these individual missions coordinate with each other as a program? Note that missions cannot be considered in isolation, as a consequence of the data continuity issues mentioned before, but also because they are linked through a common budget.

We have provided an overview of the three major groups of architectural decisions for the NEOSS system: instrument selection, instrument packaging, and satellite scheduling. In the next section, we will show that the structure behind these decisions appears often in architecture problems for other systems, which leads to the definition of a set of Patterns for use in system architecture decisions.

16.4 Patterns in System Architecting Decisions

Going from a list of decisions and alternatives (such as the one presented in the previous section for the NEOSS example) to their formulation as an architecture optimization problem can be a hard task in itself. However, we have observed that a number of Patterns appear over and over when system architecture optimization problems are formulated. For example, the problem of partitioning the Saturn V architecture into stages is fundamentally the same as the instrument-packaging problem defined in the previous section. We essentially have to allocate elements to bins in both cases, and the optimization algorithm doesn't care whether it is assigning remote sensing instruments to orbits or assigning delta-V requirements to stages. This section describes a set of Patterns that will help the system architect formulate system architecture optimization problems.

The idea of Patterns in design is often attributed to Chris Alexander, an Austrian civil architect who in 1977 published *A Pattern Language: Towns, Buildings, Construction*. [5] This book describes a set of problems that appear recurrently when buildings are designed, and it also discusses the core of a solution that is known to work well for that problem in different situations. For example, Alexander describes the Pattern of "Light on two sides of every room" as follows:

> *When they have a choice, people will always gravitate to those rooms which have light on two sides, and leave the rooms which are only lit from one side unused and empty.*

This is the descriptive part of the Pattern. Alexander provides a prescriptive part as well, which explains how to achieve this welcoming effect in different cases: small buildings (simply put four rooms, one in each corner of the house); medium-size buildings (wrinkle the edge, turn corners); and large buildings (convolute the edge further, or have shallow rooms with two windows side by side). The intent is thus to identify the Pattern when one comes across a similar problem, so that one can reuse the solution that is known to work, potentially saving resources in redesign.

This practical approach to architecture was adapted in other fields, especially in computer science, with the book *Design Patterns: Elements of Reusable Object-Oriented Software*. [6] Just as Alexander described a set of recurrent problems and solutions in civil architecture, the "Gang of Four" created a compendium of over twenty patterns in object-oriented programming, with solutions in pseudocode that are ready to be reused. These patterns include such abstract ideas as Singletons, Abstract Factories, Iterators, Interpreters, and Decorators. Most experienced object-oriented programmers are familiar with these concepts and use them daily to synthesize and communicate system architectures. [7]

Our discussion of Patterns goes beyond programmed decisions and optimization. Studying these Patterns will lead us to discuss typical architectural tradeoffs, as well as the main options for those tradeoffs, which we will call architectural "styles" (such as monolithic versus distributed architectures and channelized versus cross-strapped architectures). The Patterns effectively provide a common vocabulary for communicating and discussing trade-offs and corresponding styles. [8] Thus these Patterns are useful for formulating optimization problems but also, at a broader scale, as a framework for organizing architectural decisions.

From Programmed Decisions to Patterns

In general, optimization problems that result from programmed decisions in system architecture are instances of combinatorial optimization problems. Furthermore, most of them are similar to one or more "classical" optimization problems that appear over and over again in operations research, [9] such as the traveling salesman* and the knapsack problem† shown in Appendix D.

---

*The traveling salesman problem is formulated as follows: Given a list of cities, and the distances between them, find the shortest route that visits each city exactly once and returns to the point of origin. Its mathematical formulation is usually attributed to W. R. Hamilton.

†The knapsack problem is formulated as follows: Given a set of items, each with a mass and a value, and a rucksack that can hold a certain maximum mass, determine the number of each item to put in the rucksack so as to maximize value without breaking it. Its mathematical formulation is usually attributed to T. Dantzig.

TABLE 16.2 | The six Patterns of architectural decisions

Pattern	Description
DECISION-OPTION	A group of decisions where each decision has its own discrete set of options
DOWN-SELECTING	A group of binary decisions representing a subset from a set of candidate entities
ASSIGNING	Given two different sets of entities, a group of decisions assigning each element from one set to any subset of entities from the other set
PARTITIONING	A group of decisions representing a partitioning of a set of entities into subsets that are mutually exclusive and exhaustive
PERMUTING	A group of decisions representing a one-to-one mapping between a set of entities and a set of positions
CONNECTING	Given a set of entities that are nodes in a graph, a group of decisions representing the connections between those nodes

We now introduce our six Patterns of programmed decisions in system architecture: DECISION-OPTION, ASSIGNING, PARTITIONING, PERMUTING, DOWN-SELECTING, and CONNECTING. Some of the Patterns that we will discuss in this section are variations or generalizations of classical problems. For instance, our DOWN-SELECTING Pattern looks a lot like the 0/1 version of the knapsack problem, with the added twist that we account for interactions between elements. Table 16.2 contains a list of the Patterns and a short description of each.

Each of these Patterns enforces a different underlying formulation of decisions that can be exploited to gain insight into the system architecture (such as the architecture styles presented later in this section) and also to solve the problem more efficiently by using more appropriate tools. As we introduce the Patterns, it will become apparent that most problems can be formulated using more than one Pattern. Thus the Patterns should be seen not as mutually exclusive, but rather as complementary. That having been said, one Pattern typically is more useful than the others, because it provides more insight and/or leads to more efficient optimization.

The DECISION-OPTION Pattern

The DECISION-OPTION Pattern appears when there is a set of decisions, each with its own discrete (and relatively small) independent set of options. More formally, given a set of n generic decisions $D = \{X_1, X_2, \ldots, X_n\}$, where each decision X_i has its own discrete set of m_i options $O_i = \{z_{i1}, z_{i2}, \ldots, z_{im_i}\}$, an architecture in the DECISION-OPTION problem is given by an assignment of options to decisions $A = \{X_i \leftarrow z_{ij} \in O_i\}_{i=1,\ldots,n}$.

The DECISION-OPTION Pattern is the most direct representation of our Apollo example, in which each decision had a small number of available options (for example, EOR = {yes, no}, moonArrival = {orbit, direct}). An architecture, in the DECISION-OPTION problem, can be represented by an array of values, where each position in the array represents the value chosen for a particular decision. For example, in the Apollo example, the actual Apollo architecture can be represented by

$$A = \{\text{yes; orbit; yes; orbit; orbit; 3; 2; storable; storable}\}$$

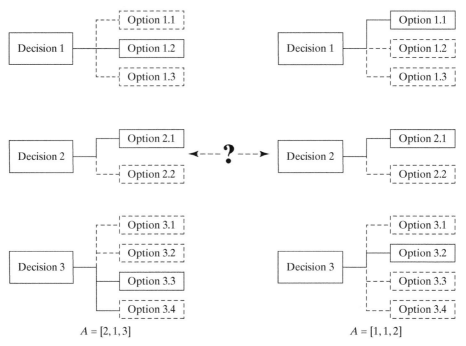

FIGURE 16.3 Pictorial representation of the DECISION-OPTION Pattern. Two different DECISION-OPTION architectures are shown for a simple case with 3 decisions and 24 possible architectures.

A pictorial representation of a generic DECISION-OPTION problem is provided in Figure 16.3, which shows 3 different decisions with 3, 2, and 4 options, respectively. Two different architectures and their representations as arrays of integers are also shown.

DECISION-OPTION problems can be readily represented using decision trees, although in practice, the tree will often be too big to represent on a single sheet of paper. They can also be represented using morphological matrices, as shown in Table 16.3.

The size of the tradespace of a generic DECISION-OPTION problem is simply given by the product of the number of options for each decision. Hence, in the example shown in Figure 16.3, there are $3 \times 2 \times 4 = 24$ different architectures. Thus the number of architectures grows very quickly with the number of decisions and options for each decision.

TABLE 16.3 | DECISION-OPTION problems can be represented using morphological matrices.

Decision/Option	Option 1	Option 2	Option 3	Option 4
Decision 1	Option 1.1	Option 1.2	Option 1.3	
Decision 2	Option 2.1	Option 2.2		
Decision 3	Option 3.1	Option 3.2	Option 3.3	Option 3.4

CHAPTER 16 • FORMULATING AND SOLVING SYSTEM ARCHITECTURE PROBLEMS   **369**

The set of options for each decision is defined as a discrete set. For problems with continuous values (such as in vehicle suspension, where the spring rate can take on any value), we would need either to provide a finite set of acceptable values (such as spring rates of 400, 500, and 600 lb/in.) or to define the boundaries and a discretizing step to construct this set.

The Apollo case is a clear instance of a "pure" DECISION-OPTION problem, because all decisions were given a discrete set of mutually exclusive alternatives. Other examples follow.

- **EXAMPLE 1:** Consider the architecture of an autonomous underwater vehicle (AUV; see Figure 16.4). A simple representation of the architecture consists of the following decisions: the configuration ("torpedo," "blended wing body," "hybrid," "rectangular"); the ability to swim (yes, no); the ability to hover like a helicopter (yes, no); the navigation method (dead reckoning, underwater acoustic positioning system); the propulsion method (propeller-based, Kort nozzles, passive gliding); the type of motors (brushed, brushless); the power system (rechargeable batteries, fuel cells, solar); and different sensor decisions: sonar (yes, no), magnetometer (yes, no), thermistor (yes, no).

(a)

(b)

(c)

FIGURE 16.4 The Bluefin 12 BOSS, the Nereus, and the SeaExplorer, three different AUV architectures. (Source: (a) National Oceanic and Atmospheric Administration (b) Image courtesy of AUVfest 2008: Partnership Runs Deep, Navy/NOAA, OceanExplorer.noaa.gov (c) Photo courtesy of ALSEAMAR)

These 10 decisions with 4, 2, 2, 2, 3, 2, 3, 2, 2, and 2 options, respectively, yield a total of 4,608 different AUV architectures before constraints are considered. [10]

- **EXAMPLE 2:** Providing communication services in remote or developing areas for commercial or military applications requires the use of dedicated assets such as satellites, drones, [11] or even balloons. Consider an aerial network of balloons hosting military communications payloads acting as communications relays between ground vehicles. [12] Major architectural decisions include deciding between two different types of radios (where the type of radio determines the range of communication) and between two balloon altitudes (where altitude drives the coverage of the network). These two decisions are made at each of 10 preselected sites. If we include the option not to have a balloon in a site, this yields $3 \times 2 \times 3 \times 2 \ldots = 6^{10} \simeq 60$ million architectures. The reader may note that this formulation enumerates the two different altitudes for the trivial case where no balloon is assigned to the site, which is unnecessary. These options could be eliminated using a constraint. Note also that this formulation assumes that we can mix and match different types of radios for different sites, which would impose an interoperability requirement between them. In this example, decisions concerning different sites are identical and have the same set of options. We will see later that a more natural formulation of this problem includes a mix of a DECISION-OPTION Pattern and an ASSIGNING Pattern.

The fundamental idea in DECISION-OPTION problems is that the options for each decision are mostly independent (except for constraints that may preclude some combinations) and, in general, differ across decisions. For example, the value "16GB" could never be an option for the decision "type of processor." We will see that this independence distinguishes the DECISION-OPTION Pattern from all the other Patterns, where decisions can share options. Moreover, in the DECISION-OPTION problem, one must choose exactly one option for each decision (for example, EOR cannot be yes and no at the same time), whereas in the other Patterns, one must usually choose a combination of options. For these reasons, the DECISION-OPTION Pattern is best represented by a morphological matrix.

The DECISION-OPTION Pattern is the most intuitive and flexible of all Patterns. No sequence, pre-conditions, or relationships are implicitly assumed between decisions. If present, such features need to be modeled using constraints (such as the invalid mission-mode options from the Apollo example). DECISION-OPTION problems appear most often for Tasks 3 and 4 of Table 16.1, the specialization and characterization of function and form.

DECISION-OPTION is a very general Pattern that can be used to represent pretty much any programmed decision problem, which is why we start with it. However, we saw in the examples that some problems have an underlying structure (coupling between decisions) that makes them easier to formulate using other Patterns.

The DOWN-SELECTING Pattern

The DOWN-SELECTING Pattern arises when we have a set of candidate elements and need to choose a subset of them. More formally, given a set of elements $U = \{e_1, e_2, \ldots, e_m\}$, an architecture in the DOWN-SELECTING Pattern is given by a subset of the elements in the set: $A = S \subseteq U$.

An architecture in the DOWN-SELECTING Pattern can be represented by a subset of elements or by a binary vector. For example, if we choose from our set of 8 candidate instruments {radiometer, altimeter, imager, sounder, lidar, GPS receiver, synthetic aperture radar (SAR), spectrometer} in the NEOSS example, two different architectures would be

A_1 = {radiometer, altimeter, GPS} = [1, 1, 0, 0, 0, 1, 0, 0]

A_2 = {imager, sounder, SAR, spectrometer} = [0, 0, 1, 1, 0, 0, 1, 1]

The DOWN-SELECTING Pattern can be seen as a set of *binary* decisions, where each decision concerns whether we choose an element or not. The size of the architectural tradespace of a DOWN-SELECTING problem is thus simply given by 2^m, where m is the number of elements in the candidate set.

A pictorial representation of the DOWN-SELECTING Pattern is provided in Figure 16.5, where two different subsets of the same set of 8 elements are shown. The subset on the left chooses 5 of the 8 elements, whereas the subset on the right chooses 3 of the 8 original elements. Selected elements are shown in black inside rectangles indicated by solid lines, whereas discarded elements are shown in gray italics inside rectangles indicated by dashed lines.

The reader may wonder why we define a new Pattern for problems that can be expressed using the DECISION-OPTION Pattern. The reason is that there is an underlying structure in the DOWN-SELECTING Pattern that is absent in the DECISION-OPTION Pattern: Architectures are defined as a subset of candidate elements. Changing a decision in the DOWN-SELECTING Pattern always means adding an element to, or removing an element from, the selected subset, which has two effects: adding or subtracting some benefit, and adding or subtracting some cost. This underlying structure allows us to choose the best set of heuristics to solve the corresponding optimization problem. For example, single-point crossover, a heuristic commonly used in genetic algorithms, is usually an appropriate choice for DOWN-SELECTING problems, because it tends to keep good combinations of elements together, and the structure of the problem is smooth enough.

The DOWN-SELECTING Pattern appears in situations where resources are limited. The typical example is a limited budget forcing an organization to choose between competing systems or projects. The instrument selection problem from our NEOSS example is an instance

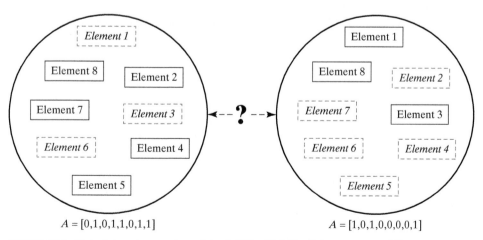

FIGURE 16.5 Pictorial representation of the DOWN-SELECTING Pattern. Two different DOWN-SELECTING architectures are shown for a simple case with 8 elements and 256 possible architectures.

of the DOWN-SELECTING Pattern, because we have to choose a subset of the candidate instruments. A key feature of this problem is that the value of choosing an instrument depends on what other instruments are selected. In other words, there are synergies and conflicts between instruments. For example, if we choose the radar altimeter but not the microwave radiometer, then the accuracy of the radar altimetry measurement will not be as good because we will not be able to correct for the effect of humidity in the air. A few more examples follow.

- **EXAMPLE 1:** IBM's Watson is a complex cognitive software system known for beating human contestants in the TV show *Jeopardy*. Watson uses multiple natural language processing (NLP) strategies to parse a question stated in English and searches an extremely large knowledge database to find the most likely correct answers. [13] The problem of selecting among candidate NLPs can be seen as a DOWN-SELECTING problem. Adding more strategies has benefits (for example, more diversification usually means more flexibility and the ability to handle more general problems), but it also has costs (development time, and computational resources for each strategy). Moreover, there might be redundancies, synergies, and interferences between some NLPs. For example, deep and shallow NLP strategies complement each other and can be combined in hybrid systems. [14]

- **EXAMPLE 2:** RapidEye is a constellation of five satellites that provide daily high-resolution optical imagery. In RapidEye's first constellation, the satellites provide images in the red, green, blue, and near-infrared bands of the electromagnetic spectrum. For the next-generation constellation, RapidEye could presumably decide to incorporate more bands into the satellites, even in the microwave region of the spectrum, since more information (such as atmospheric temperature, chemical composition, or vegetation state) can be obtained from such multi-spectral measurements. Different bands have different capabilities and applications, and some of them are more redundant than others. For example, there are several bands to measure atmospheric concentration of carbon monoxide (a powerful pollutant), including 2.2 μm and 4.7 μm. If we choose one of these two, then having the other would arguably add less value than the first one. There are also synergies between different bands. For example, adding a band at 1.6 μm can increase the value of most other bands by providing an atmospheric correction for the presence of clouds.

The DOWN-SELECTING Pattern is the most natural Pattern for Task 6 of Table 16.1, goal selection, where a subset of goals or requirements must be chosen from a candidate set. It can also be useful in Task 3 of Table 16.1 (specialization of form or function) when several options can be chosen from a group of similar elements.

The DOWN-SELECTING Pattern is similar in nature to a classical optimization problem called the 0/1 integer knapsack problem, with an important caveat. In the standard formulation of the 0/1 integer knapsack problem, we are given a set of items, each of them with a given cost and benefit. The goal is to find the number of items of each type that maximizes benefit at a given cost.* In this classical formulation, the benefit and cost of the items are independent of the other items selected. In other words, there are no interactions between the elements. But in reality, the

---

*Note that there is a more general formulation of the knapsack problem that allows choosing more than one copy of each item.

CHAPTER 16 • FORMULATING AND SOLVING SYSTEM ARCHITECTURE PROBLEMS

value of selecting a certain item will depend on the other items selected, because there are synergies, redundancies, and interferences between elements. These interactions are very important in DOWN-SELECTING problems because they can drive architectural decisions.

Consider the trivial example of a backpack and a set of available items that contains, among other things, three tubes of toothpaste of different brands (with benefits B_1, B_2, B_3), a toothbrush (B_4), a hot sandwich (B_5), an ice-cold can of beer (B_6), a towel (B_7), and soap (B_8). In a classical knapsack problem formulation, all B_i would be immutable, and thus the benefit of choosing all three tubes of toothpaste would be $B_1 + B_2 + B_3$. In reality, the benefit of having the three tubes of toothpaste will arguably be much smaller than $B_1 + B_2 + B_3$, as a consequence of *redundancy* between the items.

Similarly, in the classical formulation, the value of having the toothpaste does not depend on whether the toothbrush is selected or not. In reality, the value of having toothpaste without a toothbrush is much smaller, potentially zero. This is the effect of *synergies*, which, as we have noted before, is very important to capture.

Finally, one may argue that the value of choosing the ice-cold beer may slightly decrease if we pack the hot sandwich with it, because some of the heat in the sandwich will be transferred to the beer, ruining it. This is an example of negative interactions that we call *interferences*.

These ideas can be generalized to more complex systems. Our radar and radiometer instruments from the NEOSS example are highly synergistic elements, because the value that we get out of the combination is greater than the sum of their individual measurements. Conversely, the synthetic aperture radar and the lidar have negative interactions; both are high-energy instruments, and they are likely to have conflicting orbit requirements. Furthermore, there might be some redundancy between the radar altimeter and the lidar, since both can be used to do topographic measurements.

The importance of interactions between elements for DOWN-SELECTING problems is highlighted in Box 16.1.

Box 16.1 Insight: Interactions between System Elements

- Good architectures in DOWN-SELECTING problems contain subsets of elements that "work well together" compared to others, either because they are very synergistic or because they minimize interferences, or both. These dominating subsets are sometimes called **schemata**, especially in the context of adaptive systems. [15]
- Finding a good architecture in a DOWN-SELECTING Pattern is all about finding good schemata that have high synergies, low interferences, and low redundancy, as well as avoiding poor schemata with low synergies, high interferences, and high redundancy.

Because of the presence of redundancies, synergies, and interferences between elements, DOWN-SELECTING problems are a lot harder to solve than classical knapsack problems, since the value of an element depends on all the other selected elements. One way of approaching the problem is to enumerate all possible subsets of selected elements. This can become quite tedious because it requires pre-computing 2^N values. More sophisticated models of these interactions allow explicit modeling of the nature of the interactions and traceability of the value. [16]

The ASSIGNING Pattern

The ASSIGNING Pattern arises when we have two sets of elements (which we will simply call the "left" set and the "right" set) and we need to assign elements of the left set to any number of elements of the right set. For example, in the NEOSS example, one might predefine a set of instruments (imager, radiometer, sounder, radar, lidar, GPS receiver) and a set of orbits (geostationary, sun-synchronous, very low Earth polar orbit) so that each instrument can be assigned to any subset of orbits (including all or none of the orbits). A possible architecture is shown in Figure 16.6, where the sounder and radiometer are assigned to the geostationary orbit; the imager, sounder, radiometers, and radar are assigned to the sun-synchronous orbit; the lidar and radar are assigned to the polar orbit; and the GPS receiver is not assigned to any orbit.

A particular architecture in the ASSIGNING problem can be readily represented as an array of subsets. We can also construct the ASSIGNING Pattern as a DECISION-OPTION with only binary decisions (Do we assign element i to element j, yes or no?). Thus an architecture in the ASSIGNING problem can be represented as a binary matrix of size $m \times n$, where m and n are the number of elements in the two sets.

Figure 16.7 provides an illustration of a generic ASSIGNING problem and illustrates two alternative architectures, with their corresponding representations as binary matrices.

Given the formulation as a binary matrix, the size of the tradespace of a generic ASSIGNING problem is simply given by 2^{mn}, since there are 2^m possibilities for each decision, where m is the number of options (elements in the matrix on the right-hand side of Figure 16.9) and n is the number of decisions (elements in the matrix on the left-hand side of Figure 16.9). Hence, in the example of Figure 16.6, there are $2^{3 \times 6} = 262{,}144$ architectures. Note that the number of architectures grows exponentially with the product of the number of decisions and options.

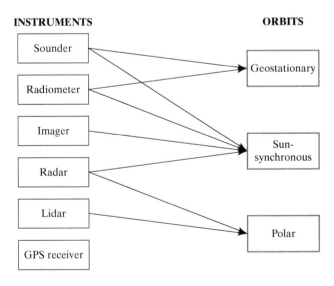

A = [geostationary ← {sounder, radiometer}, sun-synchronous ← {imager, sounder, radiometer, radar}, polar ← {radar, lidar}]

FIGURE 16.6 An example of the ASSIGNING Pattern for the NEOSS instrument-packaging problem.

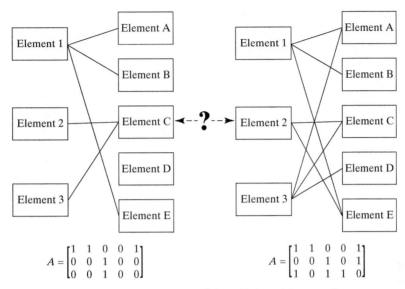

FIGURE 16.7 Pictorial representation of the ASSIGNING Pattern. Two different ASSIGNING architectures are shown for a case with 3 elements in the "left" set and five elements in the "right" set, for a total of 32,768 possible architectures.

The ASSIGNING Pattern is one of the most prominent ones in system architecture programmed decisions. Some examples follow.

- **EXAMPLE 1:** Every autonomous vehicle (be it a MQ-9 Reaper UAV, a DISH network communications satellite, a robotic vacuum cleaner, or the Google car) has a guidance, navigation, and control (GNC) subsystem that gathers information about the position and attitude of the system (navigation), decides where to go next (guidance), and changes its position and attitude to achieve the "go there" (control). We saw in Chapter 15 that a GNC system can be viewed as a set of sensors, computers, and actuators that can have connections between them. [17] These connections can be seen as two layers; sensors are connected to computers, and computers to actuators. Focusing on the first layer (sensors -> computers), if we have a predefined set of sensors and a predefined set of computers, we have to make a decision (or many decisions) about how to connect them to each other. This is a clear instance of the ASSIGNING Pattern: There are two predefined sets of elements (sensors and computers), and each element of one set (sensor) can be connected or assigned to any number of elements of the other set (computers). The same is true for the computer-to-actuator part of the problem.

- **EXAMPLE 2:** Recall our military communications aerial network example from the DECISION-OPTION Pattern. The decisions concerning which balloon type to allocate to each site can be seen as an instance of the ASSIGNING Pattern, because each balloon type can be assigned to any subset of sites (including none or all of them). Note that this assumes that we can deploy multiple balloons in a site, which may or may not be a realistic assumption.

The ASSIGNING Pattern can be thought of as a special case of the DECISION-OPTION Pattern where a set of identical decisions share a common set of options, and each decision can be assigned to any subset of those options, including the empty set. For example, one can assign workers to tasks (any worker can be assigned to any number of a pool of common tasks), instruments to orbits (any instrument can be assigned to any number of orbits), subsystems to requirements (any subsystem can be assigned to satisfy any number of requirements, or vice versa), or processes to objects (any process can be assigned to any subset of existing objects). Moreover, the ASSIGNING Pattern can also be thought of as a set of DOWN-SELECTING decisions, each decision corresponding to the assignment of an element of the left set to any subset of elements of the right set, with the additional constraint that each element of the left set can be assigned to only one subset of elements of the right set.

However, the ASSIGNING problem has an underlying structure that is absent in both the DECISION-OPTION problem and the DOWN-SELECTING problem—the fact that we are assigning elements of one set to another. If we were to formulate an ASSIGNING problem as a set of DOWN-SELECTING problems, it would not be implicit in the formulation that the candidate sets of elements for all decisions are in fact the same set, in particular what we called the "right set."

Recall that a problem does not inherently belong to any given Pattern, but rather can usually be expressed in several Patterns. It will often be the case that a problem is most naturally formulated in a particular Pattern. If we have a DECISION-OPTION problem where all decisions are identical and have the same candidate sets, then it is more appropriate (simpler, more informative, and more elegant) to formulate it as a single ASSIGNMENT problem. Moreover, formulating it as an ASSIGNMENT problem will allow us to use heuristics that take advantage of the structure of the Pattern, such as heuristics based on style or balance considerations.

The ASSIGNING Pattern often appears in Tasks 2 (function-to-form mapping) and 5 (connecting form and function) of Table 16.1. In function-to-form mapping, the ASSIGNING Pattern is essentially about choosing how coupled we want our architecture to be. Suh's principle of functional independence proposes that each function should be accomplished by one piece of form, and each piece of form should perform only one function. [18] On the other hand, having a more coupled mapping of function to form, where one element of form performs several functions, can sometimes reduce number of parts, weight, volume, and (ultimately) cost.

In the context of connectivity of form and function, the ASSIGNING Pattern is fundamentally about deciding how connected we want our architecture to be. In the NEOSS example, we assign instruments to orbits; every time we assign an instrument to an orbit, we add a "connection" in the architecture, which in this case takes the form of a copy of the instrument. In this example, connections (instruments in orbit) are costly. In the GNC example, we assign or connect sensors to computers (and computers to actuators); every time we assign one sensor to one computer, we add a connection in the architecture, which in this case can take the form of a cable, interfaces, and software to support the interfaces. In both cases, increasing the number of connections between the "left" set and the "right" set can improve system properties such as data throughput or reliability, but it comes at a price: increased complexity and cost.

Even though function-to-form mapping and the connectivity of form and function are fundamentally different problems, we see that they are decisions with similar features. In both cases, if we enforce that each element from the left set be assigned to at least one element of the right set (an additional constraint that is not present in the most general formulation of the pattern), we can define two "extreme" architectures at the boundaries of the architectural tradespace.

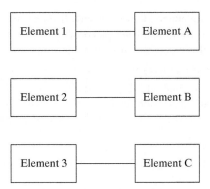

FIGURE 16.8 Channelized style of architecture in the ASSIGNING Pattern.

These are a "channelized architecture," where each "left" element is matched to exactly one "right" element (see Figure 16.8), and a "fully cross-strapped" architecture, where every "left" element is connected to every "right" element (see Figure 16.9).*

The Saturn V is a good example of a channelized architecture in a function-to-form mapping task, because the system was neatly decomposed into subsystems that performed a single function. An example of a channelized architecture in a connectivity task is the GNC system of the NASA X-38 Crew return vehicle, which has two fully independent redundant buses.

An example of a fully cross-strapped architecture in mapping function to form in an organizational context is the idea of "Total Football" in soccer, in which all ten field players play both defensive and offensive roles. The Space Shuttle avionics system is an example of a fully cross-strapped architecture in a connectivity task, because all inertial measurement units are connected to all general-purpose computers, and all computers are connected to all rudder actuation systems.

The channelized versus fully cross-strapped tradeoff applies to function-to-form mapping, and connectivity of form and function. In the case of form-to-function mapping, we can re-state

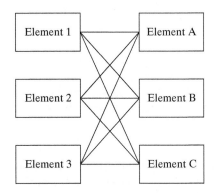

FIGURE 16.9 Fully cross-strapped style of architecture in the ASSIGNING Pattern.

---

*The names for the channelized and the fully cross-strapped architectural styles of the ASSIGNING Pattern are arguably more appropriate for the case of connectivity of form and function, but the definition and discussion apply to the case of function-to-form mapping.

Suh's principle of functional independence as a channelized architecture, where each function is accomplished by one piece of form, and each piece of form performs only one function. In the connectivity case, the tradeoff is basically throughput and reliability versus cost.

These extremes of the ASSIGNING Pattern can be seen as architecture *styles*—that is, soft constraints or driving principles that simplify the architecture (typically by making a series of similar decisions identical) and that often result in more elegant architectures. The channelized versus fully cross-strapped trade-off and corresponding styles are discussed in Box 16.2.

Box 16.2 Insight: Channelized versus Fully Cross-strapped Architecture Styles

The channelized versus fully cross-strapped tradeoff concerns the degree of coupling or connectivity between two sets of elements (called the left set and the right set) and therefore is related to system properties such as throughput and reliability. In channelized architectures, each "left" element is matched to exactly one "right" element, whereas in fully cross-strapped architectures, all elements are assigned or connected to each other.

The "channelized" style of architecture:

- Leads to a decoupled matching between the two sets of elements.
- Leads to less complex architectures, due to a reduction in the number of connections.
- May lead to less costly architectures, if the cost of the connections is important.
- May lead to lower-performance architectures, especially when performance has to do with throughput (due to bottlenecks or lack of sufficient resources) or reliability (due to lack of redundancy).
- May be less vulnerable to some "common-cause failures" thanks to the decoupling of elements. Common-cause failures occur when a failure in one element of the system triggers one or more failures in other elements of the system. Going back to the GNC example, if a sensor breaks, it could potentially kill all the computers that are connected to it, but it will not affect those that are in an independent chain.

The "fully cross-strapped" style of architecture:

- Leads to more coupled architectures.
- Leads to more complex architectures, due to an increased number of connections.
- May lead to more costly architectures, if the cost of the connections is important.
- May lead to more reliable architectures thanks to redundancy.
- May be more vulnerable to common-cause failures because of the strong coupling that allows for failures to propagate through the architecture.

The choice between these architectural styles, or of any hybrid option in between, is driven by:

- The relative cost of connections compared to the cost of the elements.
- Our cost–reliability utility curve (How much are you willing to spend to gain an extra nine in reliability?*).
- The relative likelihood of common-cause failures versus independent failures.

* The reliability of a system is commonly reported as a number of nines: R = 0.999999 corresponds to six nines, and R = 0.999 corresponds to three nines.

The PARTITIONING Pattern

The PARTITIONING Pattern appears when we have a single set of N elements and we need to partition them into a number of non-empty and disjoint subsets (from 1 to N). Each element must be assigned to exactly one subset. In other words, we cannot "repeat" elements (assign them to more than one subset) or leave elements out (assign them to no subset). More formally, given a set of N elements $U = \{e_1, e_2, \ldots, e_N\}$, an architecture in the PARTITIONING problem is given by a partition P of the set U, which is any division of U into a number of non-overlapping subsets $P = \{S_1, S_2, \ldots, S_m\}$, where $S_i \subseteq U, S_i \neq \{\phi\} \forall i$, $1 \leq m \leq N$, and the subsets are mutually exclusive and exhaustive. In other words, P is a valid architecture if:

1. The union of all subsets in P is equal to U: $\cup_{i=1}^{m} S_i = U$
2. The intersection of all elements in P is empty: $\cap_{i=1}^{m} S_i = \phi$

The instrument-packaging problem from our NEOSS example can be seen as an instance of the PARTITIONING Pattern: We have a set of instruments, and we are looking for different ways of partitioning this set into satellites. For example, all instruments can be assigned to a single large satellite, or they can all have their own dedicated satellites, or anything in between.

An architecture in the PARTITIONING Pattern is given by a partition $P = \{S_1, S_2, \ldots, S_m\}$. In the NEOSS example, given the set of instruments {imager, radar, sounder, radiometer, lidar, GPS receiver}, one possible partition is given by {{radar, radiometer}, {imager, sounder, GPS receiver}, {lidar}}, and another one is given by {{radar}, {lidar}, {imager, sounder}, {radiometer, GPS receiver}}, as shown in Figure 16.10.

Such a partition can be represented in different ways—for example, an array of integers where position i indicates the index of the subset to which element i is assigned, so that entries with the same value designate elements that are in the same subset.

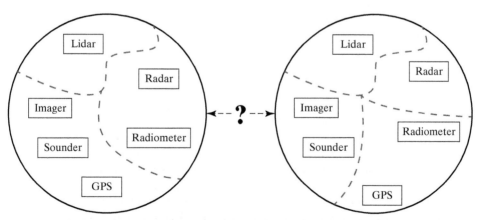

FIGURE 16.10 Illustration of the PARTITIONING Pattern in the NEOSS instrument-packaging problem.

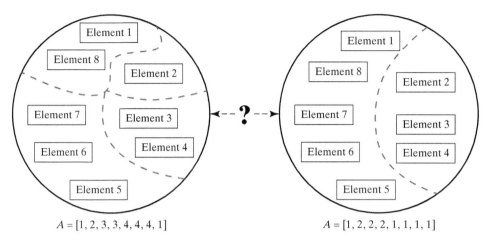

FIGURE 16.11 Pictorial representation of the PARTITIONING Pattern. Two different PARTITIONING architectures are shown for a simple case with 8 elements and 4,140 possible architectures.

A general pictorial representation of the PARTITIONING Pattern is provided in Figure 16.11, where two different partitions for a set of 8 elements are shown, together with their representations as arrays of integers. The partition on the left side divides the 8 elements into four subsets, and the partition on the right side divides the 8 elements into two subsets.

The number of possible partitions in a set grows very quickly with the number of elements in the set. For reference, there are 52 ways of partitioning 5 elements and over 115,000 ways of partitioning 10 elements. Because of the two constraints provided in the definition, partitions are harder to count than simple assignments or subsets. [19]

One might think that the PARTITIONING Pattern does not appear very often, since it has such "hard" constraints. In practice, though, many programmed decisions in system architecture are naturally formulated as PARTITIONING problems, especially when one is mapping function to form (Task 2 in Table 16.1) or decomposing function or form (Task 1 in Table 16.1).

- **EXAMPLE 1:** If one considers a set of underground oil reservoirs, the problem of how many facilities to build, and where, can be formulated as a PARTITIONING problem if subsets of reservoirs are implicitly identified with facilities. For example, if we consider three reservoirs A, B, and C, there are five possible partitions: (1) {A}, {B}, {C}, which would assign one facility to each reservoir; (2) {A, B}, {C}, which would assign reservoirs A and B to one facility and reservoir C to another facility; (3) {A, C}, {B}, which would assign reservoirs A and C to one facility and reservoir B to another facility; (4) {B, C}, {A}, which would assign reservoirs B and C to one facility and reservoir A to another facility; and (5) {A, B, C}, which would assign reservoirs A, B, and C to a single facility. Note that this formulation is based on abstract facilities. In other words, the formulation would tell us only which reservoirs are connected to which facilities, not the exact position of the facility, which would need to be determined afterwards.

- **EXAMPLE 2:** Consider an IT network for a large bank with 1,000 branches and 3,000 ATMs across the United States. These banks need to be interconnected through a number of routers. Deciding how many routers to use can be seen as a PARTITIONING problem, assuming that each branch and ATM needs to be connected to exactly one router. For example, one could imagine a solution where a single router provides service to all 4,000 nodes.

When it appears, the PARTITIONING Pattern is fundamentally about how much centralization we want in our architecture. Hence, it entails a discussion about two new architecture styles: centralized (or monolithic) and decentralized (or distributed) architectures. These styles are discussed in Box 16.3.

Box 16.3 Insight: Monolithic versus Fully Distributed Architecture Styles

The monolithic versus distributed tradeoff concerns the degree of centralization of an architecture, so it is also related to system properties such as evolvability, robustness, and flexibility. In monolithic architectures, all elements are grouped in a central "large" subset, whereas in fully distributed architectures, all elements are assigned to dedicated subsets.

The "monolithic" style of architecture:

- Leads to architectures with very few, but very complex, elements.
- Captures positive interactions (synergies) between elements, but may also capture negative interactions (interferences) between elements.
- Minimizes redundancy where there is common functionality needed by all elements.
- May lead to decreased performance of individual elements because of competition for common resources (such as bandwidth and power).
- May lead to longer development and increased likelihood of schedule slippage in system development, when an entire project must wait for integration of an unfinished element.
- May lead to less costly architectures, if the relative cost of replicating common functionality (such as solar panels, batteries, communications, and frame) required in each subset is high compared to the cost of the rest of the elements and the cost due to interferences.
- May lead to less reliable architectures due to the presence of single points of failure.
- May be less evolvable, as a consequence of longer development times.

The "fully distributed" style of architecture:

- Leads to architectures with many elements that are sometimes simpler and less costly.
- Leads to more decoupled architectures.
- May not capture important synergies between elements or avoid important interferences between elements.
- May lead to redundant replication of common functionality.
- May lead to more costly architectures, if the relative cost of replicating common functionality required in each subset is high compared to the cost of the rest of the elements and the cost due to interferences.

(continued)

- May lead to more reliable architectures, thanks to fewer single points of failure.
- May lead to shorter development time and lower risk of schedule slippage.
- May be more evolvable, thanks to shorter development times and lower costs of individual elements.

The choice between these architectural styles, or for any hybrid option in between, is driven by the resultant of all "attracting" (synergies) and "repulsive" forces (interferences):

- The relative effect of synergies and interferences between elements on system performance and cost.
- The dependence of the magnitude of synergies and interferences on the physical distance between elements in the same subset.
- The amount of functionality that needs to be replicated on every element in a distributed architecture, and the relative cost of this common functionality with respect to the cost penalty due to interferences.
- Our willingness to pay to improve programmatic aspects such as reducing development risk, shortening development time, and maintaining a flat expense profile over time.
- The cost of making a monolithic architecture that has reliability similar to that made possible by the built-in redundancy of distributed systems.

Interactions between elements—namely, the synergies and interferences introduced in the context of the DOWN-SELECTION Pattern—play a key role in the PARTITIONING Pattern. In the DOWN-SELECTION Pattern, it was assumed that interactions between elements $\{e_i, e_j\}$ occur as long as both elements are selected. In the PARTITIONING Pattern, we acknowledge that physical interactions usually require that the elements be connected. In the NEOSS example, if the radar altimeter and the radiometer are both selected in the DOWN-SELECTION problem, but are put on different spacecraft and on different orbits in the PARTITIONING problem, it will very hard to obtain the benefits of their synergistic interaction! To capture this synergy, the instruments need to be either on the same spacecraft or close enough to allow for cross-registration. Similarly, instruments can interfere electromagnetically with each other only if they are on the same spacecraft or close enough.

For example, the Envisat satellite is the largest civil Earth observing satellite ever built. It weighs about 8 mt and carries 10 remote sensing instruments. Because the instruments are on the same satellite, they can look at the same spot on Earth's surface simultaneously. Thus scientists can make the most of the synergies between these instruments by combining their measurements to generate rich data products. [20] Furthermore, Envisat had only one solar panel, one set of batteries, one set of communication antennas, and one frame for all its payload; and only one launch was needed. All of these things would have had to be replicated 10 times if the instruments had been broken down into 10 smaller satellites. However, this monolithic approach also has disadvantages. In Envisat, the synthetic aperture radar could work for only 2% of the orbit, because the solar panel of the satellite did not produce enough power for all the instruments to work at the same time; synthetic aperture radars consume a lot of power. Moreover, if the launch vehicle or the solar panel had failed, all the instruments would have been lost. In another example, the Metop satellite had a large conically scanning instrument that induced vibrations on the platform that affected a very sensitive sounder. [21] This is an example of a negative interaction or interference between elements in a monolithic architecture. Interactions can also be programmatic in nature: Envisat and Metop could not be launched until all their instruments, including the least technologically mature, were ready for launch. Envisat cost over 2 billion euros and took over 10 years to develop. This means that by the time it was launched, some of the technologies were almost obsolete.

The PERMUTING Pattern

The PERMUTING Pattern appears when we have a set of elements, and each element must be assigned to exactly one position. Choosing these positions is often equivalent to choosing the optimal ordering or sequence for a set of elements. For example, 123, 132, 213, 231, 312, and 321 are six different permutations of the digits 1, 2, and 3. Similarly, if we have three satellite missions, we have six different orders in which they can be launched. More formally, given a set of m generic elements of function or form $U = \{e_1, e_2, \ldots, e_N\}$, an architecture in the PERMUTING problem is given by a permutation O—that is, any arrangement of the elements in U into a particular order*:

$$O = \{x_i \leftarrow i \in [1; N]\}_{i=1, \ldots, N} | x_i \neq x_j \forall i, j$$

An architecture in the PERMUTING Pattern is thus readily represented by an array of integers with two possible interpretations: element-based or position-based. For example, the sequence {element 2, element 4, element 1, element 3} can be represented by the array $O = [2, 4, 1, 3]$ (element-based representation) or by the array $O = [3, 1, 4, 2]$ (position-based representation).

The pictorial representation of the PERMUTING Pattern is provided in Figure 16.12, where 2 out of 120 different permutations, and their representations as element-based arrays of integers, are shown for a generic set of 5 elements.

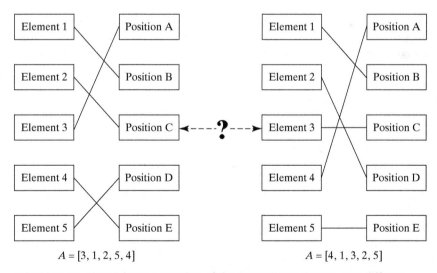

FIGURE 16.12 Pictorial representation of the PERMUTING Pattern. Two different PERMUTING architectures and their element-based representations are shown for a simple case with 5 elements and 120 possible architectures.

---

*Mathematicians call this a bijection of U onto itself.

The size of the tradespace of a PERMUTING problem of m elements is given by the factorial of m: $m! = m(m - 1)(m - 2) \ldots 1$. Note that the factorial grows extremely fast—faster than exponential functions. For 15 elements, there are over 1 trillion different permutations!

PERMUTING problems often concern the geometric layout of elements or the sequence of a set of processes or events. Recall from Parts 2 and 3 that operations are part of the architecture of the system, and they need to be considered early in the architecture process, especially because we can derive goals and metrics directly from the concept of operations during the stakeholder analysis. For example, one could try to optimize the sequence of destinations or (more broadly) tasks to do for an unmanned vehicle. The PERMUTING Pattern is also prominent when one is architecting a portfolio of systems and needs to decide the order in which the systems should be deployed. In what order should we launch the satellites in the NEOSS example? In this case, a key consideration will be related to ensuring that missions are launched in such a way as to ensure data continuity of existing measurement records. Other examples of PERMUTING problems include:

- **EXAMPLE 1:** Very large-scale integration (VLSI) circuits have on the order of 10^{10} transistors per die. Therefore, it is extremely important to minimize the total length of connections, because small variations in length can have a big impact on fabrication cost at the scale at which these chips are fabricated. One way of achieving this is to optimize the placement of interconnected cells on the chip, so that the total length of connections is minimized. This is an instance of the PERMUTING Pattern, where elements are gates or even individual transistors, and positions are positions on the board.

- **EXAMPLE 2:** A few years ago, NASA canceled the flagship space exploration program "Constellation" because of budgetary restrictions. To replace Constellation, a senior advisory committee of experts led by former Lockheed Martin CEO Norm Augustine generated several alternative strategies, one of which was known as the Flexible Path. In the context of the Flexible Path, an important architectural decision is the optimal sequence of destinations on the path to Mars (for example, Lagrange Points, then Moon surface, then Near-Earth Asteroid, and then Mars, or reverse order of Moon Surface and Near-Earth Asteroid). This is a complex decision that needs to take into account scientific, technological, and programmatic considerations.

We have seen that there are two major classes of PERMUTING problems: those that deal with time (scheduling) and those that deal with topology and geometric considerations. However, the PERMUTING Pattern is more general than it may appear at first. The PERMUTING Pattern is essentially a matching of a set of N elements to the set of integers from 1 to N, where no two elements can be matched to the same integer. Note that this formulation does not necessarily imply an ordering between the elements. The only condition is that the options be exclusive; if an element is assigned to an option, then no other element can be assigned to it. For example, if a circuit is assigned to a position in an electrical board, no other circuit can be assigned to that position.

Given this generalization, PERMUTING problems are most useful for Task 5 in Table 16.1, connectivity of function to form, as well as for defining the system deployment and/or concept of operations. Typical tradeoffs in time-related PERMUTING problems involve balanced value delivery over time and careful consideration of resource budgets. In these cases, the system architect often has to choose between front-loaded ("greedy") system deployment and incremental system deployment. This is discussed in Box 16.4.

Box 16.4 Insight: Greedy versus Incremental System Deployment

The PERMUTING Pattern often has to do with system deployment. In a system consisting of several independent systems to be deployed, the deployment strategy may be an important architectural decision to consider, because it drives value delivery to stakeholders over time. The main architectural styles related to system deployment are greedy deployment and incremental deployment.

In greedy deployment, the goal is to deploy the high-value systems as quickly as possible, even if that takes a long time. It is assumed that high-value elements are more costly and therefore take longer to develop than low-value elements. On the other hand, the incremental deployment approach seeks to deliver some value to the stakeholders as soon as possible and then progressively deploy the high-value elements of the system.

Systems that use the greedy deployment approach:

- Have high programmatic risk because they may not survive years of spending without providing value to stakeholders.
- May have lower overall deployment cost, eliminating redundancies sometimes present in incremental deployment.
- Offer fewer opportunities for changing the design.

Systems that use the incremental deployment approach:

- Have lower programmatic risk.
- May have higher deployment cost.
- Offer more flexibility.

The CONNECTING Pattern

The CONNECTING Pattern appears when we have a single, fixed set of elements, and we want to decide how to connect them. These connections may or may not have a sense of "direction." More formally, given a fixed set of m generic elements or nodes $U = \{e_1, e_2, \ldots, e_m\}$, an architecture in the CONNECTING problem is given by a graph that has U as its nodes and has a list of $1 \leq N \leq N^2$ vertices on U: $G = \{V_1, V_2, \ldots, V_N\}$ where each vertex connects two nodes: $V = \{e_i, e_j\} \in U \times U$.

The CONNECTING Pattern can be readily represented by a square binary matrix called the adjacency matrix. More precisely, the adjacency matrix A of a graph G is an $n \times n$ matrix

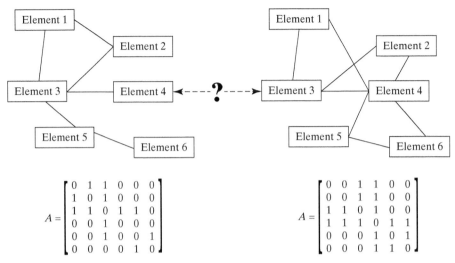

FIGURE 16.13 Pictorial representation of the CONNECTING Pattern. Two different CONNECTING architectures and their corresponding representations are shown for a simple case with 6 elements and 32,768 possible architectures.

where $A(i, j) = 1$ if node i is directly connected by an edge to node j, and $A(i, j) = 0$ otherwise. Note that in undirected graphs—that is, graphs in which edges have no orientation—$A(i, j) = 1$ implies $A(j, i) = 1$, and therefore the adjacency matrix of undirected graphs is symmetric. This is not the case for directed graphs, where edges do have a sense of direction.

The pictorial representation of the CONNECTING Pattern is provided in Figure 16.13, which shows two different ways of connecting a set of six nodes, and the corresponding binary matrices.

The size of the tradespace of a CONNECTING problem is given by the size of the set of all possible adjacency matrices on a set of m elements. The number of different adjacency matrices depends on whether the graphs being considered are undirected or directed, and on whether nodes are allowed to be connected to themselves. This leaves the four different cases that are summarized in Table 16.4.

The foregoing formulas assume that there can be a maximum of one connection between any two given nodes or between a node and itself. In other words, it assumes that the adjacency matrix is a Boolean matrix.

TABLE 16.4 | Size of the tradespace for the CONNECTING Pattern in four different cases.

	Directed Graph (non-symmetric)	Undirected Graph (symmetric)
Self-connections Allowed (diagonal meaningful)	2^{m^2}	$2^{\frac{m(m+1)}{2}}$
Self-connections Not Allowed (diagonal not meaningful)	$2^{m^2-m} = 2^{m(m-1)}$	$2^{\frac{m(m-1)}{2}}$

Examples of CONNECTING problems follow.

- **EXAMPLE 1:** Say we are in charge of architecting the water distribution network of a small region. The region is divided into areas that differ in various characteristics (such as population density and natural resources), and each area needs water. In each area, there may or may not be a water generation/treatment plant (a well or a desalination plant). The problem of figuring out which connections to make between areas, given a certain layout of water generation/treatment plants, is most naturally formulated as an instance of the CONNECTING Pattern.

- **EXAMPLE 2:** The architecture of a power grid also fits the CONNECTING Pattern very naturally. Energy is produced and/or stored at nodes, and it is transported between nodes through connections. Having more connections costs a lot of money, but it allows more balanced management of the electricity produced.

The CONNECTING Pattern is useful to address Task 5 of Table 16.1, which we labeled as "Connecting Form and Function." The two previous examples essentially deal with the definition of physical interfaces between system elements, but the nature of these interfaces can be informational, as in data networks, or can simply indicate logical dependencies (for example, software systems).

We argued in Section 16.1 that the connectivity task appears to some extent in all kinds of systems, because all systems have internal and external interfaces. One way to think about the CONNECTING Pattern is to think of the system at hand as a network and to look at different ways of connecting the set of nodes in a network. All systems can be represented as networks, but this representation is trivial for systems that are networks, such as data networks, transportation networks, power networks, and satellite constellations.* It is also easy to think about systems of systems as networks, where each system is a node and the edges are the interfaces between the individual systems.

The most natural commodity that flows through the nodes of these networks is arguably data, but vehicle traffic (ground, air, or space), electrical power, water, natural gas, and food are also possibilities. The amount and distribution of commodity that flows through the network are important drivers of the architecture. For example, centralized architectures that have one or more nodes through which most traffic flows can suffer from delays due to the appearance of bottlenecks.

More generally, the relevant tradeoffs in CONNECTING problems usually have to do with the degree of connectivity of nodes, and with their effect on performance metrics such as latency and throughput, as well as on other emergent network properties such as reliability and scalability. [22] Note, in some of these aspects, the similarity to the ASSIGNING Pattern, which can be seen as an instance of the CONNECTING Pattern where we have two types of nodes in the network, which we called the "left" set and the "right" set.

The main architectural styles in the CONNECTING Pattern come from network topologies: bus, star, ring, mesh, trees, and hybrid. These different styles are compared in Box 16.5.

---

*Note that at some level, every system can be modeled as a network, because it can be represented as a set of interconnected elements in the form domain, or as a set of connected functions in the functional domain.

Box 16.5 Insight: Architectural Styles in the CONNECTING Pattern

Architectural styles in the CONNECTING Pattern can be borrowed from network topologies (see Figure 16.14).

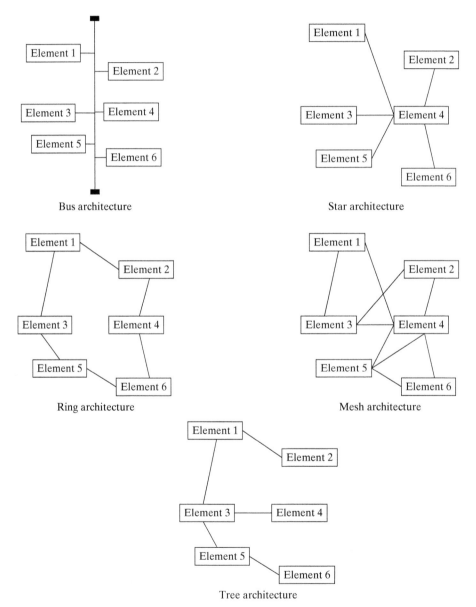

FIGURE 16.14 Architectural styles in the CONNECTING Pattern, borrowed from network topologies.

- In **bus architectures,** every node is connected to a single common interface called the bus. Advantages include low cost, and scalability. The main disadvantage is that the bus is a single point of failure. The Ariane 5 launch vehicle uses a bus architecture for the avionics system, where the bus complies with the MIL-STD-1553 standard. [23]
- **Star architectures,** also known as **hub and spoke architectures,** have a central node—the hub—to which all the other nodes—spokes—are connected. The hub acts as a relay of flow (for example, data) through the network, just like the bus in the bus architecture, but unlike the bus, it is a real node, and thus it can also perform other functions (such as acting as a source or sink of flow). Hence, the same advantages and disadvantages of bus architectures apply for star architectures. Most airlines nowadays employ hub and spoke architectures for their air traffic systems, using a handful of airports as hubs through which most flights connect.
- In **ring architectures**, every node is connected to exactly two different nodes, in such a way that all traffic through the network flows in a closed circuit, the ring. The main advantage is good performance under heavy network load. The main disadvantage is that every node in the network is a single point of failure. The computers and other devices in a bank branch are often connected in a "token ring" architecture. [24]
- **Mesh architectures** are distributed and collaborative architectures in which all nodes cooperate to act as relays. Their main advantages with respect to the previous architectures are increased fault tolerance and scalability. Their main disadvantage is that if the cost of connections is high, it can lead to costly architectures. Mesh architectures are being developed to provide Internet access to rural regions in Africa. [25] They are also at the core of peer-to-peer and agent-based software systems.
- **Tree architectures** are organized according to a hierarchy, with a root node at the top of the hierarchy. One or more child nodes are connected to the root node and form the second level of the hierarchy. In the next level of the hierarchy, each child node may in turn have zero, one, or more child nodes. The main advantages are scalability and maintainability. The main disadvantage is that if an element fails, it may affect all child elements. Commonly cited examples of tree architectures are the telephone and television networks of apartment buildings, where the telephone and television signals arrive from the outside network to a gateway node, typically located either on the first or the last floor, and are then distributed through the apartments of successive floors.

16.5 Formulating a Large-scale System Architecture Problem

We have examined six Patterns that appear when one is considering programmed decisions in system architecture: the DECISION-OPTION Pattern, the DOWN-SELECTING Pattern, the ASSIGNING Pattern, the PARTITIONING Pattern, the PERMUTING Pattern, and the CONNECTING Pattern. These Patterns can be used to formulate programmed decisions that appear when the system architect is conducting the recurring tasks that were identified earlier in the chapter and summarized in Table 16.1. Table 16.5 indicates which Patterns are usually more applicable for the various tasks.

One might think that one and only one of these Patterns will be appropriate for any particular system architecture problem and that choosing this Pattern will be straightforward. But that is not the case for real-life large-scale architecture optimization problems.

TABLE 16.5 | Relationships of system architecture tasks and Patterns. X indicates the Pattern(s) that are usually more appropriate for each task.

	Decision-Option	Down-Selecting	Assigning	Partitioning	Permuting	Connecting
Decomposition of Form and Function				X		
Mapping of Function to Form			X	X		
Specialization of Form and Function	X	X				
Characterization of Form and Function	X					
Connectivity of Form and Function			X		X	X
Definition of Scope and Selection of Goals		X				

First, as suggested by the fact that Table 16.5 is not a one-to-one mapping, the Patterns are not mutually exclusive and there is sometimes ambiguity in the choice of a Pattern for a given task. For example, in the NEOSS example, the problem of choosing which instruments go into which spacecraft can be formulated either as an ASSIGNING problem, where instruments are assigned to spacecraft or orbits, or as a PARTITIONING problem, where instruments are simply partitioned into subsets, which then are the input to an orbit and spacecraft design. Note that the two formulations are different in that the ASSIGNING Pattern allows assigning instruments to zero or more than one spacecraft or orbit, which is not allowed in the PARTITIONING Pattern. One formulation will probably make more sense than the other, depending on the context; for example, budget constraints may preclude repeating instruments.

Second, real system architecture optimization problems usually have many decisions that are rarely homogeneous in terms of their structure. Often, a large problem can be effectively decomposed into sub-problems that fit one of the Patterns. This decomposition reduces the computational complexity of the overall problem, at the expense of a loss of global optimality if couplings exist between sub-problems.

These two points are discussed in the rest of this section.

Overlap between Patterns

Relationships between these Patterns were unveiled as the Patterns were introduced in the previous section. These relationships are summarized in Table 16.6.

The relationships in Table 16.6 would seem to suggest that one can switch between some of these problem formulations but not others. In reality, although doing so requires more effort, it can

TABLE 16.6 | Outstanding relationships between Patterns. Elements in the diagonal are not meaningful. Note that the matrix is symmetric, so only the upper triangular part is shown.

	Down-Selecting (DS)	Assigning (AS)	Partitioning (PT)	Permuting (PM)	Connecting (CN)
Decision-Option (DO)	DS ~ DO with N binary decisions (choose element i yes or no)	AS ~ DO with N binary decisions (assign element i to bin j yes or no)	PT ~ DO with 1 decision containing all possible partitions, or N integer decisions (1 through N) with hard constraints to enforce partitioning structure	PM ~ DO with 1 decision containing all possible permutations, or N integer decisions (1 through N) with hard constraints to enforce permutation structure	CN ~ DO with N^2 binary decisions (connect element i to element j yes or no)
Down-Selecting (DS)		AS ~ N DS problems, one for each element	PT ~ DS with one decision for each possible subset of elements, and constraints to enforce partitioning structure.	PM ~ DS with one decision for each possible assignment of an element to a position, with hard constraints to enforce permutation structure	CN ~ DS problem where we choose a subset from a set of candidate connections
Assigning (AS)			PT ~ AS with N elements and N indistinguishable bins, and each element must be assigned to 1 bin (empty bins are OK)	PM ~ AS with same number of elements and bins, and each element must be assigned to a different bin	AS ~ CN with two distinct groups of nodes, and each node from one set can be connected only to nodes from the other set
Partitioning (PT)				PM can be expressed as a traveling salesman problem, which is an NP-complete problem. PT can be expressed as a set-partitioning problem, also NP-complete. Since it has been proved that NP-complete problems are all interchangeable, PM can be expressed as PT.	PT ~ CN where connected nodes are considered to be in the same subset, but this is a degraded formulation because multiple CN architectures correspond to the same PT architecture
Permuting (PM)					PM ~ CN with two distinct groups of N nodes, and each node from one set must be connected to exactly one node from the other set without repetition

be proved that one can switch between *any* two problem formulations. Intuitively, it is easy to see that every architectural problem can be formulated as a DECISION-OPTION problem. However, this requires explicitly enumerating what may be a large number of options for decisions that are more naturally expressed as ASSIGNING or PARTITIONING decisions. For instance, in the NEOSS example, it would be possible—though cumbersome—to express the instrument-packaging problem for 5 instruments as a single DECISION-OPTION decision with 52 options.

Proving that any architectural problem can be expressed as a DOWN-SELECTING problem is equally easy, because it suffices to assign binary codes to the options of each decision. For instance, if for a DECISION-OPTION decision we have m options, we can formulate that decision in the DOWN-SELECTING Pattern using $\log_2 m$ binary decisions. This is similar to what was done in the Apollo example with the mission-mode decision, which was effectively decomposed into multiple decisions (such as EOR yes or no, LOR yes or no). Although this is possible in general, it often requires adding a large number of constraints to rule out binary combinations that do not correspond to any option, which will happen for any m that is not an exact power of 2.

It is also relatively easy (though less intuitive) to prove, using graph theory, that any architectural problem can be expressed as a CONNECTING problem, because any DECISION-OPTION problem can be represented as a bipartite graph* in which we have one set of nodes representing the decisions and another set representing the options. Each architecture can thus be uniquely associated with an adjacency matrix that connects these two partitions.

While this is possible in general, it also comes at the price of some extra constraints, because we have to ensure that options corresponding to a certain decision are not linked to other decisions, and that each decision is assigned to exactly one option.

Proving that any architecture optimization problem can be expressed by means of exclusively PERMUTING decisions or exclusively PARTITIONING decisions is a harder task. The idea for a possible proof is that two known NP-complete problems, namely the traveling salesman problem and the set partitioning problem, can be reduced to the PERMUTING and PARTITIONING problems. Since every problem in NP (including all NP-complete problems) is reducible to every other problem in NP-complete, this means that we can express permuting problems as partitioning problems, and vice versa. Moreover, any other architecture optimization problem (DOWN-SELECTING, ASSIGNING, CONNECTING) can also be reduced to an NP complete problem, and therefore can be reduced to a PARTITIONING or PERMUTING problem.

Although every architecting problem can be expressed using any Pattern, it is clear that some Patterns are more appropriate than others to express certain problems (for instance, expressing the NEOSS instrument-packaging problem as a DECISION-OPTION problem would be awkward). Furthermore, the Pattern that we choose will affect the time it takes to solve the problem with an optimization algorithm. For example, DOWN-SELECTING and ASSIGNING problems usually are easier to solve by optimization algorithms than PARTITIONING and PERMUTING problems are.

Decomposition into Sub-problems

It is important to note that the complete set of architectural decisions for a given system will naturally not entirely fall in any one of these Patterns. Instead, different parts of the problem (subsets

---

*A bipartite graph is a graph where the nodes are divided into two groups such that there are no connections between nodes of the same group. All connections go from a node of one group to a node of the other.

of decisions) are likely to fit different Patterns. For example, the NEOSS example can effectively be decomposed into three sub-problems (instrument selection, packaging, and mission scheduling). These three sub-problems map directly to three Patterns: DOWN-SELECTING, PARTITIONING (or ASSIGNING if we assign instruments to orbits and allow for repetition of instruments), and PERMUTING.

This does not mean that a single optimization problem with all these different types of decisions cannot be written. However, in many cases, a global optimization problem will be extremely large in size, which will in general decrease the performance of the optimization algorithms.

Moreover, these different subsets of decisions are often relevant to different sets of metrics, or are relevant to the same set of metrics but operate at different scales (that is, the effect of one is much larger than the effect of the other). For example, in the NEOSS case, both the instrument selection problem and the instrument-packaging problem are fundamentally tradeoffs between cost and performance, but most of the cost and most of the performance are driven by the instrument selection problem. The instrument-packaging problem has a noticeable and important impact on both cost and performance, but this impact is still small compared to that of the instrument selection problem. Hence, in practice, decomposing the global architecture optimization problem into sub-problems that fit one of the Patterns presented in this chapter is often—but not always—the best way of decomposing the problem.

Note, however, that in the case of NEOSS, the three sub-problems are indeed clearly coupled. One cannot, for example, make an optimal decision in the instrument selection problem without thinking about the packaging of the selected instruments. This could lead to inefficiencies in the use of a certain satellite bus or launch vehicle. Similarly, it is hard to make the best packaging decision without thinking about the scheduling of the missions, since an instrument that closes an important data gap should perhaps fly in the spacecraft that is to be launched first. This is illustrated in the Figure 16.15.

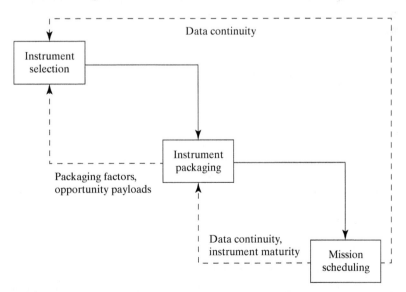

FIGURE 16.15 Decomposition of the NEOSS architecture problem into three coupled sub-problems that map to the Patterns defined earlier.

As a consequence of this coupling, simply solving the individual sub-problems and combining the results will in general lead to suboptimal solutions. This problem can be alleviated by modeling the coupling between sub-problems explicitly—for example, in the form of constraints. For instance, in the instrument selection problem, a constraint can be used to ensure that an instrument that covers an important data gap is selected. These rules will guide the search process to favor architectures that are likely to be good when the other sub-problems are solved.

In summary, when formulating a large-scale system architecture problem, the architect must decompose the problem into sub-problems that are as loosely coupled as possible, using the six Patterns to formulate the sub-problems.

16.6 Solving System Architecture Optimization Problems

Introduction

The previous sections explained how to formulate a system architecting optimization problem—that is, how to represent a complex architecture as a set of decisions using an encoding scheme (such as an array of binary variables) and write an optimization problem with the goal of finding the combinations of choices that optimize the tradeoffs between different stakeholder needs. In particular, we presented a set of six Patterns that can help the system architect turn most architectural decisions into programmed decisions. In this section, we will discuss how to solve these architecting optimization problems.

The content of this section will thus necessarily be more oriented toward tools and methods than the rest of the text. Nonetheless, our goal is not to provide an exhaustive survey of tools and methods that exist to tackle parts of these problems, but rather to present a minimum set of tools and methods and to focus on how to use them in real-world architecting problems. [26]

Full-Factorial Enumeration

We start with a very simple strategy called full-factorial enumeration, in which all possible architectures are enumerated. Each architecture is then evaluated, so that one can simply choose from the best architectures. Full-factorial enumeration is very simple and transparent, and (as we argued in Chapter 15) simulating an entire tradespace, rather than just the Pareto front, provides additional useful insight into the features of the problem.

"Small enough" architecture optimization problems can be solved by brute force—that is, by simply enumerating and evaluating all architectures. Whether a problem is small enough for full-factorial enumeration depends on the size of the tradespace, the average time it takes to evaluate one architecture on the machine(s) available, and the total computation time tolerable. The longer it takes to evaluate an architecture, the fewer architectures we can evaluate in a given amount of time. Thus there is a tradeoff between modeling breadth (the number of architectures) and modeling depth (the time it takes to evaluate an architecture).

The system architect must be careful when creating the architectural model, in particular the value functions, to ensure that it will be feasible to explore a reasonable portion of the tradespace in the allotted time. Once we have determined that a problem is small enough for full-factorial enumeration, the next step is to actually enumerate all valid architectures. Let us start with a simple

```
Architectures = {} % This array will contain the architectures
NumberOfArchs = 0 % This is a counter for the number of architectures
foreach val1 in EOR = {yes, no}
 foreach val2 in earthLaunch = {orbit, direct}
  foreach val3 in LOR = {yes, no}
   ...
      foreach val9 in ImFuel = {cryogenic, storable, N/A}
       % Add an architecture to the array and increase counter
       Architectures (NumberOfArchs++) = [val1 val2 ... val9]
      end
   ...
  end
 end
end
```

FIGURE 16.16 Pseudocode for full-factorial enumeration of architectures in the Apollo example.

example: How do we enumerate all architectures for our Apollo example from Chapter 14? Recall that in the Apollo example we had nine decisions, each with two or more options. In this case, we can simply have a set of nine nested loops, where each loop loops over all options for a decision, as shown in the pseudocode contained in Figure 16.16.

This approach is simple enough to implement for instances of DECISION-OPTION problems, because the discrete set of possible values for the decisions is usually small and available in a form (such as an array) that can readily be iterated upon. Using this approach for other types of decisions may be challenging and will in general require a preprocessing step in which all options for a particular decision (such as a PARTITIONING decision) are explicitly enumerated up-front and put in a data structure that contains one entry for each option. This can be an extremely large number of options.

Other advanced tools for full-factorial enumeration employ programming paradigms entirely different from those familiar to many engineers, such as declarative programming. Mainstream programming languages such as C and Java are procedural as opposed to declarative: The programmer has to tell the computer how to solve a problem, step by step. Conversely, in declarative programming, the programmer has to provide the goals of the problem, but not the procedure to solve it. Declarative languages have an inference algorithm that takes care of specifying the steps to solve the problem. Rule-based systems are examples of declarative programs. The use of rule-based systems for enumeration of architectures is presented in Appendix C.

Heuristics for Architecture Optimization

We showed in the previous sections how the size of the architecture tradespace can grow to billions of architectures for relatively small numbers of decisions, which implies that full-factorial enumeration is often not feasible. In these cases, we will have to resort to algorithms that explore the architecture tradespace in a more efficient and effective way.

We mentioned earlier that most architecture optimization problems are instances of non-linear, non-smooth combinatorial optimization problems. This means that algorithms that exploit strong properties of the problem (such as gradient-based algorithms and dynamic programming), are often not an option.

For this reason, meta-heuristic optimization algorithms (such as genetic algorithms) are often used to solve these kinds of problems. Meta-heuristic optimization algorithms use very abstract and general search heuristics (mutation, crossover) that assume very little about the features of the problem at hand. As a consequence, they are very flexible and can be applied to a wide variety of problems. Heuristic optimization cannot guarantee that the global optimum is found, and it is less computationally efficient than other combinatorial optimization techniques, [27] But its strength resides in its simplicity, flexibility, and effectiveness in solving a wide variety of optimization problems, especially those with the characteristics of architecture optimization problems (categorical decision variables; multiple non-linear, non-smooth value functions and constraints). These algorithms are our tool of choice in this text for solving architecture optimization problems.

Figure 16.17 represents a generic iterative optimization algorithm. It has three main modules: an architecture generator or enumerator, an architecture evaluator, and a central search agent that takes information from the enumerator and evaluator and then decides which regions of the tradespace to explore next by providing search directions and steps in these directions.

Whereas enumeration, evaluation, and down-selection (such as by means of a Pareto filter) are clearly defined tasks in the simple full-factorial enumeration case, the boundary that separates enumeration, search, and down-selection becomes blurry in the partial-search case. Indeed, in most optimization or search algorithms, generation of the search directions is intimately

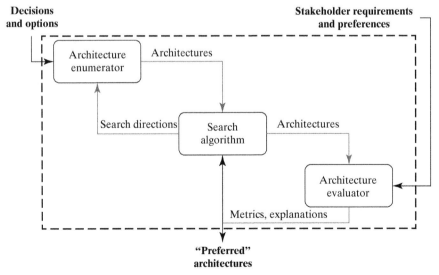

FIGURE 16.17 Typical structure of an optimization algorithm. For smaller tradespaces, the search function becomes unnecessary. The architectures are fully enumerated and evaluated, and then non-dominated architectures are chosen.

coupled with both the down-selection and the enumeration of new architectures. Architectures are not enumerated up-front, but rather are generated on the fly by applying the search heuristics to the existing architectures. Hence, a population of architectures evolves over time, ideally approximating the Pareto frontier.

A Generic Population-Based Heuristic Optimization

We'll begin with a class of meta-heuristic optimization algorithms called population-based algorithms. In population-based methods, a population of architectures is evolved and maintained at each iteration, instead of iteratively changing just one architecture. This has some advantages for architecture optimization problems, because the inherent goal is to find a few classes or families of good architectures, rather than finding the best architecture. The flow diagram for a population-based search algorithm is provided in Figure 16.18.

In a population-based search algorithm, the search and enumeration functions are performed by a set of selection and search operators or rules. Selection rules select a subset of the population that will be the basis for the construction of the new population of architectures in the next generation. For example, an intuitive selection rule is to choose architectures that are on the

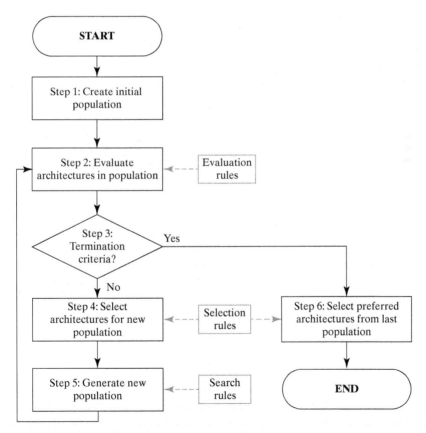

FIGURE 16.18 Population-based search algorithm.

non-dominated Pareto frontier. This is known as elitism. In practice, it has been shown that introducing a few dominated architectures that bring diversity to the population can lead to higher-quality Pareto frontiers with fewer gaps. [28]

Then, a set of search rules will combine information from the architectures in the selected subset to generate a new pool of architectures. For example, in a genetic algorithm, new architectures are generated by mutating some of these architectures (that is, randomly changing the value of a particular decision) and by crossing over pairs of good architectures (combining the values of the decisions of two good architectures to create one or two new architectures).

The new population is evaluated according to a set of metrics (these are labeled evaluation rules in Figure 16.18), and the algorithm continues until a set of termination criteria are met (such as a maximum number of generations).

Generating the Initial Population

The algorithm described in the previous section requires defining an initial population of architectures—in other words, a sample of the architectural tradespace. A simple way of performing the sampling is to generate random architectures. Mathematically, this requires obtaining random numbers and transforming those numbers into values for the architectural decisions. For example, for an unconstrained DECISION-OPTION problem, one can simply take a random value from the set of alternatives for each architectural decision, and for DOWN-SELECTING problems, one can simply obtain N random Boolean numbers (0, 1).

Obtaining random architectures is more difficult if there are constraints, because one has to check that the randomly generated architecture is consistent with all constraints. In cases with few constraints, a simple bypass is to reject infeasible architectures and keep randomly generating architectures until the desired population size is reached. However, this approach might fail in highly constrained cases, because most random architectures will be infeasible. In this case, it will be necessary to resort to more advanced techniques that either guarantee that every architecture generated is feasible by construction, or apply a repair operator that can turn an infeasible architecture into the most similar feasible architecture.

Finally, note that in some situations, we may want to bias the sampling process to drive it toward regions of interest. Consider, for example, an instance of the PARTITIONING problem with 10 elements. There are more than 115,000 different architectures, but 88% of these consist of 4, 5, or 6 subsets. This is simply because there are many more ways of partitioning a set of 10 elements into 5 subsets (45,000+) than into 2 subsets (511). If we suspect that architectures with 2 or 3 subsets may actually dominate the tradespace, it is beneficial to bias the sampling process to ensure that a relatively large fraction of the architectures will have 2 or 3 subsets. This can be done by first sampling from the number of subsets from a biased distribution, and then obtaining random samples of the desired number of subsets.

Completing an Initial Population with Deterministic Architectures

One may argue that pure randomness is not the smartest way to sample the tradespace. For example, we may want to guarantee that a few particular architectures are present in the initial population. We call these architectures "architectures of interest." In particular, if one or more baseline architectures exist for the project at hand, it is always wise to include it (or them) in the initial population.

Furthermore, it is good practice to include architectures that are in the "extreme points" of the architecture space. For example, in PARTITIONING problems, it is good practice to make sure that the monolithic architecture and the fully distributed architecture are both part of the initial population. Similarly, the channelized and fully cross-strapped architectures should be present in the initial population of an ASSIGNING problem.

In some cases, architectures in the extreme points can be obtained by using deterministic sampling schemes from the theory of Design of Experiments. Latin hypercubes and orthogonal arrays are examples of these sampling schemes. [29] It is not our intention to cover these in detail, but the main idea is that if we represent the selected architectures as vectors in the N-dimensional architecture space, we select a subset of architectures (vectors) that test each combination of values exactly once. For example, for 3 binary decisions, instead of enumerating the 8 possible architectures, we note that the following 4 architectures form an orthogonal* array: [0, 0, 0], [1, 1, 0], [0, 1, 1], and [1,0, 1]. If we look at the first and second decisions, each of the 4 possible values [0, 0], [0,1], [1,0], and [1,1] appears exactly once. The same is true for the first and third decisions, and for the second and third decisions. From a practical standpoint, for a given number of decisions N and options m, there exist tables that have pre-computed the optimal orthogonal vectors for different sample sizes. There exists a naming convention for such tables, so that table $L_P(m^N)$ contains the $P < N$ orthogonal arrays of length N containing elements from 1 to m. Thus, for instance, for 4 decisions and 3 options per decision, we can readily find what the 9 out of 81 architectures to select are by looking at the table called $L_9(3^4)$.[†]
In summary, good initial populations come from combining architectures from these three sources: architectures of interest, architectures in the extremes of the architectural region (such as those obtained through Design of Experiments), and random architectures sampled with care (taking into account how interesting certain regions of the architecture space are *a priori*).

General Heuristics and Meta-Heuristics for Efficient Search

Any optimization algorithm needs to accomplish two basic functions: exploration and exploitation. Exploration consists of finding new interesting regions of the architectural space. Exploitation consists of hill-climbing—that is, trying to find the best local architecture in a certain region of interest. Both are needed to obtain good architectures. Exploitation without exploration leads to getting "stuck" in local optima, and perhaps missing a better architecture in a different region, and exploration without exploitation leads to locally suboptimal architectures.

Population-based optimization algorithms abound in the literature. They include numerous variants of genetic algorithms, particle swarm optimization, ant colony optimization, the bees algorithm, and harmony search, among others. [30] Generally speaking, all of them have some way of accomplishing the goals of exploration and exploitation.

Hence, we focus in this chapter on two high-level strategies, or meta-heuristics, that are used or adapted in many of these algorithms and that are known to work well for wide ranges of problems: genetic algorithms [31] and local search.

---

*Note that orthogonal arrays are not orthogonal in the standard algebraic sense (that is, the standard scalar product of these vectors is not necessarily zero).

[†]Orthogonal arrays were used here to define an initial population in the context of heuristic optimization. But orthogonal arrays are actually used in different parts of the systems-engineering process. For example, they are extensively used in testing, especially software testing, to decide which subset of tests to do when full-factorial testing is impossible.

Heuristics in Genetic Algorithms

Genetic algorithms were initially proposed by Holland around 1975. [32] The basic idea is to imitate evolution through the exchange of genetic material between parent chromosomes, and this is accomplished by using two heuristics: crossover and mutation.

The flow of the algorithm is shown in Figure 16.18. An initial population of architectures is evaluated. Then a subset of architectures is selected on the basis of their fitness (the value they deliver to stakeholders) in the single-dimensional case, or based on Pareto ranking in the multi-objective case (architectures that are close to the Pareto front are more likely to be picked). The next population is generated by applying the two heuristics, crossover and mutation, to the selected architectures. The algorithm iterates until the termination criteria are satisfied.

In crossover, the selected architectures are grouped in pairs, and from each pair of parent architectures, two new children architectures are generated that have part of the genetic material (that is, the decisions) from one of their parents (the "father") and the rest from the other parent (the "mother").

There are several strategies for deciding what part of the genetic material to take from each parent. The simplest strategy, which is called single-point crossover, is illustrated in Figure 16.19. In single-point crossover, a point in the chromosome (the string of decisions) is chosen randomly. The first child takes all the decisions up to that point from the father and the rest from the mother, whereas the second child takes all the decisions up to that point from the mother and the rest from the father.

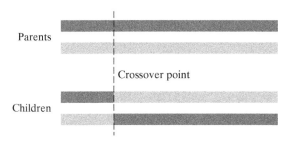

FIGURE 16.19 Single-point crossover.

Variations of this strategy use multiple random crossover points (for example, two-point crossover) or simply decide randomly whether each chromosome is taken from the father or from the mother (uniform crossover).

Holland showed in his foundational book that the effectiveness of genetic algorithms relies on adequate sampling of fragments of chromosomes called schemata. Let us start with a binary chromosome of length N or, equivalently, a simple DOWN-SELECTING problem with N binary decisions. In this case, a schema is any string of length N formed by elements from the set [{0, 1, x}, where x indicates "don't care" (that is, 0 or 1). Conceptually, schemata represent groups of architectures that share common features. For example, $[1, x, x, ... , x]$ is the schema representing all 2^{N-1} architectures that have the first binary decision set to 1. In the NEOSS example, this could be all architectures that carry the first instrument (such as the radiometer). Similarly, the schema $[x, 1, x, x, 0, x, ... , x]$ represents all architectures that carry the second instrument and do not carry the fifth instrument.

Essentially, genetic algorithms use individuals (architectures) as a means to sample a space of schemata. Indeed, each architecture can be seen as an instance of 2^N schemata. If this is hard to see, consider a trivial architecture given by $[1, 1, 1, \ldots, 1]$. Such an architecture is an instance of the schema $[1, x, x, \ldots, x]$, but also of $[x, 1, x, \ldots, x]$, $[1, 1, x, \ldots, x]$, and so forth. In total, there are 2^N of these.

Good schemata are those for which the fitness is better than the average fitness of the population. The fitness of a schema is given by the average fitness of the architectures that are instances of that schema. Because selection of architectures for crossover is done on the basis of fitness, good schemata are more likely to be selected than poor schemata. Moreover, note that crossover generates new architectures that have many schemata in common with their parents, plus a few new ones. This argument qualitatively explains why genetic algorithms converge to good architectures: Good schemata are more likely to be selected for crossover, which preserves some of them and creates some new ones.

In addition to crossover, genetic algorithms also have a mutation heuristic, which applies a small random change to a small fraction of the architectures. The goal of mutation is to decrease the likelihood that the algorithm will get stuck in a suboptimal region. In other words, mutation is enhancing the exploration function of optimization.

Current implementations of genetic algorithms include many other, more sophisticated heuristics (such as sub-populations and migration operators) and some mechanism to ensure that the Pareto front is evenly populated. The interested reader can find thorough descriptions of these mechanisms in specialized books. [33]

Augmenting the Genetic Algorithm with More Heuristics

The initial version of the genetic algorithm formulated by Holland was based largely on crossover and mutation. However, one can adapt the population-based heuristic algorithm presented earlier by adding more heuristics to it. These heuristics can be domain-independent, as are crossover and mutation, or they can use domain-specific knowledge.

An example of another domain-independent heuristic is Tabu Search, which simply consists of maintaining a list of "bad" schemata. The fundamental idea is that this list can then be used to avoid or correct "bad moves" by the other operators. [34]

Domain-specific heuristics can also be used. In the NEOSS instrument-packaging problem, the elements to be partitioned are more than just abstract elements; they are remote sensing instruments, such as radars, lidars, and radiometers. In real life, the system architect has a lot of expert knowledge about these instruments that may facilitate the search process. For example, the system architects may know that the radar altimeter and the radiometer are very synergistic instruments, so it may be obvious to them that these instruments should be placed together on the same satellite. It might take substantial computational time for the algorithm to find out that putting those instruments together is good, because the algorithm is relying to a certain degree on randomness.

In summary, our choice of heuristics for the population-based optimization algorithm can start with crossover and mutation from genetic algorithms, which are two good and simple heuristics. Other domain-independent heuristics (such as Tabu Search) and domain-specific heuristics that capture expert knowledge about the problem at hand can then be added on top of crossover and mutation.

Hence, by diversifying our portfolio of heuristics, we can hope to obtain good performance for a wider range of system architecture optimization problems. Note, however, that this diversification, like the diversification of a financial portfolio, may be subject to the effect of dilution. In other words, the effect of good heuristics is partially compensated by the effect of bad heuristics. [35] This is why it is important to develop another layer of complexity in the tool that monitors the performance of different heuristics. For example, we could imagine that, instead of dividing the population evenly among heuristics, we could assign more architectures to heuristics that have had better performance so far. The interested reader can survey the literature on hyper-heuristics and machine learning. [36]

16.7 Summary

In this chapter, we have presented a set of tools to support the system architect in the formulation and resolution of system architecture optimization problems. We began by showing that several tasks of the system architect, such as mapping function to form and choosing a system decomposition, can be formulated as architecture optimization problems. Because the step from real system architecture optimization problems to a mathematical formulation may not be self-evident, we introduced a set of Patterns of decisions that appear often when one is formulating system-architecting problems. These range from the simple DECISION-OPTION Pattern to more complex Patterns such as PARTITIONING and PERMUTING. These Patterns are not completely unrelated; rather, they overlap, in the sense that some problems can be formulated using several Patterns.

We argue that having this "library" of Patterns more closely matches the practice of system architecture: If it is closer to the architect's mental model, it will be easier to represent, and interpreting the results will be easier as well.

We then moved on to discuss how to use heuristics to solve system architecture optimization problems. Full-factorial enumeration of architectures should be used when possible, although it is often *not* possible because of combinatorial explosion. A simple algorithm (procedural nested loops) was presented, and a more sophisticated strategy based on declarative programming is provided in Appendix C.

We described our tool of choice for solving large system architecture optimization problems—namely, population-based heuristic optimization. We discussed how to generate an initial population from a combination of random and deterministic architectures. We then introduced genetic algorithms and proposed crossover and mutation as two domain-independent heuristics that can be applied to most problems. Finally, we noted that this set of heuristics can and should be customized with additional domain-independent heuristics (such as Tabu Search) and domain-specific heuristics that capture expert knowledge about the problem at hand.

References

[1] See G.A. Miller, "The Magical Number Seven, Plus or Minus Two: Some Limits on Our Capacity for Processing Information," *Psychological Review* 63, no. 2 (1956): 81–97.

[2] D. Selva and E. Crawley, "VASSAR: Value Assessment of System Architectures Using Rules." In Aerospace Conference, 2013 IEEE (Big Sky, MN: IEEE, 2013).

[3] The NEOSS example is based on substantial work by the authors on the architecture of Earth observing satellite systems. See, for example, D. Selva, B.G. Cameron, and E.F. Crawley, "Rule-based System Architecting of Earth Observing Systems: The Earth Science Decadal Survey," *Journal of Spacecraft and Rockets,* 2014.

[4] See discussion in D. Selva, "Rule-based System Architecting of Earth Observation Satellite Systems," PhD dissertation, Massachusetts Institute of Technology (Ann Arbor: ProQuest/UMI, 2012), pp. 172–188.

[5] C. Alexander, *A Pattern Language: Towns, Buildings, Construction* (Oxford University Press, 1977).

[6] E. Gamma, R. Helm, R. Johnson, and J. Vlissides, *Design Patterns: Elements of Reusable Object-Oriented Software* (Addison-Wesley Professional, 1994).

[7] For a more extensive discussion of Patterns and styles, the interested reader can refer to F. Bushmann, R. Meunier, and H. Rohnert, *Pattern-oriented Software Architecture: A System of Patterns* (New York: Wiley, 1996) or M. Fowler, *Patterns of Enterprise Application Architecture* (Addison-Wesley Professional, 2002).

[8] Note that in other references, architectural "Pattern" and "style" are considered synonyms designating the common structure seen in many problem instances. This is different from our definition, which reserves the word "Pattern" for the common problem structure and reserves the word "style" for the solutions. See, for example, N. Rozanski and E. Woods, *Software Systems Architecture* (Reading, MA: Addison-Wesley, 2012).

[9] See, for example, M. Ehrgott and X. Gandibleux, "A Survey and Annotated Bibliography of Multiobjective Combinatorial Optimization," *OR Spectrum* 22 (November 2000): 425–460.

[10] Information on different AUV architectures can be found at http://oceanexplorer.noaa.gov/ and http://auvac.org/

[11] Facebook announced in 2014 that it intends to deploy a network of 11,000 drones to provide Internet services to Africa.

[12] Adapted from S.P. Ajemian, "Modeling and Evaluation of Aerial Layer Communications System Architectures," Master's thesis, Engineering Systems Division, Massachusetts Institute of Technology, 2012.

[13] See, for example, D. Ferrucci, A. Levas, S. Bagchi, D. Gondek, and E.T. Mueller, "Watson: Beyond Jeopardy!" *Artificial Intelligence*, 2013, 199–200, 93–105. doi:10.1016/j.artint.2012.06.009.

[14] See Schäfer's dissertation: U. Schäfer, "*Integrating Deep and Shallow Natural Language Processing Components: Representations and Hybrid Architectures,* " Universitat des Saarlandes, 2007.

[15] See J.H. Holland, *Adaptation in Natural and Artificial Systems* (Cambridge, MA: MIT Press, 1992).

[16] See, for example, the authors' VASSAR formulation: D. Selva, and E. Crawley, "VASSAR: Value Assessment of System Architectures Using Rules." In Aerospace Conference, 2013 IEEE (Big Sky, MN: IEEE, 2013).

[17] See A. Dominguez-Garcia, G. Hanuschak, S. Hall, and E. Crawley, "A Comparison of GN&C Architectural Approaches for Robotic and Human-Rated Spacecraft," AIAA Guidance, Navigation, and Control Conference and Exhibit, 2007, pp. 20–23.

[18] See N. Suh, "Axiomatic Design Theory for Systems," *Research in Engineering Design* 10, no. 4 (1998): 189–209.

[19] As a side note, the number of partitions in a set of m elements with exactly k subsets is given by the Stirling numbers of the second kind, and the total number of partitions of a set is given by the Bell numbers, which are equal to the sum of all the Stirling numbers of the second kind for all possible values of k. Any combinatorics text will provide formulas for these numbers. See, for example,

R. Grimaldi, *Discrete and Combinatorial Mathematics: An Applied Introduction* (Addison-Wesley, 2003), pp. 175–180. Also, one can find most integer sequences in the online encyclopedia of integer sequences: http://oeis.org

[20] See D. Selva and E. Crawley, "Integrated Assessment of Packaging Architectures in Earth Observing Programs," in Aerospace Conference, 2010 IEEE (Big Sky, MN: IEEE, 2010), pp. 3–12, for a discussion on monolithic versus distributed architectures for Earth observing satellites, including Envisat.

[21] The interested reader can find more information about the Metop mission and the sensitive sounder in D. Blumstein, "IASI Instrument: Technical Overview and Measured Performances," *Proceedings of SPIE* (Spie, 2004), p. 19.

[22] See M. Chiang and M. Yang, M. "Towards Network X-ities from a Topological Point of View: Evolvability and Scalability," 42nd Annual Allerton Conference on Communication, Control and Computing, 2004.

[23] See N.S. Haverty, "MIL-STD 1553—A Standard for Data Communications," *Communication and Broadcasting* 10 (1985): 29–33.

[24] See N.C. Strole, "The IBM Token-Ring Network—A Functional Overview." *Network*, IEEE 1.1 (1987): 23–30.

[25] See K.W. Matthee et al., "Bringing Internet Connectivity *to Rural Zambia Using a Collaborative Approach,"*. International Conference on Information and Communication Technologies and Development, 2007. ICTD 2007 (IEEE, 2007).

[26] For a more exhaustive methods-based text, see G. Parnell, P. Driscoll, and D. Henderson (Eds.), *Decision Making in Systems Engineering and Management* (Wiley, 2011) or D.M. Buede, *The Engineering Design of Systems: Models and Methods* (Wiley, 2009).

[27] Typical combinatorial optimization algorithms include branch and bound, cutting planes, approximation algorithms, network optimization algorithms, and dynamic programming, among others. Several texts contain exhaustive introductions to these techniques. See, for example, A. Schrijver, *Combinatorial Optimization* (Springer, 2002), p. 1800, or D. Bertsimas and R. Weismantel, *Optimization over Integers* (Belmont, MA: Dynamic Ideas, 2005).

[28] See, for example, K. Deb, A. Pratap, S. Agarwal, and T. Meyarivan, "A Fast and Elitist Multiobjective Genetic Algorithm: NSGA-II," *IEEE Transactions on Evolutionary Computation* 6, no. 2 (2002): 182–197. doi:10.1109/4235.996017

[29] For an introduction to these other techniques from design of experiments, see F. Pukelsheim, *Optimal Design of Experiments* (SIAM, 2006).

[30] A good review of several meta-heuristics is given in F. Glover and G.A. Kochenberger, *Handbook in Metaheuristics* (New York: Kluwer Academic Publishers, 2003).

[31] A more exhaustive discussion of heuristics and meta-heuristics is provided in F. Glover and G.A. Kochenberger, *Handbook in Metaheuristics* (New York: Kluwer Academic Publishers, 2003). It includes a discussion of hyper-heuristics—that is, heuristics used to choose between heuristics.

[32] See J.H. Holland, *Adaptation in Natural and Artificial Systems*, 2nd ed. (Cambridge, MA: MIT Press, 1992).

[33] See, for example, D.E. Goldberg, *Genetic Algorithms in Search, Optimization, and Machine Learning* (Addison-Wesley Professional, 1989) or K. Deb, A. Pratap, S. Agarwal, and T. Meyarivan, "A Fast and Elitist Multiobjective Genetic Algorithm: NSGA-II," *IEEE Transactions on Evolutionary Computation* 6, no. 2 (2002): 182–197.

[34] Tabu search was first introduced by Fred Glover as a stand-alone meta-heuristic in F. Glover, "Tabu Search—Part I," *ORSA Journal on Computing* 1, no. 3 (1989): 190–206 and F. Glover, "Tabu Search: Part II," *ORSA Journal on Computing* 2, no. 1 (1990): 4–32. Its use within genetic algorithms has also been studied. See F. Glover, J.P. Kelly, and M. Laguna, "Genetic Algorithms and Tabu Search: Hybrids for Optimization," *Computers & Operations Research* 22, no. 1 (1995): 111–134.

[35] See an example in the context of the NEOSS problem in D. Selva, "Experiments in Knowledge-intensive System Architecting: Interactive Architecture Optimization." In 2014 IEEE Aerospace Conference, Big Sky, Montana.

[36] E. Burke, G. Kendall, J. Newall, and E. Hart, "Hyper-heuristics: An Emerging Direction in Modern Search Technology." In *Handbook of Metaheuristics*, Springer US, pp. 457–474.

Appendices

Appendix A: Effect of the Choice of Architecture Set on the Decision Sensitivity Metric

Chapter 15 introduced the sensitivity of a metric to a decision, a numerical value that tells us how sensitive a metric is to a given decision, or, equivalently, the average change in a metric when we change the value of a decision. We provided two formulations for the sensitivity metric. The first is based on main effects and is valid only for binary decisions:

$$\text{Main effect (Decision } i, \text{ Metric } M) \equiv \frac{1}{N_1} \sum_{\{x|x_i=1\}} M(x) - \frac{1}{N_0} \sum_{\{x|x_i=0\}} M(x)$$

where N_0 and N_1 are the number of architectures for which $x_i = 0$ and 1 respectively.

The second formulation extends this concept to $k > 2$ decisions:

$$\text{Sensitivity (Decision } i, \text{ Metric } M) \equiv \frac{1}{|K|} \sum_{k \in K} \left| \frac{1}{N_{1,k}} \sum_{\{x|x_i=k\}} M(x) - \frac{1}{N_{0,k}} \sum_{\{x|x_i \neq k\}} M(x) \right|$$

where K is the set of options for decision i, and $N_{0,k}$ and $N_{1,k}$ are the number of architectures for which x_i \neq k and $x_i = k$ respectively.

Both equations require the definition of a set of architectures of interest, namely $\{x\}$, and the choice of this group may lead to different results. The following numerical example supports this claim.

Consider the dummy dataset provided in Figure 1. We have three binary decisions (D1 = {Y, N}, D2 = {1, 2}, and D3 = {A, B}) and two metrics (M1 and M2).

Five out of eight architectures are non-dominated.

We computed the main effects of the three decisions for the two metrics with two different sets of architectures: the entire tradespace and the Pareto front. The results are given in Table 1.

D1	D2	D3	M1	M2
Y	1	A	1	120
Y	1	B	5/6	115
Y	2	A	7/6	130
Y	2	B	1	125
N	1	A	2/3	20
N	1	B	1/2	15
N	2	A	5/6	30
N	2	B	2/3	25

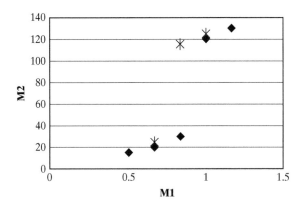

FIGURE 1 Dummy dataset: Numerical data and tradespace plot.

TABLE 1 | Main effects for the three decisions and two metrics for the entire tradespace and for the Pareto front only

Main Effects	Entire Tradespace		Front Only	
	M1	M2	M1	M2
D1	0.33	100.00	0.42	103.33
D2	−0.17	−10.00	−0.28	−28.33
D3	0.17	5.00	0.17	5.00

In the case of the entire tradespace, the main effect for metric M1 and decision D1 (0.33) is greater in absolute value than the main effects for metric M1 decisions D2 and D3 (0.17), which are identical. Therefore, D2 and D3 have identical sensitivities for metric M1. However, when we restrict the analysis to the Pareto front, the main effect for metric M1 and decision D2 (−0.28) is greater in absolute value than the main effect for D3 (0.17), which suggests that in this case, M1 is more sensitive to D2 than to D3.

This means that although the sensitivity of metric M1 to D2 and D3 is the same when all architectures are considered, it is actually greater to D2 than to D3 when only good architectures are considered.

Thus, as explained in Chapter 15, it is important to choose the group of architectures that most closely matches our goals.

Appendix B: Clustering Algorithms and Applications to System Architecture

Clustering is the general task of dividing or decomposing a set of elements into subsets or categories, or, equivalently, aggregating groups of elements into modules. Clustering plays a central role in system architecture; this has been emphasized throughout the text. Chapters 8 and 13 argued that choosing a system decomposition, essentially a clustering task, is one of the most important architectural decisions. In Chapter 15, we explained how clustering algorithms can be used to find the optimal sequence of decisions from a tradespace. Finally, the PARTITIONING problem presented in Chapter 16 can be seen as a clustering problem, as can the CONNECTING problem. In this appendix, we describe succinctly some of the most common algorithms that can be used to accomplish these tasks. Several texts exist that provide good introductions to machine learning and data mining, of which clustering is a key topic. [1]

General-Purpose Clustering Algorithms

Let us start with general-purpose clustering. Two major types of clustering algorithms are centroid-based clustering algorithms and hierarchic clustering algorithms.

K-means is one of the simplest and best-known centroid-based clustering algorithms. In K-means, we have a set of elements represented in a certain space where a distance metric can be defined (or example, $\mathbf{R}^2$ and Euclidean distance), and we need to divide those elements into K clusters. The algorithm works by iteratively refining the centroids of the K clusters. The centroids can be first initialized by choosing K random elements from the set. Then each element is assigned to the closest centroid. This defines the first set of clusters. Next, the centroids are updated by simply computing the average of the elements in each cluster, and the assignment step is repeated, yielding a new set of clusters. The algorithm iterates until the clusters don't change for two iterations.

Hierarchic clustering is also very simple and does not require predefining a number of clusters K. Instead, the algorithm is initialized with as many clusters as points, where each cluster contains a single point. Then the two clusters that are "the closest," as defined by a certain distance function, are merged into a single cluster. This distance function can, for example, compute the average Euclidean distance between the samples in the two clusters. Other distance metrics (such as Manhattan distance) and linkage criteria (such as minimum or maximum distance instead of average) can also be used.

K-means and hierarchic clustering are two of the most commonly used clustering algorithms, but they are by no means the only ones. In particular, heuristic algorithms, such as genetic algorithms, are also commonly used to solve clustering problems. [2] Clustering is a very active field of research, and new algorithms are continuously being proposed. [3]

Finding the Optimal System Decomposition

Clustering algorithms can be used to find the optimal decomposition for a system. This requires having a set of elements and some quantitative assessment of the interactions between each pair of elements—for example, in the form of a connectivity matrix C_{ij}. More precisely, given a set of N elements, C_{ij} is a DSM, i.e. an $N \times N$ matrix, where each value in the matrix is a non-negative integer representing the interaction between elements i and j. Then the algorithm can find the clustering x^* of the elements that maximizes the sum of interactions within the cluster and minimizes the sum of interactions between clusters. A penalty term that increases as the number of clusters deviates from a target value can also be added to avoid solutions with too few or too many clusters.

$$x^* = \min_x \sum_{i=1}^{N} \sum_{j=1, j\neq i}^{N} \left(1 - \delta_{x_i - x_j}\right) \cdot C_{ij} - \lambda \sum_{i=1}^{N} \sum_{j=1, j\neq i}^{N} \delta_{x_i - x_j} \cdot C_{ij} + \gamma \left(\frac{N_{\text{cluster}}}{\frac{N}{2}} - 1\right)^2$$

$$\delta_{x_i - x_j} = \begin{cases} 1, & x_i = x_j \\ 0, & x_i \neq x_j \end{cases}$$

where N_{cluster} is the number of clusters of the decomposition. A simple algorithm to solve this problem takes a random initial clustering, randomly moves elements from one cluster to another, and accepts better clusterings with a certain probability, as is done in simulated annealing. This algorithm is sometimes called the IGTA algorithm. [4] Other approaches based on genetic algorithms have also been successfully applied. [5] Note that several implementations of these algorithms are openly available online. [6]

Figure 2 shows the result of applying a genetic algorithm to a random DSM of 10 elements with the objective function described above. It is interesting to see that in the absence of negative interactions between elements (all elements of the connectivity matrix are non-negative), the

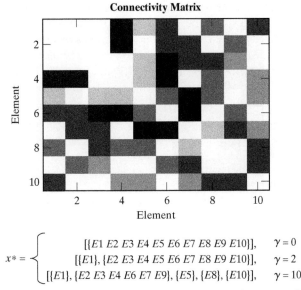

FIGURE 2 Application of a genetic algorithm to find the optimal clustering of a randomly generated DSM. The chart at the top is a graphical representation of the randomly generated connectivity matrix (darker shade means stronger coupling). The optimal clusterings for different values of the penalty for the number of clusters are shown below the chart. These values were generated with $\lambda = 0.5$.

number of clusters in the optimal decomposition strongly depends on the values of the parameters chosen for γ and λ. The optimal solution for large values of γ puts together elements 2, 3, 4, 6, 7, and 9 and decides to separate the others in different subsets.

A more interesting version of this problem appears when we allow for negative interactions between elements. Figure 3 shows an example where such negative interactions are allowed, pushing the algorithm to separate elements with strong negative interactions even for small values of γ.

These two "toy problems" were presented to illustrate how clustering algorithms can be used to find a system decomposition that is optimal in some sense. The metrics chosen to define this optimality were very simple in both cases. In some applications, it may be more interesting to construct more sophisticated metrics that, for example, transform the non-dimensional "couplings" between elements into actual costs.

Finding the Optimal Sequence of Decisions

Variants of clustering algorithms can also be used to find the optimal sequence of architectural decisions, as proposed in Chapter 15. One possible formulation of this problem follows. Begin with a set of *NDEC* architectural decisions and *NMET* metrics, a $NDEC \times NDEC$ matrix C_{ij} of non-negative integers representing the connectivity between decisions, and a $NDEC \times NMET$ matrix S_{im} of non-negative integers containing the sensitivity of all decisions to all metrics. Use the three inputs to find an ordering of decisions x (where $x(i) = j$ indicates that decision i is made in the jth position) that minimizes the connectivity between decisions that don't need to be

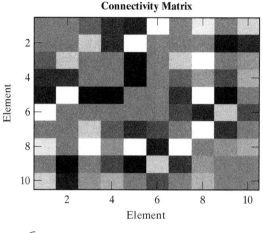

$$x^* = \begin{cases} [\{E1\ E3\ E4\ E5\ E7\ E9\}, \{E2\ E6\ E8\ E10\}], & \gamma = 0 \\ [\{E1\ E3\ E4\ E5\ E7\ E9\}, \{E2\ E6\ E10\}, \{E8\}], & \gamma = 6 \\ [\{E1\ E4\ E5\ E7\ E9\}, \{E2\ E10\}, \{E3\}, \{E6\ E8\}], & \gamma = 10 \end{cases}$$

FIGURE 3 Optimal clustering when negative interactions between elements are allowed. Black cells represent strong positive interactions, and white cells represent strong negative interactions. Gray cells such as those in the diagonal are non-interactions.

made at the same time, and minimizes the sensitivity of decisions made late in the process. More precisely, the problem can be formulated as follows:

$$x^* = \min_{x} \sum_{i=1}^{NDEC} \sum_{j=1, j\neq i}^{NDEC} |x_i - x_j| \cdot C_{ij} + \alpha \sum_{i=1}^{NDEC} \sum_{m=1}^{NMET} S_{im}(1+r)^{x_i}$$

Two parameters need to be chosen in this objective function: $\alpha \geq 0$, representing the relative importance of sensitivity versus connectivity of decisions; and $r > 0$, a discount rate that can be interpreted as the relative importance given to long-run decisions versus short-run decisions.

Stochastic hill climbing or genetic algorithms can be used to solve this problem. Figure 4 shows the result of applying a genetic algorithm to a randomly generated problem with 10 decisions.

The algorithm finds that it is always optimal to make decision D4 first; this is because D4 is a highly sensitive and connected decision (a Type I decision according to Simmons's framework presented in Chapter 15). Similarly, Decision D3 is consistently found at the end, because it has low sensitivity and coupling (it is a Type IV decision in Simmons's framework). The optimal decisions of Types II and III depend much more strongly on the value of the parameters—that is, on whether we give more weight to sensitivity or to coupling. Hence, for example, Decision D7 has low sensitivity but high coupling to more sensitive decisions, so it is made at the beginning, together with the sensitive decisions, except for very high values of α.

Again, for illustrative purposes, the example shown here is very simple, especially in the modeling of sensitivity and connectivity, and in the definition of the objective function, but the general principles hold for larger, real-life architecture problems.

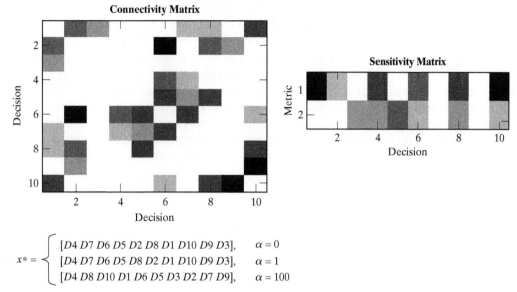

FIGURE 4 Using a genetic algorithm to find the optimal sequence in which architectural decisions should be made. The connectivity matrix containing the coupling between decisions is graphically represented at the upper left corner of the chart (darker shade means stronger coupling). The sensitivity matrix containing the sensitivity of metrics to decisions is graphically represented in the upper right corner (darker shade means stronger sensitivity). At the bottom, three optimal sequences of decisions are shown for different values of α (alpha), a parameter controlling the relative importance of sensitivity versus connectivity.

Appendix C: Rule-based Systems and Applications to System Architecture

Knowledge-based systems are computer programs that solve complex problems by "reasoning," just as humans do. [7] Knowledge-based systems have a knowledge base containing knowledge about the world (for example, a database of components, lessons learned from past systems designs, or a system architecture). Knowledge is organized in "sentences" represented in some knowledge representation language, which is typically not plain English, because that would make reasoning by computer very difficult.

A knowledge representation language consists of a syntax describing how to construct valid sentences, semantics mapping the meaning of symbols in the language with the real world, and a proof theory (some mechanism to infer new knowledge that follows logically from existing knowledge. For example, propositional logic is a very simple language in which symbols such as X, Y, and Z represent entire facts (such as "the reliability of this system is better than 0.99" or "the system consists of three components connected in series") that are either true or false. Valid sentences can be created by combining symbols through Boolean operators such as AND and OR—for example, (X AND Y OR Z).

Knowledge representation languages also have an inference mechanism that can deduce new knowledge from the existing knowledge in the database. In propositional logic, an example of a rule of inference is "If p implies q, and p is true, then I can infer that q is true." [8]

Knowledge-based systems are often classified in logic programming languages, such as Prolog; rule-based systems, also known as production systems, such as CLIPS; frame systems and semantic networks, such as OWL; and description logic systems, such as KL-ONE. This appendix will focus on rule-based systems, because they are appropriate to support the system architect in some specific tasks, such as the simulation of emergent behavior (for example, synergies and interferences) or the automatic enumeration of architectures. The interested reader can refer to one of the several excellent texts that address knowledge-based systems. [9]

Rule-based Systems

Rule-based systems are knowledge-based systems wherein short-term knowledge is represented using lists of *attribute–value* pairs called facts, and long-term knowledge is represented using logical rules (IF-THEN statements). Rules have a left-hand side (LHS) containing a list of conditions, where each condition makes reference to one or more facts, and a right-hand side (RHS) containing a list of actions, which can include adding more facts, modifying existing facts, or deleting existing facts.

Rule-based systems use a forward-chaining inference algorithm called the Rete algorithm. [10] The knowledge base is initialized with a set of rules and an initial set of facts, and the inference algorithm creates an activation record every time all the conditions of a rule are satisfied by a combination of facts in the knowledge base. At every iteration, the algorithm chooses an activation record and executes the actions in the corresponding rule. This process is done iteratively until there are no more activation records left.* This algorithm is illustrated in Figure 5.

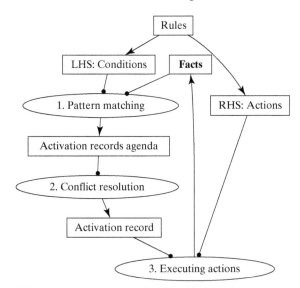

FIGURE 5 Inference algorithm in rule-based systems.

---

*Forward chaining is in opposition to backward chaining, in which the algorithm looks for facts that match the actions of a rule and asserts the conditions, working its way from the goal state to an initial state that will lead to the goals, given the knowledge in the database.

Research on rule-based systems began in the 1960s with work by Newell and Simon [11] and continued in the 1970s with Buchanan and Feigenbaum. [12] By the 1980s, rule-based systems were a mature technology applied in many disciplines, including medicine, engineering, and finance. [13]

In the rest of this appendix, we offer an overview of using rule-based systems to enumerate and evaluate architectures. For a more exhaustive discussion of these aspects, the reader is referred to related work by the authors. [14]

Declarative Enumeration of Architectures

In Chapter 16, we presented a simple algorithm to perform full-factorial enumeration of architectures based on simple nested loops. This simple algorithm is what computer scientists call an imperative program or *procedural program*. This means that the algorithm contains the procedure to solve a problem (in this case, enumerate the architectures) step by step. This approach works well for decoupled problems, where either all of the decisions are of the same type (such as all DECISION-OPTION decisions or all ASSIGNING decisions) or decisions are decoupled.

However, these conditions often do not prevail, which makes imperative enumeration challenging. For example, in the aerial network example introduced in Chapter 16, decoupling the radio type decision from the others implies that a given architecture uses either Radio A or Radio B, but not both. This eliminates architectures where both types of radios are used.

Similarly, in our NEOSS example, we had three main types of decisions: the instrument selection problem (DOWN-SELECTION), the instrument-packaging problem (PARTITIONING or ASSIGNING, depending on the formulation chosen), and satellite scheduling (PERMUTING). In this case, one cannot express the options available for the instrument-packaging problem without making a decision about the instrument selection.

Often, an architectural decision needs to be made only if another architectural decision takes a particular value. For instance, deciding which wheels on a hybrid car will be driven by the internal combustion engine (ICE) is dependent on choosing an architecture where the ICE is connected to the drivetrain.

More generally, there is sometimes an implicit sequence in which architectural decisions need to be made. Non-trivial order constraints can be very hard to encode in procedural programs and are easiest to express in a different programming paradigm called declarative programming.

In declarative programming, programs do not specify the steps to be taken in order to solve a problem, but rather just the conditions or goals in the form of rules or constraints. Declarative languages have an inference algorithm that takes care of specifying the steps to solve a problem. Rule-based systems and constraint programming are examples of declarative programs that can be used for architecture enumeration in complex cases.

Rule-based systems can be a very efficient and elegant way of enumerating architectures. For instance, Table 2 shows the only two rules that are needed to enumerate all architectures in a DOWN-SELECTING problem with N elements, followed by a PARTITIONING problem to partition the $m \leq N$ selected elements into $1 \leq k \leq m$ subsets.

In practice, using the Patterns presented earlier, it is often possible to enumerate any architectural tradespace by using just one rule per class of decision. For instance, in the guidance,

TABLE 2 | Two rules that enumerate all down-selecting and partitioning architectures using a divide-and-conquer approach

---

DEFINE-RULE down-selection-enumerator
IF EXISTS DECISION D1=[type="DOWN-SELECTING", options = {e1, e2, ..., eN}] % Decision D1 is a down-selecting decision over N elements
  AND EXISTS ARCHITECTURE A = [down_selecting=x] % a generally incomplete architecture A with down-selection x
  AND length(x) < length(options) % i.e., down-selection decision has not been finalized e.g. x = [1 0 1] for N = 6

THEN
  duplicate A [down_selecting = [x 0]] % creates new architecture fact that does not carry the next element
  duplicate A [down_selecting = [x 1]] % creates new architecture fact that carries the next element
  delete A % eliminate previous incomplete architecture lacking the selection of element

DEFINE-RULE partitioning-based-on-down-selecting-progressive-enumerator
IF EXISTS DECISION D1=[type="DOWN-SELECTING", options = {e1, e2, ..., eN}]
  AND EXISTS DECISION D2=[type="PARTITIONING", options = D1] % Decision D2 is a partitioning decision where the set of elements is given by the result of the down-selection decision D1
  AND EXISTS ARCHITECTURE A=[x, y] % a generally incomplete architecture A with down-selection x and partitioning y,
  AND length(x) == N % i.e., down-selection decision has been made e.g. x = [1 0 1 0 1 1]
  AND length(y) < sum (x) % i.e., partitioning decision is not complete, e.g. for given x, y = [1 2 2 3] is complete and y = [1 2] is not

THEN
  for i = 1 .. (max(y) + 1) % e.g. if y = [1 2], we can assign third element to existing subset (1 or 2) or new subset (3)
  duplicate A [partitioning = [y i]] % creates new architecture fact
  delete A % eliminate previous incomplete architecture lacking the assignment of element i

---

navigation, and control (GNC) example used throughout Part 4, one rule can take care of enumerating the sensors, computers, and actuators; one rule can take care of enumerating the different types of sensors, computers, and actuators; and one rule can take care of enumerating all possible ways of connecting sensors, computers, and actuators.

Declarative enumeration is also a very convenient framework for expressing constraints, since constraints can naturally be expressed as rules. For instance, in our NEOSS example, Table 3 shows how to add a constraint to avoid the radar and the lidar being put together in the same satellite.

Expressing Synergies and Interferences as Rules

Rules are also a simple and elegant way of encoding interactions between elements, such as the synergies and interferences that, as we argued in Chapter 16, can drive DOWN-SELECTING, ASSIGNING, or PARTITIONING problems. Consider the NEOSS example mentioned in Chapter 16: The radar altimeter and the microwave radiometer are synergistic instruments—that is, they have a positive interaction that results in an increase in scientific value that none of the

TABLE 3 | A rule showing how to encode a hard constraint in the NEOSS example

DEFINE-RULE hard-constraint-radar-and-lidar-separate
IF EXISTS ARCHITECTURE A=[x,y] % an architecture A with down-selection x and partitioning y,
        AND are-in-same-spacecraft(y,"radar","lidar") % radar and lidar are in the same spacecraft (i.e., their corresponding values in array y are equal)
THEN
       delete A % eliminate architecture

individual instruments can achieve. This is because the radiometer can provide atmospheric measurements that help decrease the error in the altimetry measurement, in particular the component due to the presence of water vapor in the atmosphere. This so-called wet atmospheric correction can be expressed in the form of a rule:

DEFINE-RULE radar-altimeter-microwave-radiometer-synergy
IF EXISTS MEASUREMENT M1=[parameter = sea level height, taken by = "altimeter", wet atmospheric correction = "no"]
        AND EXISTS MEASUREMENT M2=[parameter = atmospheric humidity, taken by = "radiometer"]
    AND are-in-same-spacecraft(y,"radar altimeter","microwave radiometer")
THEN
       modify M1 [atmospheric correction = "yes"]

This rule is an example of one type of emergent behavior, in which a property (a component of the error budget) of a system capability (an altimetry measurement) is modified by the interaction (data processing) with another system capability (an atmospheric humidity measurement).

Another type of emergent behavior is that in which a new capability is created from the interaction of existing capabilities. For example, the SMAP satellite mission uses what is called a disaggregation data processing algorithm to combine the soil moisture measurement taken by a radar, which has high spatial resolution but low accuracy, with the soil moisture measurement taken by a radiometer, which has low spatial resolution but high accuracy, to obtain a data product with medium spatial resolution and good accuracy. [15] This new data product emerges from the interaction of the two instruments and their capabilities.

DEFINE-RULE R1-spatial-disaggregation-synergy
IF EXISTS MEASUREMENT M1=[parameter = soil moisture, taken by = "radar", spatial resolution = "High", accuracy = "Low"]
        AND EXISTS MEASUREMENT M2=[parameter = soil moisture, taken by = "radiometer", spatial resolution = "Low", accuracy = "High"]   **AND** are-in-same-spacecraft(y,"radar ","radiometer")
THEN
       create M3= =[parameter = soil moisture, taken by = "radar+radiometer", spatial resolution = "Medium", accuracy = "Medium-High"]

The behavior encoded in the previous rules is extremely simple, but a knowledge base with even a handful of such rules can simulate extremely complex behavior. Consider that, in addition to the *R1-spatial-disaggregation-synergy* rule, we have an *R2-space-averaging* rule that trades accuracy against spatial resolution by simply averaging a measurement in space, and an *R3-time-averaging* rule that does the same in the time domain. The iterative application of these three simple rules on an initial database containing two measurements of soil moisture, taken by a radar and a radiometer, can generate a plethora of new data products: Both measurements can be averaged in space ($M1 + R2 \rightarrow M3$, $M2 + R2 \rightarrow M4$) and time ($M1 + R3 \rightarrow M5$, $M2 + R3 \rightarrow M6$). The newly created measurements can be averaged in time and space again (for example, $M4 + R3 \rightarrow M7$). The original measurements can be combined in the disaggregation scheme ($M1 + M2 + R1 \rightarrow M8$), but also many other pairs of measurements can be combined with the disaggregation scheme ($M3 + M6 + R1 \rightarrow M9$). The new measurements created are in general different from each other in spatial resolution, temporal resolution, and accuracy, so it is perfectly possible that at some point in the process, a data product is created that satisfies the needs of multiple stakeholders.

It is easy to see that this process can rapidly explode unless some stopping criteria are enforced. Sometimes, checking for the satisfaction of stakeholder needs (which can be done with more rules), can tell us when to stop simulating; there is no more value in continuing the simulation process if the current capabilities are already fully satisfying all stakeholder needs. In general, it may be necessary to simply stop the process after a certain number of emergence rules have fired.

The use of simple rules for simulating complex emergent behavior is not new. Cellular automata (such as Conway's Game of Life) and agent-based simulation are all based on the premise that extremely complex behavior can be simulated by means of simple interactions. [16]

Appendix D: Classical Combinatorial Optimization Problems

We explained in Chapter 16 that system architecture problems can be framed as combinatorial optimization problems. When we identified the six patterns of architecture problems, we mentioned that some of them look superficially like classical combinatorial optimization problems. For example, we argued that down-selecting problems look like 0/1 knapsack problems, with the difference that the value of elements is not additive—there are synergies, interferences, and redundancies.

In this appendix, we provide a list of some of the most typical combinatorial optimization problems. The list is partially adapted from Gandibleux's classification of combinatorial optimization problems according to their combinatorial structure. [17] For each problem, a source is provided. The source referred to is always representative of the foundational work on methods to solve the problem, but it is not necessarily the first time the problem was introduced. Most of these problems arise from graph theory and network analysis, and all of them are NP-complete problems. Very informally, this means that computer scientists think that a great many of these problems cannot be solved quickly (in polynomial time), so they are "hard" problems.† Note that

---

†Slightly more formally, P is the class of problems for which we can find a solution in polynomial time. NP is the class of problems for which we can check a solution in polynomial time. Most computer scientists believe that P is smaller than, and contained in, NP. NP-hard problems are at least as hard as the hardest problems in NP (that is, any NP problem can be reduced to an NP-hard problem in polynomial time), but they do not necessarily belong to NP. And NP-complete problems are simultaneously NP and NP-hard.

this is by no means an exhaustive list of combinatorial optimization problems, but it contains most of the problems relevant to system architecting.

- **Assignment problem and generalized assignment problem.** [18] In the original assignment problem, there are a number of agents and a number of tasks, and each agent has a different cost to realize each task. The goal of the problem is to assign each task to exactly one agent, in such a way that all tasks are performed and the total cost is minimized. In a generalized version of the assignment problem, agents can perform more than one task, and each has a given budget that cannot be exceeded. Furthermore, when performing a task, each agent produces a certain profit, and the goal of the problem is thus to maximize profit subject to not exceeding any agent's budget. The assignment problem is similar to our assigning pattern, but reusing our notation from Chapter 16, it allows for only channelized solutions—that is, architectures where each element of the left set is connected to exactly one element of the right set.
- **Traveling salesman problem.** [19] In the traveling salesman problem, there is a list of cities and a matrix containing their pairwise distances. The goal of the problem is to find the shortest path that passes through each city exactly once and returns to the city of origin. Because every city needs to be visited exactly once, this problem is similar to permuting problems.
- **0/1 knapsack problem.** [20] In the original knapsack problem, there is a list of items, each with a certain value and a certain weight. The goal of the problem is to determine the optimal number of items of each type to choose in order to maximize value for a certain maximum cost. In the 0/1 version of the same problem, the number of each item can take only the values $\{0, 1\}$. As we have already noted, the knapsack problem is similar to the down-selecting problem, with the important exception that the values of the items are clearly not additive in down-selecting problems, which precludes the application of very efficient algorithms.
- **Set-partitioning/set-covering problem.** [21] In the original set-covering problem, there are a list of elements referred to as the universe, and a number of predetermined sets of elements whose union is the universe. The goal of the set-covering problem is to identify the minimum number of these predetermined subsets for which the union is still the universe. The set-partitioning problem is a constrained version of the set-covering problem where the selected subsets need to be disjoint or mutually exclusive. In other words, each element can appear in only one set in the set-partitioning problem, whereas it may appear in more than one set in the set-covering problem. Clearly, our partitioning pattern is very similar to the set-partitioning problem.
- **Job-shop scheduling problem.** [22] In the simplest version of the job-shop scheduling problem, there is a list of jobs that needs to be assigned to a set of available resources (such as machines). Each job has a certain duration on each machine. The goal of the problem is to find the sequence of assignments of jobs to machines that minimizes the combined duration of the tasks. Variations of the problem include constraints between tasks (for example, a task needs to occur before another task), costs for running the machines, and interactions between the machines (for example, two machines cannot be running simultaneously). Our permuting pattern is similar to a single-machine instance of the job-shop scheduling problem, because it is truly optimizing over a space of permutations.

- **Shortest-path problem.** [23] In a shortest-path problem, there is a graph defined by a list of nodes and a list of edges, and each edge has a distance associated with it. The goal of the problem is to find the path between two given nodes that minimizes the total distance. Note that the shortest-path problem assumes that the graph is already fully defined, and therefore it is substantially different from a CONNECTING Pattern, where the goal is to find the optimal set of edges for a given set of nodes.
- **Network flow problem.** [24] In the most generic formulation of a network flow problem, there is a graph defined by a list of nodes and a list of edges. Nodes can be sources or sinks. Sources have a positive flow (supply), and sinks have a negative flow (demand). Each edge has a capacity associated with it, so that the flow through that edge cannot exceed that capacity. Each edge also has a cost associated with it. The goal of the network flow problem is to transport all the flow from the supply nodes to the demand nodes at minimum cost. This problem also assumes that the graph is fully defined and therefore is also substantially different from all our Patterns.
- **Minimum spanning tree.** [25] In the minimum spanning tree problem, there is a graph defined by a list of nodes and a list of edges. Each edge has a cost associated with it. A spanning tree is defined as a subset of the edges that form a tree that contains all nodes. The goal of the problem is to find the spanning tree of minimum cost. This problem could be seen as an instance of a connecting problem where only tree architectures are being considered.
- **Maximum satisfiability (MAX-SAT) problem.** [26] In the maximum satisfiability problem, there are a set of Boolean variables and a list of logical clauses that use that set of variables. The goal of the problem is to find the assignment to that set of Boolean variables that maximizes the number of clauses that are satisfied. Although the MAX-SAT problem does not immediately look like any of our canonical architecture problems, several problems can be reduced to MAX-SAT problems and solved efficiently using SAT solvers. For example, it is possible to reduce any DECISION-OPTION problem to a problem with only binary variables, with two types of variables: variables that determine whether a certain option is chosen for a certain decision, and variables that determine whether a certain metric is at a certain level. Then the optimization problem can at least theoretically be reduced to a satisfiability problem by encoding value functions as logical relationships between binary variables, and adding constraints so that the value of certain metrics is above a certain level. [27]

References

[1] See, for example, T. Hastie, R. Tibshirani, and J. Friedman, *The Elements of Statistical Learning: Data Mining, Inference, and Prediction* (New York: Springer, 2011).

[2] See U. Maulik and S. Bandyopadhyay, "Genetic Algorithm-Based Clustering Technique," *Pattern Recognition* 33, no. 9 (2000): 1455–1465.

[3] See, for example, R. Xu, and D. Wunsch II, "Survey of Clustering Algorithms," *IEEE Transactions on Neural Networks* 16, no. 3 (2005): 645–678.

[4] IGTA is an acronym for the Idicula-Gutierrez-Thebeau Algorithm; the first threeletters stand for the names of the people who proposed it. See R.E. Thebeau, "Knowledge Management of System Interfaces and Interactions for Product Development Processes." MS thesis, Systems Design and

Management, Massachusetts Institute of Technology; and F. Borjesson and K. Hölttä-Otto, "Improved Clustering Algorithm for Design Structure Matrix." In *Proceedings of the ASME 2012 International Design Engineering Technical Conferences.*

[5] See T.L. Yu, A. Yassine, and D.E. Goldberg, "An Information Theoretic Method for Developing Modular Architectures Using Genetic Algorithms," *Research in Engineering Design* 18, no. 2 (2007): 91–109.

[6] See http://www.dsmweb.org/

[7] This is in opposition to other problem-solving paradigms, such as searching, planning, and game playing. See S. Russell and P. Norvig, *Artificial Intelligence: A Modern Approach,* 3rd ed. (Edinburgh, Scotland: Pearson Education Limited, 2009).

[8] This is called the "modus ponens" axiom. Propositional logic has other axioms, such as "and elimination," "and introduction," "or introduction," and "resolution." Any text that provides an introduction to logic will describe these axioms. See, for example, R. Fagin, J. Halpern, Y. Moses, and M. Vardi, *Reasoning about Knowledge* (Cambridge, MA: MIT Press, 1995).

[9] See, for example, J.C. Giarratano and G.D. Riley, *Expert Systems: Principles and Programming,* Course Technology (2004).

[10] See C.C.L. Forgy, "Rete: A Fast Algorithm for the Many Pattern/Many Object Pattern Match Problem," *Artificial Intelligence* 19, September (1982): 17–37.

[11] See A. Newell and H.A. Simon, *Human Problem Solving* (Englewood Cliffs, NJ: Prentice Hall, 1972).

[12] See R. Lindsay, B.G Buchanan, and E.A. Feigenbaum, "DENDRAL: A Case Study of the First Expert System for Scientific Hypothesis Formation," *Artificial Intelligence* 61, June (1993): 209–261; and B.G. Buchanan and E.H. Shortliffe, *Rule-based Expert Systems: The MYCIN Experiments of the Stanford Heuristic Programming Project* (Addison-Wesley, 1984).

[13] See J. Durkin, "Application of Expert Systems in the Sciences," *Ohio Journal of Sciences* 90 (1990): 171–179.

[14] See D. Selva, "Rule-based System Architecting of Earth Observation Satellite Systems," PhD dissertation, Massachusetts Institute of Technology (Ann Arbor: ProQuest/UMI, 2012).

[15] N.N. Das, D. Entekhabi, and E.G. Njoku, "An Algorithm for Merging SMAP Radiometer and Radar Data for High-Resolution Soil-Moisture Retrieval," *IEEE Transactions on Geoscience and Remote Sensing* 49 (2011): 1–9.

[16] Cellular automata such as Conway's Game of Life are a good example of this. See also S. Wolfram, "*A New Kind of Science* (Champaign, IL: Wolfram Media, 2002) and M. Mitchell, *Complexity: A Guided Tour* (Oxford University Press, 2009).

[17] See M. Ehrgott and X. Gandibleux, "A Survey and Annotated Bibliography of Multiobjective Combinatorial Optimization," *OR Spectrum* 22, no. 4 (2000): 425–460. doi:10.1007/s002910000046

[18] See G.T. Ross and R.M. Soland, "A Branch and Bound Algorithm for the Generalized Assignment Problem," *Mathematical Programming* 8 (1975): 91–103.

[19] See G. Dantzig and R. Fulkerson, "Solution of a Large-scale Traveling-Salesman Problem," *Journal of the Operations Research Society* 2, no. 4 (1954):393-410.

[20] See A. Drexl, "A Simulated Annealing Approach to the Multiconstraint Zero-One Knapsack Problem," *Computing* 40, no. 1 (1988): 1–8. doi:10.1063/1.3665391

[21] See R.S. Garfinkel and G.L Nemhauser, "The Set-Partitioning Problem: Set Covering with Equality Constraints," *Operations Research* 17, no. 5 (1969): 848–856. doi:10.1287/opre.17.5.848

[22] See A.S. Manne, "On the Job-Shop Scheduling Problem," *Operations Research* 8, no. 2 (1960): 219–223.

[23] S.E. Dreyfus, "An Appraisal of Some Shortest-Path Algorithms," *Operations Research* 17, no. 3 (1969): 395–412.

[24] See J. Edmonds, "Theoretical Improvements in Algorithmic Efficiency for Network Flow Problems," *Computing* 19, no. 2 (1972): 248–264.

[25] See R. Graham, "On the History of the Minimum Spanning Tree Problem," *Annals of the History of Computing* 7, no. 1 (1985): 43–57.

[26] See P. Hansen and B. Jaumard, "Algorithms for the Maximum Satisfiability Problem," *Computing* 303 (1990): 279–303.

[27] See D. Rayside and H.-C. Estler, "A Spreadsheet-like User Interface for Combinatorial Multi-objective Optimization." *Proceedings of the 2009 Conference of the Center for Advanced Studies on Collaborative Research*—CASCON'09, 58 (doi:10.1145/1723028.1723037) for a SAT-based implementation of system architecture optimization.

Chapter Problems

System Architecture–Chapter 2

Learning Objectives
- Develop reference definitions in your context.
- Separate function from form.
- Think about architecture as a composite of decisions.

Problem 1
For each of the terms that follow:

- Create your own definition first.
- Find several other definitions, and cite the sources.
- Compare and critically analyze all the definitions.
- Synthesize your final definition.
- Include an illustrative example of something that fits your definition.

    1a. What is a **system**?
    1b. What does it mean to be **complex** (in the context of a technical system)?
    1c. What is **value**?
    1d. What is a **product**?

Problem 2
An alternative definition of system architecture is "the ten most influential decisions made in the definition of the system." Choose a make and model of a sedan costing less than $50,000.

    Beginning with the general concept of a car as a transportation device, list what you believe are the ten most important technical decisions. Then narrow the field of concepts from a car as a transportation device to your specific make and model, and include two to five choices for each decision.

    We will refine the methods of decision and option selection through the course of this text. The purpose of this exercise is to get you thinking about the potential, and the challenges, associated with this approach.

Problem 3
For four of the "simple" systems listed in the following table, identify the object elements or abstractions of form that you will represent.

    a. Identify the system, its form, and its function (Section 2.3).
    b. Identify the entities of the system, their form and function, and the system boundaries and context (Section 2.4).
    c. Identify the relationships among the entities in the system and at the boundaries, including their form and function (Section 2.5).
    d. Based on the function of the entities, and on their functional interactions, identify the emergent properties of the system (Section 2.6).

422 CHAPTER PROBLEMS

Sector	Simple System
Aeronautics	Model balsa glider with propeller
Electrical	Simple crystal radio
Software	Prime number search code (see pseudocode below)
Optical	A simple refracting telescope
Judicial	Jury in a U.S. court
Ocean	A basic fiberglass canoe
Computer	Simple 1-bit full adder
Medical Services	Simple gender-neutral office exam with physician (see narrative below)

Prime Number Search Pseudocode

/* Comment – Pseudocode Function to test if a number is a prime number */
/* Comment – The number to test is the variable TestPrime */
/* Comment – The function returns a True or a False */

Function IsPrime (**TestPrime**)
   Initialize variable **IsPrime** = False
   Initialize variable **TestNum** = 0
   Initialize variable **TestLimit** = 0

   /* Comment - Eliminate even numbers /
   If **TestPrime** Mod 2 = 0 Then
      IsPrime = False
      Exit Function
   End If

   /* Comment - Loop through ODD numbers starting with 3 */
   Assign **TestNum** = 3
   Assign **TestLimit** = **TestPrime**

   Do While **TestLimit** > **TestNum**
      If **TestPrime** Mod **TestNum** = 0 Then
         IsPrime = False
         Exit Function
      End If

            TestLimit = TestPrime \ TestNum

        /* Comment - we only bother to check odd numbers */
        TestNum = TestNum + 2

    Loop

    /* Comment - If we made it through the loop, the number is a prime. */
    IsPrime = True

End Function

Narrative for Simple Medical Service

You enter the examining room for a simple, gender-neutral annual physical. Body readings (height, weight, blood pressure, temperature) are taken, and other standard procedures for a routine physical are performed. You receive a clean bill of health and leave the examining room at the end of the examination.

Do not consider the scheduling, billing, or additional lab work or other medical services that might follow an examination; this system is contained within the walls of the examination room for the period that you are there.

System Architecture–Chapter 3

Learning Objectives

- Distinguish between decomposition and hierarchy.
- Develop a critical analysis of system diagrams.

Problem 1

The Open Systems Interconnection (OSI) model is frequently used as a layered representation of communication systems. Does this representation use any of the other tools or concepts for reasoning about complex systems? If so, create a simple diagram showing a portion of the OSI that emphasizes the other tool(s) or concept(s) you identify.

424 CHAPTER PROBLEMS

Application
Provides access to the OSI environment for users and also provides distributed information services.

Presentation
Provides independence to the application processes from differences in data representation (syntax).

Session
Provides the control structure for communication between applications; establishes, manages, and terminates connections (sessions) between cooperating applications.

Transport
Provides reliable, transparent transfer of data between end points; provides end-to-end error recovery and flow control.

Network
Provides upper layers with independence from the data transmission and switching technologies used to connect systems; responsible for establishing, maintaining, and terminating connections.

Data Link
Provides for the reliable transfer of information across the physical link; sends blocks (frames) with the necessary synchronization, error control, and flow control.

Physical
Concerned with transmission of unstructured bit stream over physical medium; deals with the mechanical, electrical, functional, and procedural characteristics to access the physical medium.

Problem 2

A DSM of the TCP/IP suite is shown on the next page, read across the rows to the column. For example, an entry at [2, 5] would indicate a connection from 2 to 5.

Create a network representation of this layered hierarchy, showing the connections between elements, as well as which layer the elements are located in.

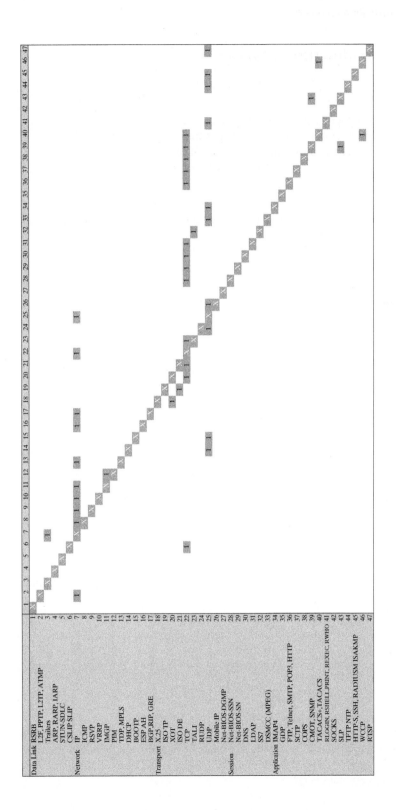

System Architecture–Chapter 4

Learning Objectives
- Learn how to identify the form of simple systems.
- Analyze and compare various approaches for analyzing form.
- Learn how to represent simple systems in OPM.

Problem 1
For four of the "simple" systems listed in the "matrix of simple systems" from Chapter 2, identify the object elements or abstractions of form that you will represent. Develop a graphical decompositional view and a graphical structural view for the *form* of the system.

Problem 2
For one of the four "simple" systems in Problem 1, develop a list-like representation of the decompositional view and of the structural view of form. How did you choose to translate the structural information into a list-like view?

Problem 3
Revisit the labeling of the lines on your graphical representation of the structural view. Are they really describing form? Can you identify the *classes* of structural relationships? How much of this information was also conveyed on your list-like view?

Problem 4
How are objects of form and structural information normally conveyed in your field or discipline? Cite two examples.

System Architecture–Chapter 5

Learning Objectives
- Analyze and compare various approaches for analyzing structure, functional interaction, and expression of function.
- Learn how to represent simple systems in OPM.

Problem 1
For four of the "simple systems" described in the Chapter 2 Problems, answer Questions 5a to 5d in Table 5.1.

Problem 2
For two of the four simple systems, develop an OPD of the value-related operand, delivered function, internal functions (operands and processes), and form. In addition to the OPD, develop another type of graphical representation of the function and form of a system. How useful is (are) this (these) representation(s)?

Problem 3

For two of the simple systems, develop a list-based way to represent the value-related operand, delivered function, internal functions (operands and processes), and form. How useful is this representation?

Problem 4

How is functional information normally conveyed in your field or discipline? Are processes indicated? Operands? Their combination into functions? In your response, cite two examples, one each from different disciplines.

Problem 5

Produce the projection of operands onto processes for the alternative functional architecture for the pump, shown in Figure 5.14.

System Architecture–Chapter 6

Learning Objectives

- Develop synthesized approaches to analyzing form and function.
- Identify the functional architecture of a system.
- Apply these approaches to systems of "medium" complexity.

In each of the following problems, choose two of the medium-complexity systems from the list that follows. (Try to pick two from different domains. For example, do not pick both the glider and the sailboat.)

Problem 1

If the system you choose has a parts list posted, begin with that. For at least one of the two systems, develop your own "Level 2" decompositional model (that is, pick at least one system that does not have a parts list provided) so that it contains no more than 30 to 50 Level 2 instrument objects of form. Identify reasonable choices for a simplified Level 2 model with 20 to 30 parts. Show the formal structure for the simplified Level 2 objects of form for each of the two systems, indicating the nature of the formal structure (connectivity, topological, etc.). Is the formal structure important in these systems? Show it in a matrix (DSM or other) or graphical (OPM or other) representation.

Problem 2

For each of the two systems, what are the internal operands that move among or are changed by the instrument objects of form? Indicate the operand interaction by showing how these pass between the instrument objects. Show this in a matrix (DSM or other) or graphical (OPM or other) representation.

Problem 3

For each of the two systems, what are the key value-related internal processes? Develop a model of the key internal value-related processes and the flow of value through the intermediate operands. Represent this with an instrument-centric and a process-centric view of the "flow" along the primary value path. Again, use either a matrix (DSM or other) or graphical (OPM or other) representation. Identify objects of form whose function is not yet identified: interfaces, other value processes, or supporting processes.

Problem 4

Identify the important steps in operation (behavior/dynamics) of the product system, versus the static functions for both systems. For one of the two systems, develop a representation of the behavior of the system, such as the line operations diagram for the corkscrew. Was the dynamic behavior of the system important to represent?

Problem 5

Represent the system from Problem 4 using the appropriate SysML diagrams.

Problem 6

For each of the two systems, what are the key supporting processes and instruments at Level 2? Develop a layered representation of the functional architecture. Did the inclusion of supporting processes and instruments reveal new external interfaces?

Problem 7

For each of the two systems, using the diagram you developed in Problem 6, produce the projection onto processes.

Problem 8

For each of the two systems, using the diagram you developed in Problem 6, produce the projection onto objects. Which of the projections (the one you produced in Problem 7 or the one you produced in Problem 8) is more helpful?

List of Medium-Complexity Systems

Sector	Medium-Complexity System
Aeronautics	Standard class glider (15-m span)
Electrical	Superheterodyne receiver
Software	Mailing list
	Externally delivered value function: Forward electronic message to a list of subscribers
	Example: grouplist@mit.edu
Optical	Sextant
Judicial	U.S. court criminal court. See the following description.
Ocean	Simple 18- or 19-foot sloop (such as Rhodes 19). See the following description.
Computer	A handheld calculator supporting 4-bit additions and subtractions
Medical Services	Medium-complexity, gender-neutral annual physical exam (See the narrative that follows.)

Glider Description

Glider (typical of a 15-m standard class sailplane)

- fuselage
- landing gear wheel
- landing gear wheel axle
- landing gear wheel retraction assembly
- landing gear wheel control push rods
- landing gear wheel actuation lever
- canopy
- canopy hinge
- canopy latch
- tow hook
- tow hook release lever
- tow hook release cable
- tail skid
- vertical tail (integral to fuselage)
- rudder
- rudder hinge
- rudder cables
- rudder pedals
- aileron push rods
- control stick (attached to aileron and elevator push rods)
- elevator push rods
- dive brake push rods
- dive brake actuation lever
- seat
- altimeter
- airspeed indicator
- airspeed indicator tubing
- airspeed indicator probe (in airstream)
- compass
- GPS navigation instrument
- yaw string (to sense direction of free stream)
- air vent (to direct air to pilot)
- competition number
- N number (registration number painted on fuselage)
- wing (2)
- wing spar (2)
- wing spar connecting pins (2)
- ailerons (2)
- aileron hinges (2)
- aileron push rods in wing
- aileron push rod coupling to fuselage push rods
- dive brakes (2)
- dive brake hinges (2)
- dive brake push rods (2)
- dive brake push rods couplings to fuselage push rods
- tip skids (2)
- winglets (2)
- horizontal tail
- elevator
- elevator push hinge
- chart
- registration certificate
- type certificate
- parachute
- barf bag
- pencil, pen
- log book
- tools
- water, food

tow plane	airport
tow pilot	landing field
tow rope	trailer
pilot	car to pull trailer

Mailing List Description

Externally delivered value function: *Forward electronic message to a list of subscribers.*

Example: grouplist@mit.edu

Assumptions: There are a few variants of this type of system, so you can make some simplifying assumptions, including (but not limited to) the following:

- *Membership is unrestricted. Each person who wants to receive messages can subscribe/unsubscribe to the list without the intervention of a moderator.*
- *Anyone can send to this list. (Spammers rejoice!)*

Do not get too far down into the details of how an e-mail system works or what an e-mail is. From our perspective, it is simply an addressed message. Feel free to demarcate hardware elements of form in a general way, but focus on elements of software form, such as the server or the application. Do not descend into algorithms or commands within functions.

Judicial Court Description

Elements in the Courtroom

judge	witness for the prosecution—arresting officer
defendant	witness for the prosecution—eye witness (2)
defendant's attorney (2)	witness for the prosecution—expert witness (2)
defendant's attorney paralegal	witness for the defense—character witness (2)
prosecutor (2)	witness for the defense—alibi witness
court reporter	witness for the defense—expert witness (2)
bailiff	law books
clerk of the court	book of faith (for swearing witnesses)
jury (12)	gavel
judge's law clerk	

Sailboat Description

Sailboat (typical of an 18-foot sloop)

- hull
- deck
- seats
- mast step (hole in which mast seats)
- centerboard well (which houses centerboard when up)
- floor boards
- centerboard
- centerboard hinge
- centerboard raising lever
- rudder
- tiller (attached to rudder for sailor input)
- rudder hinge
- mast
- pulley at mast top for main halyard
- pulley at mast top for jib halyard
- pulley at mast top for spinnaker halyard
- fore stay
- side stays (2)
- aft stay
- main sail boom
- goose neck (attaching main boom to mast)
- main sail boom vang (for applying down force to boom)
- main sail
- main sail halyard (rope for raising main sail)
- main sail halyard cleat
- main sail sheet (for controlling position of sail)
- main sail sheet cleat
- out haul (for stretching main sail along boom)
- out haul cleat
- main sail tack attachment (attaching sail at bottom to mast)
- main sail batten
- jib
- jib halyard
- jib halyard cleat
- jib sheets (for controlling position of jib) (2)
- jib sheet cleats (2)
- jib tack attachment (attaching jib at bottom to deck)
- spinnaker
- spinnaker halyard
- spinnaker halyard cleat
- spinnaker pole
- spinnaker pole downhaul
- spinnaker pole downhaul cleat
- spinnaker pole uphaul
- spinnaker pole uphaul cleat
- spinnaker sheet (2)
- spinnaker sheet cleat (2)
- paddle
- anchor
- anchor line
- anchor line cleat
- compass
- charts
- life vests
- flares
- cooler
- water, food
- horn
- documentation papers
- docking fenders
- docking lines
- bilge pump

System Architecture–Chapter 7

Learning Objectives
- Identify the solution-neutral function.
- Represent the differences between concepts.
- Develop multiple concepts.

Problem 1

For any system of your choosing, produce a hierarchy of solution-neutral statements, with multiple potential specializations at each level.

Problem 2

Identify the solution-neutral statement of the function of the device for each of the medium-complexity systems listed in the Chapter 6 Problems.

Problem 3

For each of those systems, identify the principal beneficiary and the needs of that beneficiary that are addressed by the system. Are there other important beneficiaries that can be identified? If so, what are their needs? Review the statement of the solution-neutral externally delivered function from Problem 1. Is it a good statement of functional intent? How might you make it more precise?

Problem 4

For the two medium-complexity systems you chose in Chapter 6, what is the concept (the specialized operand, process, and instrument)? Can you identify two or three principal internal functions in this rich multifunctional concept?

Problem 5

For the solution-neutral statements of function you identified in Problems 2 and 3, produce two alternative concepts. Do your alternative concepts differ in specific operand, operating process, instruments, or a combination thereof?

Problem 6

Based on the needs identified, how would you rate the performance of the concepts?

Problem 7

For both of the systems you chose, how would your representation of concept of operations or concept of services differ from the operations diagram you constructed in Chapter 6?

System Architecture–Chapter 8

Learning Objectives

- Develop familiarity with Design Structure Matrix (DSM) clustering.
- Represent decomposition as a choice.

Problem 1

Suggest a Level 1 modularization for the two systems you used in the problems for Chapters 6 and 7. How was it informed by the formal structure, operand interactions, and internal function(s)?

Problem 2

What tradeoffs or compromises did you make in selecting the modularization?

Problem 3

What internal interfaces would you have to define if you used this modularization? (Do not characterize all of them; just list them.)

Problem 4

Represent your systems at Level 2 in a DSM, and then use a clustering algorithm (such as, with available MATLAB code for download) to suggest alternative possible Level 1 modularizations.

System Architecture–Chapter 9

Learning Objectives

- To abstract individual PDPs into one generic reference PDP which can serve as a basis to derive specific PDPs (like the ones in your company). The generic PDP should be applicable to complex systems.
- To compare product development and system architecture methods and identify their relative strengths.

Problem 1

It is important to thoroughly understand the common best practices in PDP. Based on the PDP shown in Figure 9.5, consider, in teams of 3 or 4, how a PDP for a large, complex, evolving system should look. Then examine the PDPs from the companies of your team members. What do they have in common? Where and why are they different? Synthesize a generic PDP in as much detail as possible, while still keeping it generic.

Deliverables

- Comparison of the elemental steps or tasks that are common to the company PDPs (there may be 20 to 40 steps). Identify the steps that are common to all or most of your PDPs.

CHAPTER PROBLEMS

(Prepare a matrix with columns labeled Step Name, A Few, Most, and All, and list the steps with an X in the column indicating how often they occurred.)
- One synthesized reference PDP model per team (one chart, plus explanatory annotations if necessary). Show enough detail so that each of the main steps are labeled, and include some breakdown or explanation of each.
- Description of how the company-specific PDPs differ from the reference process (one or two charts). What are the factors that drive these differences?

Problem 2

Figure 9.6 presents seven questions, multiplied by three lifecycle phases, that cover many elements of PDPs and corporate initiatives. Recognizing that no method can answer all questions and still be targeted, complete the table below, indicating only those questions that are answered. For each, make a short statement about the contribution of that method to the question, or the expected output.

Phase	Attribute	Method (Example: System Architecture)
Conceiving the product (or service) and operations	Why?	
	What?	
	How?	
	Where?	
	When?	
	Who?	
	How much?	
Design process	Why?	
	What?	
	How?	
	Where?	
	When?	
	Who?	
	How much?	
Implementation process	Why?	
	What?	
	How?	
	Where?	
	When?	
	Who?	
	How much?	

Sources for some potential methods follow.

Agile: Robert Cecil Martin. *Agile Software Development: Principles, Patterns, and Practices.* Prentice Hall-PTR, 2003.

Lean: Daniel T. Jones and Daniel Roos. *Machine That Changed the World.* New York: Simon and Schuster, 1990.

Rechtin and Maier: Mark W. Maier and Eberhardt Rechtin. *The Art of Systems Architecting.* Vol. 2. Boca Raton, FL: CRC Press, 2000.

Product Design and Development: Karl T. Ulrich and Steven D. Eppinger. *Product Design and Development.* Vol. 384. New York: McGraw-Hill, 1995.

System Safety: Nancy Leveson. *Engineering a Safer World: Systems Thinking Applied to Safety.* Cambridge, MA: MIT Press, 2011.

Axiomatic Design: Nam P. Suh. "Axiomatic Design: Advances and Applications (The Oxford Series on Advanced Manufacturing)." 2001.

OPM: Dov Dori. *Object-Process Methodology: A Holistic Systems Paradigm; with CD-ROM.* Vol. 1. Springer-Verlag, 2002.

Problem 3

Imrich identified six important ideas in his framework for architecture (context, content, concept, circuitry, character, and magic). How do these ideas map to rest of this text? Are they applicable only to systems with a high degree of human interaction, or are they more broadly applicable?

Problem 4

What does Imrich's discussion of concept suggest about the deliverables of the architect?

System Architecture–Chapter 10

Learning Objectives

- To identify the impact of upstream and downstream influences on architecture.
- To identify the impact of architecture on upstream and downstream influences.

Problem 1

In your current company or a past role, how is each of the four levels of corporate strategy communicated? How do you observe it impacting architecture?

436 CHAPTER PROBLEMS

Problem 2

Complete the following impact graph for strategy.

Strength of Impact	Impact *of* Architecture	Strategy Factor	Impact *on* Architecture	Strength of Impact
		Mission of business Scope of business		
		Enterprise financial goals, including market share goals		
		Enterprise goals on core competencies, internal strengths, weaknesses, competitive position		
		Resource allocation decisions and business case process		
		Corporate initiatives, action plans, functional strategies		

Problem 3

How is marketing structured in your company? Do you have separate organizations for inbound and outbound marketing, and what effect does (or could) that decision have?

Problem 4

Complete the following impact graph for marketing.

Strength of Impact	Impact *of* Architecture	Marketing Factor	Impact *on* Architecture	Strength of Impact
		Customer and stakeholder needs analysis		
		Market segmentation		
		Market size and penetration		
		Competitive analysis		
		Product and service function, features		
		Product and service pricing		
		Communications and differentiation plan		
		Distribution channels		

Problem 5
How is regulation considered in your enterprise? How does it impact architecture?

Problem 6
Complete the following impact graph for strategy.

Strength of Impact	Impact *of* Architecture	Regulatory Factor	Impact *on* Architecture	Strength of Impact
		Regulatory compliance		
		Anticipated regulation		
		Standards		
		Product liability		

Problem 7
How is technology developed and acquired? If it is acquired, how is it incorporated in your company? How does it impact architecture?

Problem 8
Complete the following impact graph for technology infusion.

Strength of Impact	Impact *of* Architecture	Technology Factor	Impact *on* Architecture	Strength of Impact
		Technology planning and assessment		
		Technology transfer and product infusion		
		Technology strategy as a competitive advantage		
		Architecting system to deal with future technology advances		
		Intellectual property and its strategy		

Problem 9
Using the operations framework presented in this chapter, complete it using a system you are familiar with. Are there standard processes established for each of the steps?

Problem 10

Choose a platform you are familiar with, one that includes shared parts or elements across products. Identify which benefits are most important to the platform. To your knowledge, were these the benefits that were originally envisioned? Which benefits were costed in the business case?

Problem 11

For the same platform, identify which drawbacks and costs are applicable. Which of these are actively managed?

Problem 12

Describe how the system architecture of the platform has evolved (along with the business case) in your enterprise. If appropriate, explain these using the ABCD framework. What factors cause your enterprise to focus on certain items in the first pass, rather than leaving them for the second pass? How is this related to the type of enterprise (technology-driven, market-driven, largely new designs, largely redesign, and so on)?

System Architecture–Chapter 11

Learning Objectives

- Identify the needs of stakeholders.
- Compare methods for identifying and prioritizing stakeholder needs.
- Develop goals for the system.

Problem 1

Answer Questions 7a, 5a, 5b, and 5c from Table 8.1 to produce a preliminary description of the system under consideration.

Problem 2

Compare and contrast the process of identifying and characterizing customer needs for two different types of customers and needs. The first type should be the reference case of the well-known or established product, with the general consumer as the beneficiary and customer. Choose a second case from among the following, based on your experience and expertise: a latent need or gap for general consumers; an OEM customer (that is, selling up the supply chain); a different organization within your enterprise; or a government or policy-driven organization. Briefly document the stakeholder and need identification, characterization, and validation process for the second case, and the process for interacting with the agent of the stakeholders. Be honest and tell it like it is, not as it should be. (For example, if chatting at the golf course is important, put it in). Discuss what similarities and differences exist between the reference consumer case and the one you chose to analyze. What are the ultimate pieces of information that both types of customers seek to produce about needs, and which are necessary in order for architecting to begin (to drive ambiguity out of the system by setting system goals)?

Problem 3

Who are the stakeholders and beneficiaries for your system, and what are their needs (including stakeholders who would represent society, direct beneficiaries, and so on)? What are the needs of your enterprise, considering it a stakeholder as well, and how is value exchanged? Which stakeholders appear to be beneficial stakeholders, which problem stakeholders, and which charitable beneficiaries? How would you group or segment the stakeholders so that your "stakeholder map" is as simple as possible yet has the important stakeholders on it?

Problem 4

Characterize the needs of the stakeholders (including your enterprise) in terms of one (or at most two) parameters that you think best fit your case (intensity, satisfaction, etc.) How does value flow from your enterprise outputs to the stakeholders and back to the necessary resource inputs of your enterprise? How does competition in supply (or importance of supplier) influence the flows? Choose the six to ten most important stakeholders (one being the producing enterprise) and make a value flow map of the system.

Problem 5

Using an analysis method of your choosing, devise a list of the prioritized needs of the principal stakeholders, both within a single stakeholder and among various stakeholders. How might you validate the prioritization of the needs within a stakeholder? How would you do so across stakeholders?

Problem 6

Interpret the needs as goals. Create a statement of the goals (the system problem statement, and the descriptive goals) at the highest level (that is, without any decomposition to goals associated with the first level of decomposition of process or form). Comment on the rigidity of the goals.

Problem 7

Prioritize the descriptive goals into critical, important, or desirable. Try to develop appropriate metrics that would go with each. If possible, compare the original (produced by someone else) goals statement and your refined version of it. Is your formulation "better"?

Problem 8

Comment if you can on the tests of the goals you have developed. Are they complete, consistent, attainable, humanly solvable, and so on? How would you validate that if the goals are met, the needs of the stakeholders will be addressed?

Problem 9

Based on your work experience and/or discussion with a mentor, reflect on the process of defining goals for systems. Is there a formal process that is (or was) used at your enterprise? Are there explicit processes for prioritization, and if so, how do they take into consideration the needs of different stakeholders? Are there explicit checks for representativeness, completeness, consistency, and human solvability? How might you implement these checks, knowing that many failures of systems are in fact due to incomplete, inconsistent, or unclear goal setting? If you are working in a group of two or three, you may want to compare and contrast the approaches at your companies.

System Architecture–Chapter 12

Learning Objectives:
- Develop alternative concepts.
- Decompose concepts into concept fragments.
- Recombine concept fragments to form new concepts.
- Identify sources of creativity.

Problem 1

For the product or system you chose in the problems for Chapter 11, start with the goals you developed, and work toward the concept definition of the system. Identify and analyze the current concept to the best of your ability, and identify at least one alternative concept. Develop several concepts. Based on the solution-neutral statement of primary functional intent, identify several alternative specializations of (potentially) operands and (certainly) processes. Then, for several of the specialized processes, identify specialized instrument objects. This will produce the highest-level concept "tree" or morphological matrix. What sources of creativity did you draw on to develop these concepts? How do these sources of creativity compare with those you read in the chapter?

Problem 2

Expand the concepts and identify concept fragments. Create a tree or morphological matrix representation of these decisions. What constraints exist among the potential concept fragments that are represented? How does this limit the space of possible concept fragment combinations?

Problem 3

Combine concept fragments to produce as many integrated concepts as you can feasibly enumerate. Did you identify any concepts that were not originally identified?

Problem 4

Now prune this tree or set of decisions to the point where you have two well-defined concepts. Again, what criteria did you use for pruning the tree? Were they related to the system problem statement? To the goals? The final result of this process should be two well-rationalized concepts for the system (the current or reference one and one good alternative) and an understanding, on your part, of the constraints and metric evaluations that led to these two solutions.

System Architecture–Chapter 13

Learning Objectives
- Distinguish complexity from complicatedness.
- Choose among multiple possible decompositions of a system.

For the product or system you chose in the problems for Chapters 11 and 12, start with the concept definition and constituent processes you identified.

Problem 1
Compare the complexity of the two concepts you generated in the problems for Chapter 12. Is one of them more complex? More complicated? Identify what decisions or entities drive the apparent complexity. Do you think that the concept description you created can reveal the complexity, or do you believe that the complexity will not be known until the system is implemented?

Problem 2
Based on your concept, create a map of the Level 1 processes, and the Level 1 decomposition of form, for both concepts.

Problem 3
What insight can you obtain about the "goodness" of the decomposition of the system at Level 1 by examining Level 1 only?

Problem 4
Decompose Level 1 to Level 2 for both concepts.

Problem 5
Produce two possible modularizations for Level 1 for each of the concepts. What planes of decomposition does each employ? Which other modularization factors could pose challenges for your modularization? Which modularization is more elegant?

Problem 6
In the process of modularizing, you have defined interfaces that may be ad hoc or may become stable features of the architecture. Choose one of the important interfaces that you defined, and describe whether it is ad hoc or stable. Then describe the form and function of the interface, how open or closed it would be, and how you would propose to control it over the product or system lifecycle.

System Architecture–Chapter 14

Learning Objectives
- Describe the context in which architectural models are useful and feasible.
- Compare related methods for representing decisions.
- Indicate which methods can be manually processed and which require computational effort.

Problem 1
A central idea in this chapter is that we can model a system architecture as a set of interconnected decisions. For a system of your choice, provide a set of architectural decisions, typically 5 to 10, along with the options for each decision.

Problem 2
Are the decisions coupled or decoupled? Create a DSM at either the decision level or the option level, and indicate which decisions are coupled, if any.

Problem 3
What are the main components of an architectural model?

Problem 4
Describe in your own words the Apollo architectural model:
a. Enumerate the decision variables and their allowed options.
b. Enumerate the constraints.
c. Enumerate the metrics.
d. How are metrics related to decisions?
e. What is the size of the tradespace?

Problem 5
Formulate an architectural model for a system of your choice:
a. Enumerate the decision variables and their allowed options.
b. Enumerate the constraints.
c. Enumerate the metrics.
d. How are metrics related to decisions?
e. Compute the size of the tradespace.

Problem 6
What is a morphological matrix? Construct a morphological matrix for a system of your choice.

Problem 7
Transform the morphological matrix that you constructed in Problem 5 into a decision tree with no chance nodes. Note that you have freedom in your choice for the sequence of the decisions.

Problem 8

How are architectural decisions different from other decisions?

Problem 9

Why is it hard to create good architectural models?

System Architecture–Chapter 15

Learning Objectives

- Discuss a tradespace as a mental model.
- Identify the primary elements of tradespace analysis.
- Use a tradespace to prioritize architectural decisions.

Problem 1

What is an architectural tradespace? Why is this a useful mental model?

Problem 2

What is a Pareto front of a tradespace? How is it computed?

Problem 3

Why is it useful to have the entire tradespace of architectures, as opposed to just the Pareto frontier?

Problem 4

Why is the concept of dominance not enough to use as the single means for selecting a system architecture?

Problem 5

What causes the appearance of clusters and stratification in a tradespace?

Problem 6

What is the fuzzy Pareto frontier? How is it computed, and why is it a useful concept?

Problem 7

Why is sensitivity analysis necessary? What can (and what can't) one learn from sensitivity analysis?

Problem 8

Describe a methodology to sort architectural decisions in order of importance. What decisions should be made first?

Problem 9

How can one use the results of the sensitivity analysis to refine an architectural model?

Problem 10

Write down your own procedure, describing the analysis you would do if you were given an architectural tradespace.

System Architecture–Chapter 16

Learning Objectives

- Distinguish between architecting problems.
- Choose which problem is most appropriate for a given application.
- Determine when formulation as an architecting problem is both feasible and useful in defining the architecture of a system.
- Describe the trades among potential algorithms for a given architecting problem.
- Identify where full-factorial enumeration is appropriate.

Problem 1

What tasks of the system architect are amenable to formulation as optimization problems? Which tasks are not? How do these map to the deliverables of the architect?

Problem 2

We are all confronted with the common problem of choosing the configuration for our new computer. Decisions to make include the choice of operating system (Windows, Mac, Linux) and processor (such as, for Windows, i7-960, i7-930, i5-760, AMD).
   a. For this example, formulate a DECISION-OPTION problem with 7+/− 2 decisions, and several options for each decision.
   b. What would be the relevant metrics?
   c. How do the metrics map to the decisions?
   d. What constraints should be taken into account?

Problem 3

Formulate a system architecture problem as an optimization problem, using one of the patterns presented.
   a. Identify the Pattern chosen, along with the architectural decisions, options, metrics, value function, and constraints.
   b. Can you think of a reformulation of the same problem using a different Pattern? What are the relevant differences between the two formulations?

Problem 4

Assign each of the following problems to the most appropriate class of Patterns discussed in this chapter (DECISION-OPTION, DOWN-SELECTING, ASSIGNING, PERMUTING, PARTITIONING, CONNECTING). Note that more than one assignment is sometimes possible.

a. Consider the system architecture of a power system, simplified to the point of being represented by a mix of renewable and conventional energies. You are given certain projections of demand, and some characteristics of different sources of energy (including efficiency, non-recurring and recurring cost, and reliability), and your goal is to find the architectures that optimize the tradeoff between these metrics.
b. Consider a small, portable system to estimate the weight and chemical composition of the food we eat so that we can infer nutritional information. We simplify the architecture of the system to be represented by the platform (smartphone or stand-alone), type of sensor (a choice among three commercial miniature spectrometers), degree of autonomy (business intelligence is embedded in the product versus on a remote server accessed via the Internet).
c. Google has a division that funds both internal and external research projects. This division has a given budget and a list of proposals for technology investments, and it has to choose a subset of these proposals that maximizes return for the given budget. The values of these investments are not independent of each other. For example, investing in a technology for control of unmanned aerial vehicles will affect the value of investing in a conceptual study for an Internet network based on the deployment of thousands of such vehicles, since the probability of success of the latter depends on the former.
d. Consider the system architecture of a small city, modeled as a 10 by 10 grid delimiting rectangular regions each of which measures 10 km by 10 km. We can make each region be industrial, residential, commercial, or green. We are given forecasts for population growth and the needs of that population, as well as economic factors in the region, and the goal is to find a good set of architectures that fosters both economic growth and the well-being of the population.
e. Consider a home automation system, including voice-activated control of lighting, temperature, doors, windows, and entertainment systems. A set of sensors and actuators for each of these functions is available, and the only remaining decision concerns the number and location of the computers. For example, you can decide to put a computer in each room and distribute the decision-making process; or you can put, in an accessible location, a single central computer that controls all the sensors and actuators in the home.

Problem 5

Find two instances of each class of problem in your own domain. Which ones were more natural, or easier to find? Why do you think that is?

Problem 6

How does the interaction between elements drive DOWN-SELECTING problems? Provide examples from your own domain.

Problem 7

What is the fundamental tradeoff in the ASSIGNING Pattern? What domain-specific information drives ASSIGNING problems? For the two ASSIGNING problems found in Problem 3, describe a channelized architecture and a fully cross-strapped architecture, and comment on their characteristics. Are they good candidate architectures? What are their advantages and disadvantages?

Problem 8

What is the fundamental tradeoff in the PARTITIONING Pattern? What domain-specific information drives PARTITIONING problems? For the two PARTITIONING problems found in Problem 3, describe a monolithic architecture and a fully distributed architecture, and comment on their characteristics. Are they good candidate architectures? What are their advantages and disadvantages?

Problem 9

Write down a procedure illustrating how to use the six classes of Patterns for a real-life large, complex system in your domain. What are the most challenging steps of the procedure?

Problem 10

Why is full-factorial enumeration not viable in many cases? What are the alternatives?

Problem 11

Assuming that evaluating an architecture takes 1 second of computational time regardless of the number of decisions, what should we set as the maximum size of the following system architecture problems to ensure that we can do full-factorial enumeration in less than 24 hours? Repeat the computation for $t = 0.01$ second.
  a. A DOWN-SELECTING problem with N decisions with 3 options each
  b. An ASSIGNING problem with N elements in the left set and $N - 1$ elements in the right set
  c. A PARTITIONING problem with N elements
  d. A PERMUTING problem with N Elements
  e. A CONNECTING problem with N nodes where edges are undirected

Problem 12

Sketch an algorithm to do full-factorial enumeration of a PARTITIONING problem.

Problem 13

What are the main steps of any population-based heuristic optimization algorithm?

Problem 14

How do you choose an initial population of architectures?

Problem 15

We generate a random population of 100 different architectures in a PARTITIONING problem of size $N = 8$.

a. What fraction of these 100 architectures is expected to have elements 1 and 2 in the same subset?
b. What fraction of these 100 architectures is expected to consist of 2 subsets?

Problem 16

What is a schema? Give examples of schemata for the NEOSS mission-scheduling problem.

Problem 17

Why do genetic algorithms work? In what contexts are they applicable? In what contexts are they not applicable?

Problem 18

Give an example of a domain-specific heuristic for a system with which you are familiar.

Index

A

A posteriori sensitivity, 345
Abstractions, 23–24, 229
Accompanying systems, 75
Activity diagram (SysML), 46
Address, 74
Address resolution protocol (ARP), 172
Adjacency matrix, CONNECTING Pattern and, 385–386
Aggregation of form, 18
Algorithms, 395–401, 436–440
   clustering, 436–440
   declarative enumeration and, 442–443
   genetic, 400–402
   inference, 441
   meta-heuristic, 396–398
   optimization, 395–401
   population-based, 397–398, 399
   procedural programs, 442–443
   rule-based systems, 441
   search, 396–398
   sequence of decisions from, 438–440
   system decomposition from, 437–438
Ambiguity
   architect role for, 178–179
   architectural decisions and, 327
   corporate functions with, 181–182
   fuzziness, 181
   principle of, 182
   principle of goals and, 252
   reduction of, 180–183
   uncertainty, 181
   unknown information, 181
   upstream process and, 178–179
Amplifier system
   abstractions in, 23
   form and function identification, 13, 15–16
   formal structure and functional relationships of, 27–29
   emergence and structure of, 32
   primary externally delivered functions, 90, 91–92
   principle value-related internal functions, 94
Anticipated regulation, 206
Apparent complexity, 36, 289–291
Architects, 178–196
   ambiguity, reduction of, 178–179, 180–183
   creativity of, 178, 179
   deliverables of, 183–184
   holistic view of, 178, 189–190, 191
   managing complexity, 178, 179
   principle of stress of modern practice, 192
   principle of the role of, 180
   product development process (PDP), 184–192
   upstream process and, 178–179
   W questions, 189–190
Architectural competition, 280–284
   dominant architecture, 282–283
   future concepts of, 283–284
   product differentiation and, 280–282
Architectural decisions, principle of, 197. *See also* Decision-making process; Decisions
Architectural optimization problems, *see* Optimization problems
Architectural tradespaces, 331–358
   architecture representation, 332–334
   clusters, 341–343
   CONNECTING Pattern and, 386
   decision space view, 352–353
   degree of connectivity, 350–351
   Design Structure Matrix (DSM) for, 350
   quadrants for, 352–353
   impact on metrics, 351–352
   impact on other decisions, 350–351
   metrics, 332–333, 351–352
   Monte Carlo analysis, 348–349
   organizing architectural decisions, 350–356
   Pareto frontier, 334–341, 352
   penalties, 345–347

point designs for, 333–334
principle of coupling and organization of architectural decisions, 354
principle of robustness of architectures, 349
quadrants for decisions, 352–353
robustness of results, 346–349
scenarios, 345–346
sensitivity analysis, 345–349
sequencing decisions, 354–356
stratification, 343–344
structure of, 341–344
Architecture Business Case Decision (ABCD), 217, 219–221
architectural framework, 220–221
customer need and, 219–220
technology and, 219–220
Architecture representation tools, 45–48
integrated 3D models, 45–48
Object Process Methodology (OPM), 45, 47–48
projections, 45
Systems Modeling Language (SysML), 45–47
Unified Modeling Language (UML), 45–46
views, 45
Architecture, 2–7. *See also* System architecture
advantages of, 2–5
analysis of form in, 58–63
analysis of formal relationships in, 63–75
benefit of, 51–52
bus style, 388–389
channelized style, 377–378
civil versus system, 193–196
complex systems, 2
cross-strapped style, 377–378
decision-making process, 313–330
decisions, as, 3, 309–310
defined, 9, 51
design and, 4–5
distributed (decentralized) style, 381–382, 399
form and, 53–57
formal context and, 75–77

formal relationships and, 63–75
function in, 83–89
mesh style, 388–389
monolithic (centralized) style, 381–382, 399
objectives for architects, 5–6
patterns in, 365–389
platforms and, 222–224
principles, 6
ring style, 388–389
set choice effects on sensitivity metric, 435–436
software system analysis, 77–82
star (hub and spoke) style, 388–389
system optimization problems, 359–405
tradespace representation of, 332–334
tree style, 388–389
ASSIGNING Pattern, 367, 374–378
Atomic parts, 41, 42–43
Attainable goals, 245, 252–253
Attribute–value pairs, 441
Attributes of objects, 56
Awareness of needs, 236

B

Balance, principle of, 244
Ballooning (expanding) concepts, 263–264, 279–280
Beginning (stakeholder identification), principle of, 229
Behavior, 126–128
Beneficial stakeholders, 228
Beneficiaries, 227–232
distinguishing stakeholders from, 227–228
identifying needs of, 230–232
operator and, 228
stakeholders and, 227–232
Benefit
architecture and, 51–52
at cost, 13
delivery, 91
intensity, 236
value and emergence of, 13

Binary decisions, 371
Bivalent architecture, 269
Block definition diagram (SysML), 46, 47, 70
Blueprints, 95
Boundaries, 18–19, 24–26
   context and, 24–26
   defined for system, 24–26
   external interface, 25, 28–29
   form and function relationships across, 28
   formal context and, 75, 76
   identification of, 18–19
   software system analysis and, 80–81
   system interfaces as, 123–125
Bounds for stakeholder identification, 229
Bread slice making
   internal functions, 96, 98–99
   projection onto form of, 134
   projection onto objects of, 131
   system architecture of, 113–114, 116–117
   SysML diagrams for mapping of, 116–117
Bubblesort software system analysis, 77–82
   See also Software system analysis
Bus architecture style, 388–389
Business unit strategy, 199

C

Centrifugal pump analysis, 58–63
   elements of form, 60
   entity identification, 60–61
   external functions of, 92–93, 96
   functional architecture of, 100–102
   hierarchy in elements of form, 61
   internal functions of, 96–97, 100–102
   medium-complexity system, 61–63
   operation of, 58–59
   principle value-related internal functions, 94, 96
   process-operand (PO) array for, 96–97, 102
   simple system, 58–61
   simplified system representation of, 129–130
   supporting layers of, 122–123
   system architecture of, 117–118, 122–123, 129–130
   system of, 59–60
   value-related operand of, 92–93
   value stream objects, 132, 133
Channelized architecture style, 377–378, 399
Charitable stakeholders, 228
Choice activity, 318
Circulatory system
   abstractions in, 23
   form and function identification, 14, 16
   principle value-related internal functions, 94–95
Civil architecture, system architecture versus, 193–196
Class/instance relationships, 43
Clustering, 173–175, 436–440
   algorithms, 436–440
   decomposition from, 437–438
   Level 1 architecture, 173–175
   optimization from, 437–440
   sequence of decisions from, 438–440
   system architecture applications, 436–440
Clusters, tradespace analysis use of, 341–343
Combinatorial optimization problems, 445–447
Combined operand and instrument object mapping, 114–115
Commissioning, 210–212
Competition, marketing and, 203
Complex systems, 2, 35–50
   approaches to thinking about, 49
   architecture of, 2, 262–263
   architecture representation tools for, 45–48
   atomic parts of, 41, 42–43
   complicated versus, 36
   creativity and, 262–263
   decomposition of, 39–43
   defined, 35–36
   design development, 36–39
   design of, 4
   features of, 48
   hierarchy of, 39–41
   integrated 3D models, 45–48
   logical relationships, 43–44
   reasoning through, 44–45
   top-down/bottom-up reasoning, 44
   zigzagging, 44–45

INDEX **451**

Complexity, 35–36, 286–308
  apparent, 36, 289–291
  architect management of, 178, 179
  complicated versus, 36, 288–290
  decomposition for management
    of, 286–308
  defined, 287
  essential, 291–295
  hierarchy and, 288
  investment in, 295
  managing, 295–302
  organization and, 291
  principle of "2 down, 1 up," 298
  principle of 2nd law, 295
  principle of apparent complexity, 292
  principle of decomposition, 296
  principle of elegance, 300
  principle of essential complexity,
    292–295
  systems and, 35–36, 286–288
  understanding, 286–295
Compliance with regulations, 205
Complicated things, complexity and,
  36, 288–291
Concept fragments, 153–154,
  269–274
  concept ballooning for combinations
    of, 279–280
  creativity applied to, 269–247
  decomposition for, 269–271
  development of, 269–274
  function expansion for, 269–271
  integrated concepts and, 153–154
  morphological matrix for, 153–154
Concepts, 137–159
  applying creativity to, 263–268
  ballooning (expanding) and recombining
    fragments, 263–264, 279–280
  creativity and generation
    of, 262–285
  defined, 142
  development of, 143–146, 268–269,
    271–274
  evolution and refinement of, 274–277
  expanded to architecture, 162–163
  fragments, 153–154, 269–274
  functional intent of, 146, 149, 153
  hierarchy of, 149, 151–152
  identifying, 266–268
  integrated, 152–156, 274–279
  intent and, 149, 151–152
  naming conventions, 147
  notion of, 142–143
  operations (conops), 157–158
  organizing alternatives, 147–150
  selection of for further development,
    277–279
  services and, 157–158
  software patterns as, 146
  solution-neutral functions and,
    137–138
  specialization of solution-neutral function,
    143–145
  system architecture and, 160–176
  system descriptions from, 137–138
CONNECTING Pattern, 385–389
Connections, 64, 66, 118
Connectivity, degree of, 350–351
Connectivity formal relationships,
  71–74, 80, 81
Consistent goal, 245, 252–253
Constraints, 314–315, 363, 443
  architecture models, 314–315
  architecture problems, 363
  declarative enumeration and, 443
Context, 24–26
  boundary definition and, 24–26
  defined, 24
  formal, 75–77
  use, 76–77
Contingency operations, 212
Corporate strategy, 198–201
  business unit, 199
  enterprise goals, 200
  executive, 199
  functional, 199–200
  initiatives and action plans, 201
  mission and scope, 200
  resource allocation decisions, 200
  shareholder annual report, 198–199
  upstream influences on, 198–201
Coupling, emergence and, 327

Coupling and organization of architectural decisions, 354
Coupling of needs, 236
Creativity, 262–285
    applying to concept, 263–268
    architect role and, 178, 179
    architectural competition, 280–284
    ballooning (expanding) concepts, 263–264, 279–280
    complex system architecture and, 262–263
    concept fragment development, 269, 271–274
    concept generation and, 262–285
    developing concepts, 268–269
    function expansion, 269–271
    identifying concepts, 266–268
    integrated concept evolution and refinement, 274–277
    principle of, 267
    Pugh matrix for concept selection, 277–279
    selection of integrated concepts for further development, 277–279
    structured, 265–266
    unstructured, 263–264
Critical goals, 250–252
Cross-strapping architecture style, 378–379, 399
Crossover points, 400–401

D

Decision-making process, 313–330
    calculated outcomes, 315–317
    constraints and metrics for architecture model, 314–315
    decision support and, 317–328
    decision trees for, 323–326
    Design Structure Matrix (DSM) for, 321–323
    formulating decision problem, 312–317
    heuristics for decisions, 312–313
    interconnection of decisions, 311–312
    morphological matrix for, 321
    properties of architectural decisions and metrics, 326–328
DECISION-OPTION Pattern, 367–370
Decision space view, 352–353

Decision support, 317–328
    choice activity, 318
    decisions and, 317–319
    defined, 318
    design activity, 318
    intelligence activity, 318
    representing layer, 319–320
    review activity, 318
    simulating layer, 320
    structuring layer, 320
    system (DSS), 319–320
    system architecture and, 326–328
    tools for, 320–326
    viewing layer, 320
Decision trees, 323–326
Decisions
    architectural, 3, 197–198, 327
    architectural tradespaces for, 350–356
    decision support and, 317–319
    defined, 317
    degree of connectivity, 350–351
    design and, 4–5
    heuristics for, 312–313
    impact on metrics, 351–352
    impact on other decisions, 350–351
    influences on, 197–225
    interconnection of, 311–312
    main effect of, 351–352
    non-programmed, 319
    organizing, 350–356
    patterns in, 365–389
    principle of architectural decisions, 197
    principle of coupling and organization of, 354
    programmed, 318–319
    properties and metrics of, 327
    resource allocation, 200
    sensitivity to metrics, 351–352, 435–436
    sequencing, 354–356
    spectrum of, 318–319
Declarative enumeration, 442–443
Decommissioning, 210–212
Decomposition, 18, 39–43, 57
    atomic parts from, 41, 42–43
    choosing, 296–298
    clustering algorithm for, 437–438

complex systems, 41–43
complexity managed with, 286–380
concept fragments from, 269–271
defined, 18, 39
distinct system elements, 19
entities into form and function, 18
form, 57, 300–301
function, 300–301
function expansion from, 269–271
hierarchic, 40–41
integral system elements, 19
medium-complexity systems, 41, 61–63
modular system elements, 19
modularity and, 298–302
multi-level, 41–42, 61
objects from form, 57
one-level, 41–42, 58
potential planes for, 301–302
principle of, 296
principle of elegance, 300
Saturn V launch vehicle, 303–305
simple systems, 41, 58–61
Space Station Freedom (SSF), 305–308
sub-problems from, 392–394
system into entities, 19
system processes of, 39–43, 437–438
two-level, 41
Degree of connectivity, 350–351
Deliverables of the architect, 183–184
Department of Defense Architecture Framework (DoDAF), 45
Deployment, PERMUTING Pattern and, 384–385
Design activity, 318
Design for X (DfX), 212–214
Design Structure Matrix (DSM)
 architectural tradespaces and, 350
 bubblesort software structure, 80, 81
 connectivity formal relationships, 73–74, 81
 decision-making process and, 321–323
 formal relationships, 70–71, 80
 function interaction notation, 106
 projecting operands onto processes using, 105

Design team
 abstractions in, 23
 complex design development, 36–39
 evolution of system thinking of, 21
 form and function identification, 14, 16
 hierarchic decomposition of, 40–41
 principle value-related internal functions, 94–95
 system boundaries of, 25
 value-related operand of, 92
Detriment of needs, 236
Diagrams
 activity, 46
 block definition, 46, 47, 70
 decomposition of form, 57–63
 formal relationships, 68–70
 function representation, 86–89
 graphical representation, 56–63, 67–68
 internal block, 46, 47, 70
 legend for SysML graphical elements, 69
 Object Process Methodology (OPM), 47–48, 56–63, 67–68, 86–87
 object representation, 56–57
 package, 46, 47
 parametric, 46, 47
 requirement, 46, 47
 sequence, 46
 state machine, 46, 88–89
 structure representation, 68–70
 Systems Modeling Language (SysML), 46–47, 68–70, 88–89
 Unified Modeling Language (UML), 45–46
Distinct system entities, 19
Distributed (decentralized) architecture style, 381–382, 399
Domain-specific heuristics, 401
Dominance, Pareto frontier and, 334–335
Dominant architecture, 282–283
DOWN-SELECTING Pattern, 367, 370–373
Downstream influences, 207–217
 Design for X (DfX), 212–214
 implementation, 207–210
 Legacy elements, 214–215
 operations, 210–212
 platforming, 216–217, 218–219

Downstream influences (*continued*)
   principle of product evolution, 216
   principle of reuse of Legacy
      elements, 214
   product and system evolution, 214–217
   product lines (families), 215–216
   reuse, 214–215
   systems attributes (ilities) as, 213–214
Duality, 78–79
Dynamic behavior, 128

E

Elegance (architect systems), principle of, 300
Emergence, 8–13, 28–33
   attributes of operation, 11–12
   coupling of architectural variables and, 327
   defined, 10
   entity combination for, 18
   experiments for, 31
   formal relationships and, 28–32
   functional interactions and, 103–104
   functional relationships and, 28–32
   functions, 10–11
   importance of, 28–33
   modeling for, 31
   performance, 11
   precedent of, 31
   predicting, 31–32
   principle of, 12
   system failure and, 30–31
   systems and, 8–13
   unanticipated and undesirable
      (emergency), 12–13
   values, 13
Emergency (undesirable emergence), 12–13
Emergency operations, 212
Emergent behavior, 444
Enterprise goals, corporate strategy, 200
Enterprise product development process
   (PDP), 184–188
Entities, 9, 17–28
   abstractions and, 23–24
   decomposition into form and function, 18
   defined, 9
   distinct, 19
   external interface, 25, 28–29

   focus on importance of, 21–23
   form of, 17–18
   function of, 18
   holistic thinking about, 19–21
   identification of, 17–26
   integral, 19
   modular, 19
   relationships among, 26–28
   system decomposition into, 19
   system boundaries and, 18–19,
      24–26, 28
Essential complexity, 291–295
Exchange, 232–234, 237–238
   stakeholder relationships
      and, 232–234
   system formation from, 237–238
   value delivery from, 232–234, 237–238
Executive corporate strategy, 199
Expansion
   ballooning concepts,
      263–264, 279–280
   concepts to architecture, 162–163
   creativity and, 269–271, 280–281
   functions, 269–271
Experiments, emergence prediction by, 31
Exploitation functions, 399
Exploration functions, 399
External function, 89–93, 107–108
   primary value-related, 89–91
   secondary value-related, 107–108
   value-related operands, 90–93
External interface, 25, 28–29

F

Flying wing aircraft architecture, 222–224
Focus, 21–23
   importance of for entities, 21–23
   principle of, 22
Form, 13–18, 53–82. *See also*
      Formal context; Formal relationships
   aggregation, 18
   analysis of in architecture, 58–63
   architecture and, 53–57
   canonical system characteristic of, 17
   decomposition, 57, 300–301
   defined, 13, 54

entities and, 17–18, 19
features of function and, 109
identification of, 13–15, 16
implementation of, 53
Level 1 architecture and, 163–165
mapping, 18, 111–121, 164–165
Object Process Methodology (OPM)
 for, 56–63
objects as, 56–57
operation of, 53
projection onto, 132–135
relationships among entities, 26–28
representation of, 54–57
software system analysis and, 77–82
system architecture and, 110–121,
 123–125, 132–135
system attribute of, 54
system interfaces in, 123–125
zigzagging, 163–164
Form-to-function mapping, 274–275
Form-to-process mapping, 117–118
Formal context, 75–77
 accompanying systems, 75
 architecture analysis using, 75–77
 system boundary, 75, 76
 use context, 76–77
 whole product system, 75–76
Formal relationships, 26–29, 32–33,
 63–75
 analysis of in architecture, 63–75
 connections, 64, 66
 connectivity, 71–74, 81
 defined, 26, 63
 Design Structure Matrix (DSM)
  representation, 70–71, 73–74, 80, 81
 diagrams for, 68–70
 emergence and, 32–33
 entity interactions, 26–28
 external interfaces, 28–29
 graphical representation, 67–68
 identification of other types of, 74–75
 mapping and, 118–121
 N-squared table representation,
  26, 29, 70
 Object Process Methodology (OPM)
  representation, 67–68, 72–73

software system analysis, 79–80
spatial/topological relationships, 64–67
structural representation, 68–70, 80
structure as, 63–64, 118–121
Systems Modeling Language (SysML)
 representation, 69–70
tables for, 70–71, 80, 81
Forward engineering, 137. *See also* Concept;
 Solution-neutral functions
Front-loaded (greedy) system
 deployment, 385
Full-factorial enumeration, 394–395
Function, 10–11, 15–16, 83–109. *See also*
 Functional interactions; Functional
 relationships
 analytical representation of, 86–89
 architecture and, 83–89
 architectural decisions and objectivity
  of, 327
 benefit of delivery, 91
 canonical system characteristics of, 17
 defined, 10, 83
 decomposition, 300–301
 emergent, 10–11
 entities, 18
 expansion of, 269–271
 external, 89–93, 107–108
 features of form and, 109
 form mapping into, 164–165
 identification of, 15–16
 internal, 94–98, 107–108
 Level 1 architecture, 164–165
 mapping, 111–121
 Object Process Methodology (OPM),
  86–87
 operands, 15, 17, 84–89
 primary value-related, 90–91
 process, 15, 17, 84–89
 process operand (PO) array, 96–97,
  99, 102
 relationships among entities, 26–28
 secondary value-related, 107–108
 standard blueprints for, 95
 system architecture and, 110–121,
  122–128
 system interfaces in, 123–125

Function (*continued*)
  Systems Modeling Language (SysML), 88–89
  value-related, 89–93, 94, 107–108
  zooming, 18
Functional architecture, 98–99, 104–107. *See also* Functional interactions
Functional intent, 139, 141, 146, 149, 153
Functional interactions, 98–107
  emergence, 103–104
  functional architecture and, 98–99, 104–107
  identifying, 99–102
  mapping and, 118–120
  Object Process Methodology (OPM), 100, 101, 104
  process operand (PO) array, 99, 102
  projecting operands onto processes, 105
  software system analysis, 104–107
  value pathway, 102–103
  zooming, 103, 104
Functional relationships, 28–32
  defined, 26
  emergence and, 32–33
  entity interactions, 26–28
  external interfaces, 28–29
  N-squared table representation, 26, 29
Functional strategies, 199–200
Function–goal reasoning, 165
Function-to-form mapping, 164–165
Fuzziness, 181
Fuzzy Pareto frontier, 338–340, 352

G

Generic product development process (PDP), 188–189
Genetic algorithms, 400–402
Global product development process (PDP), 189–192
Goals, 226–261
  attainable, 245, 252–253
  beneficiaries, 227–232
  complete, 245, 246
  consistent, 245, 252–253
  criteria for, 245–247, 254
  critical, 250–252
  defined, 245
  desirable, 250–252
  humanly solvable, 245, 246–250
  important, 250–252
  interpreting needs as, 244–250
  IT architecture requirements, 256–257
  needs-to-goals framework, 227
  principle of ambiguity and goals, 252
  principle of balance, 244
  principle of the beginning, 229
  principle of the system problem statement, 247
  prioritizing, 250–253
  representative, 245, 256
  stakeholders in IT architecture, 254–259
  stakeholders, 227–243
  system problem statement (SPS), 247–250
  translating needs into, 226–261
Goods, 16
Granularity (abstractions) for stakeholder identification, 229
Graphical representation, *see* Diagrams; Object Process Methodology (OPM)
Grouping stakeholders, 234–236
Guidance, navigation and control (GNC) system, 335–338

H

Heuristics, 395–402
  algorithms, 395–401
  architecture optimization and, 359–402
  crossover points, 400–401
  decisions using, 312–313
  deterministic architecture and, 398–399
  domain-specific, 401
  genetic algorithms, 400–402
  initial population generation and completion, 398–399
  mutation, 401
  optimization problems and, 365–402
  population-based algorithms, 397–398, 399
  search efficiency, 399
Hierarchic decomposition, 40–42

Hierarchy
    complex systems, 39–41, 61
    complexity and, 288
    concept tree, 149, 151–152
    elements of form, 61
    intents and concepts, 151–152
    principle of "2 down, 1 up," 298
    solution-neutral functions, 139–140, 149, 151–152
Holism, principle of, 20
Holistic thinking, 19–21
    architect view of, 178, 189–190, 191
    entity identification using, 19–21
    known-unknowns, 20
    product development process (PDP) framework, 189–190, 191
    unknown-unknowns, 20
Home data network architecture, 170–173
Human relationships, 74
Humanly solvable goals, 245, 246–250
    defined, 245
    difficulty interpreting, 246
    principle of the system problem statement, 247
    system problem statement (SPS), 247–250
    To-By-Using framework for, 247–249

I

"ilities" (systems attributes), 13, 213–214
Implementation, 207–210
    defined, 209
    downstream influences on, 207–210
Inbound marketing, 201, 202–203
Incremental system deployment, 385
Inference algorithm, 441
Influences on system architecture
    downstream influences, 207–217
    principle of architectural decisions, 197
    product case for, 217, 219–221
    upstream influences, 198–207
Initiatives and action plans, corporate strategy, 201
Integral system entities, 19
Integrated concepts, 152–156, 274–279
    concept fragments for, 153–154
    creativity applied to, 274–279
    evolution and refinement of, 274–277
    flexibility of, 154
    form-to-function mapping of, 274–275
    morphological matrix for, 153–156
    selection of for further development, 277–279
Integrated 3D models, 45–48
Intelligence activity, 318
Intent, 139–144
    concept development and, 143–144
    defined, 140
    functional, 139, 141, 146, 149, 153
    Level 2 architecture, 166
    regulatory, 205
    solution-neutral functions, 139–142
    specialization of solution neutral functions using, 143–144
Interferences, 373
Interferences as rules, 443–445
Internal block diagram (SysML), 46, 47, 70
Internal function, 94–98, 107–108
    identification of, 94–95
    predicting emergence of, 96
    process-operand (PO) array for, 96–97
    standard blueprints of processes, 95
    principle value-related, 94
    secondary value-related, 107–108
International Council on System Engineering (INCOSE), 45
IP packet, 172
IT architecture, 254–259
    goal requirements, 256–257
    mappings for, 257–258
    Olympic system challenge, 254–255
    reliability through redundancy, 258–259
    stakeholders and, 254–259
    system boundary of, 255–256
    utility applications, 257–258

K

Kano analysis, 236–237, 242
Knowledge-based systems, 440–441
Known-unknowns, holistic thinking and, 20

L

Layers of system architecture, 122
Legacy elements, 214–215
Level 1 architecture, 162–165
   clustering, 173–175
   concept expanded to architecture, 162–163
   development of, 162–165
   form defined for, 163–164
   mapping function to form, 164–165
   modularizing the system, 173–175
   zigzagging, 163–164
Level 2 architecture, 166–170
   development of, 166–170
   home data network architecture at, 170–173
   intent and, 166
   processes and relationships, 167–168
   recursion for, 166–167
   zooming to, 167
Liability, regulation and, 206
Logical relationships, 43–44

M

Main effect, 351–352, 435
Management
   architect role for, 178, 179
   choosing a decomposition, 296–298
   complexity, 178, 179, 286–308
   decomposition and, 295–302
   modularity and decomposition, 298–302
   potential planes for decomposition, 301–302
Many-to-many mapping, 115–117
Mapping, 18, 111–121
   ASSIGNING Pattern and, 376–377
   combined operand and instrument object, 114–115
   enabling function and performance, 120
   form, 18, 111–121
   formal structure and, 118–121
   form-to-function, 274–275
   form-to-process, 117–118
   function, 111–121
   functional interactions from, 118–120
   function-to-form, 164–165
   integrated concepts, 274–275
   IT architecture and, 257–258
   Level 1 architecture, 164–165
   many-to-many, 115–117
   no instrument, 114–115
   one-to-many, 115–116
   one-to-one, 115
   system architecture and, 111–121
   system optimization problem and, 376–377
Market penetration, 203
Market sizing, 203
Marketing, 201–203
   architectural competition, 280–282
   corporate strategy and, 200
   inbound, 201, 202–203
   outbound, 201–202, 203
   product differentiation and, 280–282
   upstream influences on, 201–203
Matrix representation
   adjacency matrix, 385–386
   architectural tradespaces, 350
   concept fragments, 153–154
   concept selection, 277–279
   CONNECTING Pattern, 385–386
   connectivity formal relationships, 73–74, 81
   decision-making process and, 321–323
   Design Structure Matrix (DSM), 70–71, 73–74, 321–323, 350
   formal relationships, 70–71, 80
   morphological matrix, 153–156, 321
   Pugh matrix, 277–279
Media access control (MAC), 172
Medium-complexity systems, 41, 61–63
Membership, 74
Mesh architecture style, 388–389
Meta-heuristic optimization algorithms, 396–398
Metrics
   architectural tradespaces, 332–333, 351–352
   architecture model constraints and, 314–315
   decision impact on, 351–352

decision-making process, 314–315, 326–328
main effect, 351–352
Pareto ranking, 341
properties of architectural decisions and, 326–328
sensitivity of decisions to, 351–352, 435–436
Mission and scope, corporate strategy, 200
Model input parameter values, 348
Modeling, emergence prediction by, 31
Modeling breadth versus depth, 327
Modular system entities, 19
Modularity and decomposition, 298–302
Modularizing the system, 173–175
Monolithic (centralized) architecture style, 381–382, 399
Monovalent architecture, 269
Monte Carlo analysis, 348–349
Morphological matrix, 153–156, 321
Multi-level decomposition, 41–42, 61
Multivalent architecture, 269
Mutation, 401

N

N-squared table representation, 26, 29, 70
Naming conventions, concepts, 147
NASA Earth Observing Satellite System (NEOSS), 363–365
National Polar-Orbiting Environmental Satellite System (NPOESS), 4–5
Needs, 226–261
  beneficiaries, 230–232
  characterizing, 236–244
  defined, 230
  exchange for stakeholder value delivery, 232–234
  interpreted as goals, 244–250
  Kano analysis, 236–237, 242
  principle of balance, 244
  prioritizing, 240–244
  stakeholder system classification of, 260
  stakeholders, 230–234
  transformative architecture and, 226
  translating into goals, 226–261
Needs-to-goals framework, 227

Network address translation (NAT), 172
Networks, home data architecture, 170–173
No instrument mapping, 114–115
Non-idealities in value pathway, 121–122
Non-programmed decisions, 319

O

Object Management Group (OMG), 45
Object Process Methodology (OPM)
  analysis of form, 58–63
  approach of, 45
  connectivity formal relationships, 72–73
  decomposition of form, 57–63
  diagram representations, 47–48, 56–57
  formal relationships, 67–68
  function representation, 86–87
  functional interactions, 100, 101, 104
  graphical representations, 67–68
  object relationships, 47–48
  object representation, 47
  value-related operands, 92–93
Objective functions, architectural decisions and, 327
Objects, 56–57, 131–132
  attributes of, 56
  decomposition of form into, 57
  defined, 56
  form representation as, 56–57
  projection onto, 131–132
  states, 56
  system representation of, 131–132
One-level decomposition, 41–42, 58
One-to-many mapping, 115–116
One-to-one mapping, 115
Operand, 15, 85
  analytical representation of, 86–89
  attribute, 91–92
  canonical system characteristic of, 17
  classes of interactions, 135
  defined, 15, 85
  external function analysis and, 90–93
  function as process and, 84–89
  projecting onto processes, 105
  value-related, 90–93
Operational behavior, 125–128

Operations
    attributes of, 11–12
    commissioning, 210–212
    concept of (conops), 157–158
    contingencies, 212
    cost, 128
    decommissioning, 210–212
    defined, 210
    downstream influences on, 210–212
    emergency, 212
    sequence of, 127
    stand-alone, 212
Operator, 126, 228
Optimization problems, 359–405
    algorithms for, 395–401
    architectural, 359, 361–364
    combinatorial, 445–447
    constraints, 363
    decomposition into sub-problems, 392–394
    deterministic architecture and, 398–399
    exploitation functions for, 399
    exploration functions for, 399
    full-factorial enumeration for, 394–395
    genetic algorithms for, 400–402
    heuristics for, 395–402
    initial population for, 398–399
    large-scale system problem formulation, 389–394
    meta-heuristic optimization algorithms, 396–398
    NASA Earth Observing Satellite System (NEOSS), 363–365
    patterns for, 365–394
    population-based algorithms, 397–398, 399
    search algorithms for, 396–398
    solving, 394–402
    system architect decisions and, 365–389
    system architecture task and pattern relationships, 389–394
    tasks for, 359–360, 389–394
    value functions, 362
Organization, complexity and, 291
Outbound marketing, 201–202, 203
Ownership, 74

P

Package diagram (SysML), 46, 47
Parametric diagram (SysML), 46, 47
Pareto frontier, 334–341, 352
    architectural tradespaces and, 334–341, 352
    dominance and, 334–335
    fuzzy frontier, 338–340, 352
    guidance, navigation and control (GNC) system, 335–338
    mechanics of, 341
    mining data on, 339–340
    Utopia point for plots, 337
Pareto ranking, 341
PARTITIONING Pattern, 367, 379–383
Patterns, 365–394
    adjacency matrix for, 385–386
    applications of, 366–367
    architectural tradespace and, 386
    ASSIGNING, 367, 374–378
    bus architecture style, 388–389
    channelized architecture style, 377–378, 399
    concept of in architecture, 365–366
    CONNECTING, 385–389
    cross-strapping architecture style, 377–378, 399
    DECISION-OPTION, 367–370
    distributed (decentralized) architecture style, 381–382, 399
    DOWN-SELECTING, 367, 370–373
    front-loaded (greedy) system deployment, 385
    incremental system deployment, 385
    interactions between system elements, 373
    mapping with, 376–377
    mesh architecture style, 388–389
    monolithic (centralized) architecture style, 381–382, 399
    optimization problems and, 365–389
    overlap between, 391, 392
    PARTITIONING, 367, 379–383
    PERMUTING, 367, 383–385
    ring architecture style, 388–389
    star (hub and spoke) architecture style, 388–389

sub-problems and, 392–394
system architectural decisions
    and, 365–389
system architecture task relationships to,
    389–394
tree architecture style, 388–389
Peer-to-peer protocol over Ethernet
    (PPPoE), 172
Penalties for architectural tradespaces,
    345–347
Performance, 11
PERMUTING Pattern, 367, 383–385
Planes for decomposition (potential),
    301–302
Platforming, defined, 216
Platforms, 216–217, 218–219, 222–224
    architecture and, 216–217, 222–224
    B-52 versus B-2 bomber architecture
        and, 222–224
    benefits of, 218
    cost sharing and, 217
    costs and drawbacks, 219
    downstream influences on, 216–217
    product development complexity and, 217
Point designs, 333–334
Population-based algorithms, 397–398, 399
Precedent, emergence prediction by, 31
Primary function, 89
Primary value-related external function,
    89–91
Principles
    ambiguity, 182
    ambiguity and goals, 252
    apparent complexity, 292
    architectural decisions, 197
    balance, 244
    beginning (stakeholder identification), 229
    benefit of delivery, 91
    coupling and organization of architectural
        decisions, 354
    creativity, 267
    decomposition, 296
    dualism, 78
    elegance (architectural systems), 300
    emergence, 12
    essential complexity, 292–295

focus, 22
holism, 20
product evolution, 216
reuse of Legacy elements, 214
robustness of architectures, 349
role of the architect, 180
2nd law (complexity), 295
solution-neutral function, 139
stress of modern practice, 192
system problem statement, 247
"2 down, 1 up," (hierarchic systems), 298
value and architecture, 111
Probability distribution, 348–349
Problem stakeholders, 228
Procedural programs, 442–443
Process, 15, 85
    analytical representation of, 86–89
    canonical system characteristic of, 17
    defined, 15, 85
    function as operand and, 84–89
    Level 2 architecture relationships and,
        167–168
    projecting operands onto, 105
Process-form (PF) array, 130
Process operand (PO) array, 96–97
    creation from OPM diagrams, 99
    functional interactions, 99, 102
    internal function identification, 96–97
Product, defined, 9
Product case combining influences, 217,
    219–221. See also Architecture Business
    Case Decision (ABCD)
Product development process (PDP),
    184–192
    enterprise, 184–188
    Generic, 188–189
    Global, 189–192
    holistic framework for, 189–190, 191
    principle of stress of modern
        practice, 192
Product evolution, 214–217
    architectural competition, 280–284
    downstream influences on, 214–217
    principle of, 216
Product lines (families), 215–216
Programmed decisions, 318–319

Projected system representation, 130–135
  classes of operand interactions, 135
  form, 132–135
  objects, 131–132
  value stream objects, 132, 133, 134
Projections, 45
Pseudo-regulation, 206
Pseudocode, 77–78
Pugh matrix, 277–279

Q

Quadrants for architectural tradespace decisions, 352–353

R

Recursion, 43–44
Recursion for level 2 architecture, 166–167
Regulation, 204–206
  anticipated, 206
  compliance with, 205
  liability and, 206
  pseudo-regulation, 206
  regulatory intent, 205
  sources of, 204–205
  standards and, 206
  upstream influences on, 204–206
Regulatory intent, 205
Reliability through redundancy, 258–259
Representative goals, 245, 256
Representing layer, decision support systems (DSS), 319–320
Requirement diagram (SysML), 46, 47
Resource allocation decisions, corporate strategy, 200
Reuse, 214–215
Reuse of Legacy elements, principle of, 214
Reverse engineering, 52. *See also* Form; Function
Review activity, 318
Ring architecture style, 388–389
Robustness of sensitivity analysis results, 346–349
Rule-based systems, 440–445
  attribute–value pairs for, 441
  constraints and, 443
  declarative enumeration and, 442–443
  emergent behavior and, 444
  inference algorithm for, 441
  interferences, 443–445
  knowledge-based systems and, 440–441
  procedural programs, 442–443
  synergies, 443–445
  system architecture applications, 440–445

S

Saturn V launch vehicle decomposition, 303–305
Scenarios, sensitivity analysis and, 345–346
Schemata, 373, 401
Search algorithms, 396–398
2nd law (complexity), principle of, 295
Segmentation of markets, 203
Sensitivity analysis, 345–349
  penalties, 345–347
  probability distribution, 348–349
  robustness of results, 346–349
  scenarios, 345–346
Sensitivity of decision to metrics, 351–352, 435–436
Sequence, 74, 79, 127
Sequence diagram (SysML), 46
Sequencing decisions, 354–356
Services, 16, 157–158
Shareholder annual report strategy, 198–199
Simple systems, 41, 58–61
Simplified system representation, 129–130
Simulating layer, decision support systems (DSS), 320
Software system analysis, 77–82
  boundaries and, 80–81
  connectivity structural relationships, 80, 81
  duality and, 78–79
  external value-related operands, 96
  formal entities and relationships, 79–80
  functional architecture in, 104–107
  internal functions, 97
  object of form for, 77–78
  principle of dualism, 78
  pseudocode for, 77–78
  sequence and, 79
  system architecture of, 118, 119
  use context for, 81

Solar system
  abstractions in, 23
  form and function identification, 15, 16
Solution-neutral functions, 137–159
  concept and, 137–138
  defined, 139
  functional intent and, 139
  hierarchy of, 139–140, 149, 151–152
  identification of, 140–142
  intent of, 140–142
  principle of, 139
  specialization of, 143–145
Space Station Freedom (SSF) decomposition, 305–308
Spatial relationships, 64
Spatial/topological relationships, 64–67, 118
Specialization relationships, 43
Stakeholder maps, 238, 239, 240
Stakeholders, 203, 227–244
  beneficial stakeholders, 228
  beneficiaries and, 227–232
  bounds for identification, 229
  charitable stakeholders, 228
  dimensions of needs, 236–237
  distinguishing beneficiaries from, 227–228
  exchange and relationships of, 232–234, 237–238
  granularity (abstractions) for identification, 229
  grouping, 234–236
  identifying needs of, 230–232
  IT architecture and, 254–259
  marketing and needs of, 203
  operator and, 228
  principle of balance, 244
  principle of the beginning, 229
  prioritizing needs, 240–244
  problem stakeholders, 228
  value delivery of, 232–234, 237–238
Stand-alone operations, 212
Standard blueprints of internal processes, 95
Standards, regulation and, 206
Star (hub and spoke) architecture style, 388–389

State machine diagram (SysML), 46, 88–89
States of objects, 56
Strategy, *see* Corporate strategy
Stratification, tradespace analysis and, 343–344
Stress of modern practice, principle of, 192
Structure, 63. *See also* Formal relationships
Structured creativity, 265–266
Structuring layer, decision support systems (DSS), 320
Subjectivity, architectural decisions and, 327
Sub-problems, optimization problems decomposed into, 392–394
Supply availability, 236
Supporting layers, 122–123
Synergies as rules, 443–445
System, defined, 9
System architecture, 110–136, 160–176, 197–225
  civil architecture versus, 193–196
  complexity of, 35–36, 286–288
  clustering, 173–175, 436–440
  concept in, 160–176
  decision support for, 326–328
  defined, 110
  downstream influences, 207–217
  form in, 110–121, 123–125, 132–135
  formal structure and, 110–111
  front-loaded (greedy) deployment, 385
  functional architecture and, 110–111
  functions in, 110–121, 122–128
  home data network architecture, 170–173
  incremental deployment, 385
  influences on, 197–225
  knowledge-based systems, 440–441
  Level 1 development, 162–165
  Level 2 development, 166–170
  mapping, 111–121
  modularizing the system, 173–175
  non-idealities, 121–122
  objects, projection onto, 131–132
  operational behavior, 125–128

System architecture (*continued*)
  optimization problems for, 359–405
  patterns in, 365–389
  PERMUTING pattern and, 385
  principle of architectural decisions, 197
  principle of elegance, 300
  principle of value and architecture, 111
  product case for, 217–221
  questions for defining a system, 54, 84, 112, 138, 161
  rule-based systems, 440–445
  supporting layers, 122–123
  system interfaces, 123–125
  system representations, 129–136
  upstream influences, 198–207
System evolution, 214–217
System failure, emergence and, 30–31
System interfaces, 123–125
System problem statement (SPS), 247–250
System representations, 129–136
  architectural tradespaces, 331–358
  form, 132–135
  objects, 131–132
  process-form (PF) array for, 130
  projected, 130–135
  simplified, 129–130
System thinking, 8–34
  canonical characteristics, 17
  concept of, 8
  emergence and, 8–13, 28–33
  entities of a system, 9, 17–28
  essential features of, 33
  evolution of, 21
  form of system, 13–18
  function of system, 10–11, 15–16
  product versus, 9
  relationships among entities, 26–28
  system failure and, 30–31
  tasks for identification, 8
Systems attributes (ilities), downstream influences of, 213–214
Systems Modeling Language (SysML), 45–47, 48, 69–70
  architectural view representation, 45–47
  diagrams, 46–47, 68–70
  formal (structural) representation, 68–70
  function representation, 88–89
  legend for graphical elements, 69
  state machine diagrams, 88–89
  Unified Modeling Language (UML) and, 45–46

T

Tables for formal relationships, 70–71, 80, 81
Tabu search, 401
Tasks, system thinking identification, 8
Technology infusion, 206–207
Technology Readiness Level (TRL), 207, 208
To-By-Using framework, 247–249
Top-down/bottom-up reasoning, 44
Topology, 64
Tradespaces, *see* Architectural tradespaces
Transformational grammar, 17
Tree architecture style, 388–389
Tube-and-wing aircraft architecture, 222–224
"2 down, 1 up" (hierarchical systems), principle of, 298
Two-level decomposition, 41

U

Uncertainty, 181
Unified Modeling Language (UML), 45–46
Unknown information, 181
Unknown-unknowns, holistic thinking and, 20
Unstructured creativity, 263–264
Upstream influences, 198–207
  corporate strategy, 198–201
  marketing, 201–203
  regulation, 204–206
  technology infusion, 206–207
Upstream process, 178–179
Urgency of needs, 236
Use case diagram (SysML), 46
Use context, 76–77, 81

Utility applications, IT architecture, 257–258
Utopia point, 337

V

Value
  attribute–value pairs, 441
  benefit at cost as, 13
  defined, 52, 91
  model input parameters, 348
Value and architecture, principle of, 111
Value delivery
  exchange as, 232–234, 237–238
  indirect, 232, 237–238
  stakeholder relationships and, 232–234
  stakeholder maps for, 237–238
Value flow, 238–240
Value functions, 362
Value loop, 239, 242
Value pathway, 102–103
  functional interactions and, 102–103
  non-idealities in, 121–122

Value-related function, 89–93, 94, 107–108
  external, 90–91
  internal, 94
  operands, 90–93
  OPM diagrams for, 92–93
  primary, 89–91
  principle of benefit delivery, 91
  secondary, 107–108
Value stream objects, 132, 133, 134
Variable types, architectural decisions and, 327
Viewing layer, decision support systems (DSS), 320
Views, 45

W

W questions, 189–190
Whole product system, 75–76
Winter Olympics scenario,
  see IT architecture
Wireless access point (WAP), 172

Z

Zigzagging, 44–45, 163–164
Zooming functions, 18, 103, 104, 167